Rethinking Global Land Use in an Urban Era

Strüngmann Forum Reports

Julia Lupp, series editor

The Ernst Strüngmann Forum is made possible through the generous support of the Ernst Strüngmann Foundation, inaugurated by Dr. Andreas and Dr. Thomas Strüngmann.

This Forum was supported by funds from the

Rethinking Global Land Use in an Urban Era

Edited by

Karen C. Seto and Anette Reenberg

Program Advisory Committee:
Eric F. Lambin, Julia Lupp, Cheikh Mbow, Charles L. Redman,
Anette Reenberg, Karen C. Seto, and Thomas Sikor

The MIT Press
Cambridge, Massachusetts
London, England

© 2014 Massachusetts Institute of Technology and
the Frankfurt Institute for Advanced Studies

This volume is the result of the 14th Ernst Strüngmann Forum, held Sept. 23–28, 2012, in Frankfurt am Main, Germany.

Series Editor: J. Lupp
Assistant Editor: M. Turner
Photographs: U. Dettmar
Design and realization: BerlinScienceWorks

All rights reserved. No part of this book may be reproduced in any form by electronic or mechanical means (including photocopying, recording, or information storage and retrieval) without permission in writing from the publisher.

MIT Press books may be purchased at special quantity discounts for business or sales promotional use. For information, please email special_sales@mitpress.mit.edu or write to Special Sales Department, The MIT Press, 55 Hayward Street, Cambridge, MA 02142.

The book was set in TimesNewRoman and Arial.
Printed and bound in the United States of America.

Library of Congress Cataloging-in-Publication Data

Rethinking global land use in an urban era / edited by Karen C. Seto and Anette Reenberg.
  pages cm. — (Strüngmann forum reports)
Includes bibliographical references and index.
ISBN 978-0-262-02690-1 (hardcover : alk. paper)
1. Land use. 2. Urbanization. 3. Sustainable development. 4. Environmentalism. I. Seto, Karen Ching-Yee II. Reenberg, Anette.
HD111.R45   2014
333.77—dc23

2013044422

10 9 8 7 6 5 4 3 2 1

Contents

The Ernst Strüngmann Forum		vii
List of Contributors		ix
Preface		xiii
1	Rethinking Global Land Use in an Urban Era: An Introduction *Karen C. Seto and Anette Reenberg*	1

Land-Use Competition

2	Trends in Global Land-Use Competition *Eric F. Lambin and Patrick Meyfroidt*	11
3	Food Production and Land Use *Peter J. Gregory and John S. I. Ingram*	23
4	Finite Land Resources and Competition *Helmut Haberl, Cheikh Mbow, Xiangzheng Deng, Elena G. Irwin, Suzi Kerr, Tobias Kuemmerle, Ole Mertz, Patrick Meyfroidt, and B.L. Turner II*	35
5	Land-Use Competition between Food Production and Urban Expansion in China *Xiangzheng Deng, Yingzhi Lin, and Karen C. Seto*	71

Distal Land Connections

6	Globalization, Economic Flows, and Land-Use Transitions *Peter J. Marcotullio*	89
7	Applications of the Telecoupling Framework to Land-Change Science *Jianguo Liu, Vanessa Hull, Emilio Moran, Harini Nagendra, Simon R. Swaffield, and B. L. Turner II*	119
8	Significance of Telecoupling for Exploration of Land-Use Change *Hallie Eakin, Ruth DeFries, Suzi Kerr, Eric F. Lambin, Jianguo Liu, Peter J. Marcotullio, Peter Messerli, Anette Reenberg, Ximena Rueda, Simon R. Swaffield, Birka Wicke, and Karl Zimmerer*	141
9	Palm Oil as a Case Study of Distal Land Connections *Birka Wicke*	163

Decision Making, Governance, and Institutions

10	Emergent Global Land Governance *Matias E. Margulis*	183

11	**Large-Scale Land Transactions: Actors, Agency, Interactions** *Carol A. Hunsberger, Saturnino M. Borras Jr., Jennifer C. Franco, and Wang Chunyu*	201
12	**Private Market-Based Regulations: What They Are, and What They Mean for Land-Use Governance** *Graeme Auld*	217
13	**Changes in Land-Use Governance in an Urban Era** *Bradford S. Gentry, Thomas Sikor, Graeme Auld, Anthony J. Bebbington, Tor A. Benjaminsen, Carol A. Hunsberger, Anne-Marie Izac, Matias E. Margulis, Tobias Plieninger, Heike Schroeder, and Caroline Upton*	239

Urbanization and Land Use

14	**Range of Contemporary Urban Patterns and Processes** *Hilda Blanco*	275
15	**How Is Urban Land Use Unique?** *Dagmar Haase*	299
16	**Reconceptualizing Land for Sustainable Urbanity** *Christopher G. Boone, Charles L. Redman, Hilda Blanco, Dagmar Haase, Jennifer Koch, Shuaib Lwasa, Harini Nagendra, Stephan Pauleit, Steward T. A. Pickett, Karen C. Seto, and Makoto Yokohari*	313

Looking Forward

17	**Ways Forward to Explore Sustainable Land Use in an Urbanizing World** *Anette Reenberg and Karen C. Seto*	333
Bibliography		341
Subject Index		387

The Ernst Strüngmann Forum

Founded on the tenets of scientific independence and the inquisitive nature of the human mind, the Ernst Strüngmann Forum is dedicated to the continual expansion of knowledge. Through its innovative communication process, the Ernst Strüngmann Forum provides a creative environment within which experts scrutinize high-priority issues from multiple vantage points.

This process begins with the identification of themes. By nature, a theme constitutes a problem area that transcends classic disciplinary boundaries. It is of high-priority interest, requiring concentrated, multidisciplinary input to address the issues involved. Proposals are received from leading scientists active in their field and are selected by an independent Scientific Advisory Board. Once approved, a steering committee is convened to refine the scientific parameters of the proposal and select the participants. Approximately one year later, the central meeting, or Forum, is held to which circa forty experts are invited.

Preliminary discussion for this theme began in 2009, and on August 18–20, 2011, the steering committee meeting was convened. This committee—comprised of Eric Lambin, Cheikh Mbow, Charles Redman, Anette Reenberg, Karen Seto, Thomas Sikor, and Julia Lupp—identified key issues for debate and selected the participants for the Forum, which was held in Frankfurt am Main, from September 23–28, 2012.

The activities and discourse surrounding a Forum begin well before participants arrive in Frankfurt and conclude with the publication of this volume. Throughout each stage, focused dialog is the means by which participants examine the issues anew. Often, this requires relinquishing long-established ideas and overcoming disciplinary idiosyncrasies that otherwise might inhibit joint examination. When this is accomplished, a unique synergism results from which new insights emerge.

This volume conveys the synergy that arose out of a group of diverse experts, each of whom assumed an active role, and is comprised of two types of contributions. The first provides background information on key aspects of the overall theme. Originally prepared in advance of the Forum, these chapters have been extensively reviewed and revised to provide current understanding on these key topics. The second (Chapters 4, 8, 13, and 16) summarizes the extensive discussions that transpired. These chapters should not be viewed as consensus documents nor are they proceedings; instead, they are intended to transfer the essence of the discussions, expose the open questions that still remain, and highlight areas in need of future enquiry.

An endeavor of this kind creates its own unique group dynamics and puts demands on everyone who participates. Each invitee contributed not only their time and congenial personality, but a willingness to probe beyond that which is evident, and I wish to extend my gratitude to all. A special word of thanks goes to the steering committee, the authors of the background papers, the reviewers of the papers, as well as the moderators of the individual working groups: Cheikh Mbow, Ruth DeFries, Thomas Sikor, and Charles Redman. To draft a report during the Forum and bring it to a final form in the months thereafter is no simple matter, and for their efforts, I am especially grateful to Helmut Haberl, Hallie Eakin, Bradford Gentry, and Christopher Boone. Most importantly, I extend my sincere appreciation to Karen Seto and Anette Reenberg. As the scientific chairpersons for this Forum, their commitment and support ensured a most vibrant intellectual gathering.

A communication process of this nature relies on institutional stability and an environment that encourages free thought. The generous support of the Ernst Strüngmann Foundation, established by Dr. Andreas and Dr. Thomas Strüngmann in honor of their father, enables the Ernst Strüngmann Forum to conduct its work in the service of science. In addition, valuable partnerships exist with the Frankfurt Institute for Advanced Studies, which shares its vibrant intellectual setting with the Forum, and the Volkswagen Stiftung, which provided financial support for this theme.

Long-held views are never easy to put aside. Yet when this is achieved, when the edges of the unknown begin to appear and gaps in knowledge are able to be defined, the act of formulating strategies to fill such gaps becomes a most invigorating exercise. It is our hope that this volume will convey a sense of this lively exercise. More importantly, we hope that it will spur further enquiry and lead to a better understanding of the challenges that must be addressed, if sustainable land use is to be achieved in this century.

Julia Lupp, Program Director
Ernst Strüngmann Forum
Frankfurt Institute for Advanced Studies (FIAS)
Ruth-Moufang-Str. 1, 60438 Frankfurt am Main, Germany
http://esforum.de

List of Contributors

Graeme Auld School of Public Policy and Administration, Carleton University, Ottawa, ON K1S 5B6, Canada
Anthony J. Bebbington Graduate School of Geography, Clark University, Worcester, MA 01610, U.S.A.
Tor A. Benjaminsen Noragric Norwegian University of Life Sciences, 1432 Aas, Norway
Hilda Blanco Price School of Public Policy, University of Southern California, Von Kleinsmid Center 367, Los Angeles, CA 90089–0626, U.S.A.
Christopher G. Boone School of Sustainability, Arizona State University, Tempe, AZ 85287–5402, U.S.A.
Saturnino M. Borras Jr. International Institute of Social Studies, The Hague, and Transnational Institute, Amsterdam, The Netherlands; China Agricultural University, Beijing, China
Wang Chunyu China Agricultural University, Beijing, China
Ruth DeFries Department of Ecology, Evolution and Environmental Biology, Columbia University, New York, NY 10027, U.S.A.
Xiangzheng Deng Institute of Geographic Sciences and Natural Resources Research, Beijing, and Center for Chinese Agricultural Policy, Chinese Academy of Sciences, Beijing, 100101 China
Hallie Eakin School of Sustainability, Arizona State University, Tempe, AZ 85287, U.S.A.
Jennifer C. Franco Transnational Institute, Amsterdam, The Netherlands; China Agricultural University, Beijing, China
Bradford S. Gentry Yale School of Forestry and Environmental Studies, Kroon Hall, New Haven, CT 06410, U.S.A.
Peter J. Gregory Centre for Food Security, School of Agriculture, Policy and Development, University of Reading, Reading, RG6 6AR and East Malling Research, East Malling, ME19 6BJ, U.K.
Dagmar Haase Department of Geography, Humboldt University Berlin, 12489 Berlin, Germany
Helmut Haberl Institute of Social Ecology, Alpen-Adria University, 1070 Vienna, Austria
Vanessa Hull Center for Systems Integration and Sustainability, Michigan State University, East Lansing, MI 48823, U.S.A.
Carol A. Hunsberger International Institute of Social Studies, 3629 AX The Hague, The Netherlands
John S. I. Ingram Environmental Change Institute, Oxford University Centre for the Environment, Oxford, OX1 3QY, U.K.

Elena G. Irwin Department of Agricultural, Environmental and Development Economics, Ohio State University, Columbus, OH 43210, U.S.A.

Anne-Marie Izac Consortium Office Agropolis International, Montpellier Cedex 5, France

Suzi Kerr Motu Economic and Public Policy Research, Wellington 6142, New Zealand

Jennifer Koch Biological and Ecological Engineering, Oregon State University, Corvallis, OR 97331, U.S.A.

Tobias Kuemmerle Geography Department, Humboldt-University Berlin, 10099 Berlin, Germany

Eric F. Lambin Georges Lemaître Centre for Earth and Climate Research, Earth and Life Institute, Université catholique de Louvain, 1348 Louvain-la-Neuve, Belgium; School of Earth Sciences and Woods Institute, Stanford University, Stanford, CA 94305, U.S.A.

Yingzhi Lin School of Mathematics and Physics, China University of Geosciences, Wuhan, 430074 China

Jianguo Liu Center for Systems Integration and Sustainability, Michigan State University, East Lansing, MI 48823, U.S.A.

Shuaib Lwasa Department of Environmental Management, Makerere University, Kampala, Uganda

Peter J. Marcotullio Department of Geography, Hunter College-CUNY, New York, NY 10065, U.S.A.

Matias E. Margulis Department of International Studies, University of Northern British Columbia, Prince George, BC V2N 4Z9, Canada; Max Planck Institute for the Study of Societies, 50676 Cologne, Germany

Cheikh Mbow World Agroforestry Centre, UN Avenue-Gigiri, 00100 Nairobi, Kenya

Ole Mertz Department of Geography and Geology, University of Copenhagen, 1350 Copenhagen K, Denmark

Peter Messerli Centre for Development and Environment (CDE), University of Bern, 3012 Bern, Switzerland

Patrick Meyfroidt Georges Lemaître Centre for Earth and Climate Research, Earth and Life Institute, Université catholique de Louvain, 1348 Louvain-la-Neuve, and F.R.S.-FNRS, 1000 Brussels, Belgium

Emilio Moran Department of Geography and Center for Global Change and Earth Observations, and Center for Systems Integration and Sustainability, Michigan State University, East Lansing, MI 48823, U.S.A.

Harini Nagendra Ashoka Trust for Research in Ecology and the Environment, Royal Enclave, Bangalore 560064, India

Stephan Pauleit Strategic Landscape Planning and Management, Technical University of Munich, 85354 Freising, Germany

Steward T. A. Pickett Cary Institute of Ecosystem Studies, Millbrook, NY 23545, U.S.A.

List of Contributors

Tobias Plieninger Berlin-Brandenburg Academy of Sciences and Humanities, Ecosystem Services Research Group, 10117 Berlin, Germany

Charles L. Redman Global Institute of Sustainability, Arizona State University, Tempe, AZ 85287–5402, U.S.A.

Anette Reenberg Department of Geography and Geology, University of Copenhagen, 1350 Copenhagen, Denmark

Ximena Rueda Department of Environmental Earth System Science, Stanford University, Stanford, CA 94305, U.S.A.

Heike Schroeder School of International Development, University of East Anglia, Norwich NR4 7TJ, U.K.

Karen C. Seto School of Forestry and Environmental Studies, Yale University, New Haven, CT 06511, U.S.A.

Thomas Sikor School of International Development, University of East Anglia, Norwich NR4 7TJ, U.K.

Simon R. Swaffield School of Landscape Architecture, Lincoln University, Lincoln 7647, Christchurch, New Zealand

Billie Lee Turner II School of the Geographical Sciences and Urban Planning, Arizona State University, Tempe, AZ 85287, U.S.A.

Caroline Upton Department of Geography, University of Leicester, Leicester LE1 7RH, U.K.

Birka Wicke Copernicus Institute of Sustainable Development, Utrecht University, 3584 CS Utrecht, The Netherlands

Makoto Yokohari Graduate School of Frontier Sciences, University of Tokyo, 5-1-5 Kashiwanoha, Chiba 277–8563, Japan

Karl Zimmerer Department of Geography, Pennsylvania State University, University Park, PA 16802, U.S.A.

Preface

The Ernst Strüngmann Forum is, in its unique format, a setting that fosters an optimal environment for creative scientific inquiry and dialog. Indeed, the idea for this book grew out of interactions at the fourth Ernst Strüngmann Forum, "Linkages of Sustainability" (Graedel and van de Voet 2008), to which we were invited. Put simply, we found this gathering to be the most exciting academic workshop that we have ever attended. We believe that it is the Forum's approach—a week-long, presentation-free intellectual retreat that is best imagined as a think tank, marked by intensive exploration and discussions—that makes the difference. This interactive process encourages novel ideas to emerge out of a dynamic dialog between scholars from wide-ranging disciplinary expertise.

As a tool, the Forum is second to none in fostering creative ideas at the intersection of disciplinary domains.

Hence, we thought that our common interest in contributing toward a rethinking of the conceptualization of global land use in the contemporary urban era could be significantly advanced if we were given the opportunity to chair a Forum on this topic. It goes without saying that we were extremely grateful when the Board approved our proposal, thus making it possible to convene forty leading researchers to discuss, formulate, and refine new visions for future research on global land use.

It is widely acknowledged that global resources are becoming scarcer due to growing populations and increasing wealth, which is creating an insatiable demand. At this Forum we focused specifically on the resource of land. In the scientific literature as well as in the general media, the pressures on land are well documented and becoming increasingly evident in a variety forms, for example, in connection with nature conservation strategies, urban expansion, new visions for use of land to produce energy to replace fossil fuel, as well as concerns about large land acquisitions.

Over the last couple of decades, the land-change science community has been successful in creating a more consolidated and coherent disciplinary platform for land-use research (Turner et al. 2007). Nonetheless, many research efforts on land-related issues are still anchored in other disciplines that are underrepresented in land-change science (e.g., disciplines that emphasize economics, governance, and ethics). The entire range of relevant research communities does not have an established forum in which to interact, and consequently, possibly important theoretical and conceptual innovations do not materialize. Although the need for this type of dialog has been expressed by research communities from both the natural and social sciences as well as by policy communities, the obstacles for co-creation of ideas across disciplines have not been overcome. This Forum has hopefully initiated a dialog that will continue and grow.

There was consensus on the steeering committee that in order to create fresh ideas, this Forum needed to convene participants from a range of perspectives, including those that are underrepresented and those that are tangential to land-change science. The final group of participants came from the disciplines of economics, history, political science, geography, sustainability science, and law. Throughout the Forum, discussions were deeply thought provoking and far ranging in scope. Numerous gaps in knowledge were immediately identified as were limitations in existing methods and conceptual frameworks; these are highlighted throughout the present book. Looking beyond the scope of the Forum and this resulting volume, one of the most rewarding outcomes we witnessed was the development of ideas that are being used to form new research projects. It is our sincere hope that these will make a positive contribution to the research agenda on global land issues in the years to come.

Crucial for the success of the Forum was the engagement of the right people. We owe a debt of gratitude to the members of the steering committee: Charles Redman, Cheikh Mbow, Eric Lambin, Thomas Sikor, and Julia Lupp. They worked with us to refine the objectives and identify the big questions that were posed to the discussion groups. Perhaps more importantly, they shared their respect and contacts in their respective networks. Without this, it would not have been possible to gather such a strong group of diverse expertise.

Equally, we are grateful to each of the invited scholars, who allocated their time and offered their commitment throughout this project.

Without generous funding, it would not have been possible to convene this Ernst Strüngmann Forum. Interestingly, among most scientific sources of funding, it seems specifically challenging to find support for interdisciplinary and explorative thinking. Hence, the financial support provided by the VW Foundation and the Ernst Strüngmann Foundation is of inestimable importance.

Finally, we would like to thank the invaluable support provided by the Forum's staff. Without the wise guidance and comprehensive editorial support provided by Julia Lupp and her colleagues, we would not have been able to develop and present the ideas that resulted from this Forum.

As you read our collective results, we hope that a bit of our enthusiasm spills over to you. For to address the challenges inherent in sustainable land use, much work remains. We hope that the perspectives offered in this book and the tools that were developed at the Forum will open up opportunities and lead to a better understanding of the numerous issues that must be addressed if we are to achieve sustainable land use in the 21st century.

—Karen C. Seto and Anette Reenberg

1

Rethinking Global Land Use in an Urban Era

An Introduction

Karen C. Seto and Anette Reenberg

Contemporary Global Change Trends

Be it conserving forests for biodiversity, preserving farmland for agricultural production, or halting urban sprawl, land issues are emerging as central to geopolitics, economics, globalization, human well-being, and environmental sustainability. With a projected 9+ billion people on the planet by the middle of the century—67% of whom will live in urban areas—there is increasing concern that there will not be enough land to meet societal and ecosystem needs. Moreover, many land uses are incongruous: land used for housing cannot be used to grow food; land for forests and carbon sequestration cannot be used for factories; biodiversity hotspots thrive only without human alteration. Beyond the physical limits imposed by a finite terrestrial surface area, there are issues of ethics and fairness. Who should have the right to convert the world's forests to timber and eventually houses? Where should land be conserved for food production versus for housing or timber products? In what regions of the world should mining take place and should those communities be compensated? Who has the right to use nontitled land?

Three important trends are currently reshaping land use locally and globally: urbanization, the growing integration of economies and markets, and the emergence of new land-use agents (Young et al. 2006; Zhu et al. 2006; Montgomery 2008; Lieser and Groh 2011). Concomitant with the transition toward a global dominance of urban livelihoods will be changes in lifestyles and thereby in diet and land use (Kastner et al. 2012). Migrant and urban diets are more energy rich than rural diets (Bowen et al. 2011). Income gains associated with urbanization increase the consumption of luxury goods such as meat products, multiplying the current indirect pressure on agricultural and pasture land. Although distant societies have been connected for centuries through trade, the twenty-first century will be characterized by an acceleration of simultaneously occurring, global-reaching changes: increases in real-time information and

communication, large-scale investments, massive rural to urban migration, climate change, and other environmental changes. At the same time, international investments in ecosystem services, agricultural production, and urban real estate are changing the nature of land use. These new agents are global in reach, large in scale, and will have huge implications for the geographic allocation of supply of and demand for land. Although the implications of such changes for sustainability of human societies' uses of land are profound, the current methods for describing and understanding resource-use intensity, and the land system in particular, are incomplete.

Over the last twenty years, land-change science has made significant progress in advancing our understanding of land-use dynamics at various temporal and spatial scales (Turner et al. 2007). The land-change science community has increasingly acknowledged the complexity of human-environment systems, improved the conceptual platform for land-change studies, and examined and documented the drivers of land-use change. Overall, the global overview of land-use development has been significantly improved and has benefited from remote sensing and modeling advancements. However, absent from current land-change discussions is a framework that examines explicitly the trade-offs between land uses and agents of change at the global scale. Such a new conceptualization would need to provide methods to measure the linkages between land uses across geographic space and across time. For example, how does urban development and associated loss of agricultural land in one region drive the expansion of agricultural production and deforestation in another area? Here, the concept of "urban land teleconnections" may be a useful point of departure to develop methods and metrics that allow for full geographic and temporal accounting of the connections among land uses, especially those linked to urbanization processes in distal places (Seto et al. 2012b).

A conceptual framework would also need to address the long-standing issue regarding the tension between the need to *use* land for societal benefits (e.g., to grow crops, to graze, to provide housing and commerce, or to provide for timber products) and the need to *conserve* land (e.g., for wildlife habitat, forests, use by future generations, or provision of ecosystem services). The challenge of balancing these opposing and normative priorities for the use of Earth's finite land resource has become larger as the pressure on land has accelerated due to pressures such as population growth and soil degradation. With a growing human population and increasing demand for resources that require land, there are also mounting conflicts among different uses of land. Although estimates of "untouched" land vary, it appears that virtually all land areas on Earth have been influenced by humans in one way or another (Kareiva et al. 2007; Turner et al. 2007). Hence, a number of questions seem to be pressing:

How much land is required to provide living space for the estimated 10 billion people that will inhabit Earth by 2100 versus the amount of land needed for agriculture to feed the world's people? How is the amount of land required for agricultural production affected by *where* agricultural land will be available? How are food production and consumption connecting distant places and globalizing pressures on land? How does need to mine land for ore conflict with the location of human settlements and cities?

The optimal allocation of land uses may be calculated economically (Lopez et al. 1994) or mathematically (Aerts et al. 2003). However, in reality, local institutions, governance regimes, and cultural context hugely influence land use. Moreover, the stakeholders for land are increasingly global and not local. The new mercantilism of the twenty-first century is characterized by sovereign wealth funds (SWFs), the state-owned investments that are often used to advance political and economic agendas (Pazarbaşioğlu et al. 2007). The last official estimate was that SWFs controlled over 3 trillion USD in assets in 2006.[1] More recent reports indicate that they are now worth more than 5 trillion USD (TheCityUK 2012). Governments around the world are beginning to recognize the value of land, especially fertile farmland, and are buying land abroad. As part of its quest for food security, Saudi Arabia is buying up farms in Ethiopia (Lippman 2010), South Korea is buying farmland in the Sudan, and China is investing in the Congo (Doriye 2010). These global forces are interacting with local conditions in ways that often have unintended social and environmental consequences, such as conflicts due to conflicting land uses and over-withdrawal of local water resources. The rising land acquisition by high-income countries—especially for farmland and rare earth minerals—in developing countries is of particular concern.

The magnitude and rate of change in recent years and simultaneous trends of urbanization, economic integration, and emergence of new land agents requires a rethinking of sustainable global land use: what it means, how it is conceptualized, how it can be achieved, as well as normative and fairness issues.

Emerging Issues for Land Systems Research in an Urbanizing Era

Feeding, sheltering, and clothing a global population in the face of global environmental, demographic, and economic changes is not uniquely a contemporary challenge. However, the current saturation of potentially productive land, accelerating teleconnections, and the growing urbanization of Earth constitute a significantly different state of affairs as compared to previous historical experiences. It raises a number of compelling challenges and questions that deserve to be addressed in a systematic fashion.

---

[1] Data from Deutsche Bank Research article, "Sovereign Wealth Funds – State Investments on the Rise" available at http://www.dbresearch.com/PROD/CIB_INTERNET_EN-PROD/PROD0000000000215270.pdf

The fundamental goal of this Forum was to reinvent land-change science by exploring new theoretical concepts which reflect contemporary trends in land use, urbanization, and integration of economies. Such a reconceptualization of land-change science needs to take into account a wide range of contemporary realities and trends in global land use. These may be illustrated by, but are not limited to, the following examples:

- Food demand through 2050 will increase three times faster than population growth (Australian Bureau of Agricultural Resource Economics Sciences 2012).
- More urban land area will be developed during the first three decades of the twenty-first century than all of human history (Angel et al. 2011; Seto et al. 2011).
- By 2050, a 140% increase in soybean production will be needed to support the feed requirements for livestock for meat consumption, a figure that does not include feedstock for biofuel production (Bruinsma 2009).
- Land grabbing has become a widespread global issue: 986 deals have been recorded since 2000, totaling 57,341,608 ha—approximately the size of half of Western Europe—and 41% of the land acquired is in Africa (http://landportal.info/landmatrix).
- CB Richard Ellis, the world's largest commercial real estate company, had USD 159 billion of transaction activity worldwide in 2011, more than the gross domestic product of Hungary (http://ir.cbre.com).
- From the takeover of indigenous land in Bangladesh (Adnan and Dastidar 2011) to forceable land acquisitions in Ethiopia, foreign land deals are being called not only violations of human rights (Center for Human Rights Global Justice 2010), but they are also resulting in violent and deadly clashes (Romero 2012).

Five key themes guided our inquiry:

1. We need to understand the growing competition for access to and use of productive land given finite land resources. What are these competitions, where do they exist, and how do they manifest themselves? How do our current analytical frameworks incorporate/omit this issue?
2. We need to identify the new forms of distal land connections in the twenty-first century (e.g., urban land teleconnections) and their implications for global land use and society. For example, how does the demand for sustainable building materials change land use and forest management practices in faraway places? What are the implications of these teleconnections for sustainability and sustainable land use? Who are the key decision makers behind these arrangements?
3. There is a need to identify the effects of increasing global land connections and competition on local land-use decisions and emergent global

land governance. What are the underlying mechanisms behind how these teleconnections get established?
4. We need to identify the new agents and practices in global land use. We know that local land-use practices are increasingly being affected by nonlocal interests. Who are the new agents? How do they appear and get access to land? How do these new land-use practices change how we conceptualize and implement land sustainability concepts?
5. We need to make explicit the normative evaluations (e.g., efficiency, equity, justice) as they are applied to land use. Who should decide on land-use issues? How would normative issues change land-use decision making and conceptual frameworks about land sustainability?

Explorative Lenses

How these issues affect the opportunities and constraints for sustainable land use must be seen in a multidimensional perspective. Figure 1.1 shows what we propose to be a set of four lenses for the exploration of the global land use and three contemporary trends that will constitute a second organizing principle for the analysis. As the figure suggests, scale, rate, and place will be used as parts of a third analytical dimension. While this graphical representation of the proposed analytical decomposition of global land-use changes attempts to

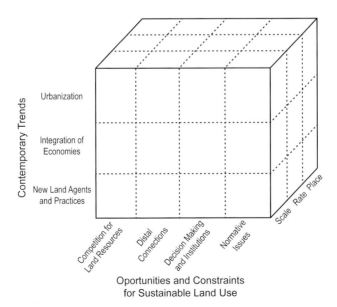

Figure 1.1 Multidimensional perspectives to explore opportunities and constraints for sustainable land use.

place themes to be explored (or of importance) into discrete boxes, we also acknowledge that the themes are closely linked together. The four lenses directly influence the way in which land-use choices can be explored with respect to sustainability. Hence, using the lenses as entrance points may provide a manageable simplification that can guide a rethinking of land-change science, as long as these entrance points are seen as perspectives that later need again to be considered in concert. This provisional, conceptual decomposition of the opportunities and constraints of the land system can be helpful to conceptualize dimensions of land sustainability, and, in turn, to identify points of intervention as well as intervention options that are critical or opportunistic.

The "competition for land resources" lens directs, for example, attention to huge alterations following in the wake of urbanization (new types of housing stock, road infrastructure, new demands for food and recreation, land-sparing urban forms, etc.); to consequences of economic interaction (leading to land grabbing, or to global harmonization of local land uses); or to new land practices and rapid transformations introduced by new land agents.

The "distal connection" lens draws attention to pressures on ecosystems services caused by urban teleconnections and the role of suburbanization of the countryside. It explores integration of economies by understanding disconnections of places of population pressures and places of production. In addition, it looks into issues such as global market places for land and labor as well as transnational coalitions and advocacy agendas.

The "decision making and institutions" lens directs attention to the new urban decision makers, such as master planning and international real estate companies. Integration of economies leads to new incentives for large-scale investors and supranational policies for restitution of land to evicted people. It also addresses issues of how international real estate companies become new urban decision makers.

The "normative issues" lens pays specific attention to important analytical implications of adopted norms, values, and classifications. For example, how do we decide what is just or fair land use in an interconnected world? With respect to the integration of economies, the definitions of systems boundaries must be considered because they have implications for assessing sustainability as well as for accounting the uses of distal land resources. Finally, the notion of who has the right to the land as well as who is setting the priorities for the use of land becomes important in relation to new land agents and practices.

Cutting across these important two sets of dimensions are the notions of scale, place, and rate. With this we want to stress that scales of phenomenon must be considered, that rates of change processes have huge implications for impacts as well as possible intervention, and that some actions at local scales and in specific places may in fact undermine land sustainability of the whole system or in other places.

The Utility of a Multidimensional Rethinking of Land Use

What do we hope to get from the proposed conceptual advancement of a multidimensional rethinking of land-change science in an urbanizing era? If successful, the results of this Forum would aid in consolidating and defining new perspectives to better understand the challenges and opportunities for sustainable land use in the twenty-first century. By addressing the planetary division of land use as well as the local dimensions of land use, we hope that this think tank will provide a conceptual platform to address the questions of how Earth's finite land resources should be used and allocated, how the possible global distribution of land use may be affected by local limits, governance, and institutions, and where the likely land resource-related conflicts are expected to arise in the upcoming decades.

The multidimensional approach proposed seeks to account for the presence of a limited number of prominent change trends, which have a significant bearing on the land-change processes. It recognizes the importance of exploring place and momentum of alterations. Finally, it suggests that the analysis of sustainable land use be organized around a set of apparently useful analytical lenses.

Land-Use Competition

2

Trends in Global Land-Use Competition

Eric F. Lambin and Patrick Meyfroidt

Abstract

This chapter identifies the main land-change trends and competitions in global land use. To produce commodities for global markets, additional land is required, and this demand competes for the most productive and accessible lands. There is less underutilized productive land than is generally assumed, and converting that land is always associated with social and ecological trade-offs. Current data provide no clear evidence in support of either of two competing views—a Malthusian and Ricardian view—on the global availability of productive land: Is there a finite stock of suitable land, leading to a strict competition between land uses, or will increases in the price of goods produced from the land make it economically feasible to bring marginal land into use? The main "friction points" in global land use are expected to be between forests and agriculture; urban land use and intensive agriculture; tree plantations and natural forests; bioenergy, feed crops, and food crops; and intensive cropland and extensive agriculture. Land changes are increasingly influenced by the multiple processes of globalization. Discussion concludes by identifying a few key research questions for land-change science.

Introduction

Despite a continuous increase in the productivity of agriculture and forestry, the pressure to convert natural ecosystems remains high. Human population growth and increases in per capita consumption of material goods is expected to continue to lead to a growing demand for commodities produced from the land. Over the last decade, commodities produced for global markets, whose production occupies vast amounts of land and has high-income elasticity (e.g., soybean, palm oil, beef, coffee, timber), have expanded rapidly. Global food demand is projected to continue to increase as a function of per capita income. The environmental impacts of meeting this demand will depend on whether current trends of greater agricultural intensification in richer nations and greater land clearing in poorer nations continue (Tilman et al. 2011). As long as the rate of increase in demand is greater than the rate of productivity increase, conversion of land under natural vegetation cover to productive uses will be unavoidable. Additional land demands often compete for the most productive and accessible lands that are least vulnerable to natural hazards and climatic variability, where a skilled labor force is available, and where political stability favors productive uses and long-term investments in infrastructures. In this chapter we provide an overview on global land use and identify the main land-change trends and competition between land uses.

Current State of Global Land Use

We begin by summarizing various estimates of global land use for the year 2000 and the 2000–2010 period (see Table 2.1, adapted and updated from Lambin and Meyfroidt 2011), retaining low and high estimates. The most important, and least known, figure for global land-use budgeting is the area of potential available cropland (PAC),[1] defined as the moderately to highly productive land that could be used in the future for rainfed farming, with low to moderate capital investments; this land is neither under intensive use, legally protected, nor under intact forest cover. Only about a quarter of the total ice-free land is suitable for rainfed agriculture given temperature, rainfall, and soil constraints. Early estimates of the amount of land available for future expansion of cultivation ranged from 1670 to 1900 Mha in developing countries, depending on the method used (Young 1999a), and around 2300 Mha globally (Ramankutty et al. 2002). More recent, global-scale PAC estimates have significantly revised this figure downward. A new global assessment of agroecological zones identified 3,330 Mha of land that are very to moderately suitable for rainfed crop production (IIASA/FAO 2012). Of the uncultivated fraction of these lands (1,800 Mha), only 445 Mha globally are nonforested, nonprotected, and populated with less than 25 persons/km^2; therefore, they are assumed to be available for potential cropland expansion if one attempts to minimize ecological costs of land conversion (World Bank 2010). Adopting a spatially explicit, country-by-country approach, a more recent study based on a limited number of regional or country case studies (including the South American Chaco, Cerrado, and Amazon arc of deforestation, the Democratic Republic of Congo, Indonesia, and Russia) suggests that there is less PAC than is generally assumed, and that converting land is always associated with social and ecological trade-offs (Lambin et al. 2013). Once social, institutional, economic, and physical constraints are taken into account, there are few remaining places with "free and easy" lands. Ecological costs of converting that remaining land would be significant, both in terms of carbon and biodiversity. This

---

[1] "Cropland" is being used here as a general term, consistent with FAO's definition of "arable land and permanent crops."

Table 2.1 Primary land uses in 2000 and 2010 in million hectares (Mha). Source for 2000 figures were obtained from Lambin and Meyfroidt (2011; see section on "Supporting Information"). Urban built-up area was updated based on Angel et al. (2011) (low) and Seto et al. (2011) (high). Data on annual change in cropland and pasture were obtained from Ramankutty (2013) (respectively, low and high) and FAOSTAT (respectively, high and low). Those for natural forests are based on FAO FRA 2010 Remote Sensing Survey[1] (data from 2000–2005) (low) and FAO FRA 2010 (high) (Lindquist et al. 2012). Information on planted forests derives from FAO FRA 2010 (includes all planted forests); urban built-up area is from Angel et al. (2011); potential available cropland is from World Bank (2010) (high) and Lambin et al. (2013) (low).

	Area 2000 (Mha)		Annual change 2000–2010 (Mha)	
	Low	High	Low	High
Cropland	1,510	1518	+0.2	+1.7
Pastures	2,800	3,410	–7.7	–7.2
Natural forests	3,143	3,871	–11.3	–10.1
Gross deforestation			–15.2	–13
Gross natural regrowth			+3.9	+2.9
Planted forests	126	215	+4.9	+4.9
Urban built-up area	60	73	+1.0	+2.7
Potential available cropland (without forest conversion)	138	445		

[1] http://www.fao.org/forestry/fra/remotesensingsurvey/en/

study found potential available cropland in the six study regions; after accounting for land accessibility constraints and ecological trade-offs, this represents only one-third, on average, compared to the PAC extracted from the widely cited IIASA/FAO (2012) study of agroecological suitability. A more realistic estimation of the availability and geographic distribution of the potential available cropland, or land reserve, is a priority for land-use planning and policy foresight as well as to inform markets and potential investors.

Future Land Use

Multiple demands for land cumulate and lead to rapid conversion. Demands for more cropland lead to an increase in food, feedstocks, and biofuel production; for industrial forestry to produce timber; for fast-growing trees for carbon sequestration; and for urban and recreational spaces to accommodate a growing urban population (see Table 2.2; the multiple sources of the data cited in this section are described in the "Supporting Information" section in Lambin and Meyfroidt 2011; see also Smith et al. 2010; Tilman et al. 2011). In addition, demands for protected areas for nature and biodiversity conservation as well as for natural or managed ecosystems to provide a range of regulation and

Table 2.2 Projected land use for 2030: total additional land-use demand compared to the 2000 baseline (in millions hectares). Source: Lambin and Meyfroidt (2011, section on "Supporting Information"); urban expansion was updated based on Angel et al. (2011) (low) and Seto et al. (2011) (high).

	Low	High
Additional cropland	81	147
Additional biofuel crops	44	118
Additional grazing land	0	151
Urban expansion	66	153
Expansion industrial forestry	56	109
Expansion of protected areas	26	80
Land lost to land degradation	30	87
Total land demand for 2030	303	845

cultural ecosystem services further contribute to potential conflicts between various land uses.

Globally, to feed a growing population may require an additional 2.7–4.9 Mha of cropland per year on average, depending on future diets, food wastages, and food-to-feed efficiency in animal production (Wirsenius et al. 2010). Most of this expansion is likely to occur in Latin America and Africa, while cropland is expected to decline in developed countries (Smith et al. 2010). In 2007, production of feedstocks for the current generation of biofuels required about 25 Mha. Meeting the current policy mandates of biofuel use would require an increase of 1.5–3.9 Mha yr^{-1}. Projections of pasture expansion range between 0 and 5 Mha yr^{-1}, depending on intensification of livestock production systems, which become increasingly decoupled from the land (Naylor et al. 2005). Expansion would occur mainly in Latin America and East Asia, while pasture area would decrease in North America and Europe. The land footprint of cities is less than 0.5% of Earth's total land area, but urban area is predicted to more than double by 2030, according to the low scenario, expanding most rapidly in sub-Saharan Africa as well as in South and Central Asia. Demand for industrial forestry will grow by 1.9–3.6 Mha yr^{-1}, mainly in Asia and subtropical regions, to meet an increase in demand for wood products of 2.8–40.3%, depending on income elasticity of demand and on fuelwood substitution (Meyfroidt and Lambin 2011). Industrial forestry may replace natural forests but will also encroach on agricultural land. Protected areas will continue to expand by 0.9–2.7 Mha yr^{-1}. Land degradation negatively affects land productivity and makes about 1–2.9 Mha unsuitable for cultivation per year, with a high rehabilitation cost. All of the above future land demands are of the same order of magnitude. Climate change will affect agriculture directly through changes in agroecological conditions, potentially opening or closing lands for cultivation. Geographic shifts in land suitability will not affect all prime lands, thanks to adaptations of farming systems; however, interannual fluctuations in crop yields will probably increase.

By comparing Tables 2.1 and 2.2, we see the following:

1. Under a scenario that already includes significant increases in land productivity, total land demands could exceed, in the coming decades, the area of productive land that is potentially available.
2. Cropland and pasture expansion are likely to accelerate, probably with a continuing expansion in the tropics and a slowing down of agricultural decline in temperate zones (less net abandonment).
3. Expansion of planted forests is likely to slow down due to increasing competition with other land uses.
4. Urban sprawl on agricultural lands is likely to accelerate compared to recent expansion.

In summary, the additional land demand for all agricultural, bioenergy, tree plantations, urban, and nature conservation uses is estimated to range from 303–845 Mha by 2030 compared to the 2000 baseline (Lambin and Meyfroidt 2011). Based on these global trends, productive land could become a scarce resource in most developing countries by 2030. Under that scenario, lands with a lower productivity will be brought into use, and forests will continue to be converted for agriculture. Indeed, market responses associated with land scarcity are likely to stimulate the future adoption of more efficient land-management practices. Innovations that could prevent a global land shortage include technological breakthroughs on genetically modified crops or second generation biofuels, investments for restoration of degraded lands, adoption of more vegetarian diets in rich countries, strict land-use planning to preserve prime agricultural land, or new industrial processes to produce synthetic food, feed, and fibers.

There are two different views—a Malthusian and Ricardian view—on the global availability of productive land (Lambin 2012). A crude "Malthusian" view refers to the assertion that resource depletion and scarcity occur when the rate of resource consumption outstrips the ability to provide new resources, thus leading to a crisis. Population equilibrates with resources at a level that is mediated by technology and a conventional standard of living. Malthus did not explicitly address the issue of land scarcity. Applied to land, this view suggests that there is a limited stock of suitable land, leading to a strict competition between land uses and, eventually, absent any intervention, an overuse of the resource which would lead to declining welfare. David Ricardo's law of rent states that the rent of a land site is equal to the economic advantage obtained by allocating the site to its most productive use, relative to the advantage obtained by using marginal land for the same purpose, given the same inputs of labor and capital. According to Ricardo's theory, under conditions of land scarcity and relatively inelastic demand, an increase in the amount of goods produced from land requires that the price of these goods increase to make it economically feasible to bring marginal land into use. Therefore, different land uses do not strictly compete against each other for space, as they can expand with few constraints on land availability thanks to investments in land conversion, restoration, and productivity enhancement. Converting marginal lands will indeed be associated with social and ecological trade-offs, as these lands are also utilized in various ways and do provide ecosystem goods and services.

Key Land-Use Competition

In this section, we discuss future trends in land change that will lead to competition between a few land uses and in a few geographic regions. These trends constitute the major "points of friction" and "geographic hot spots" in global land use (see also Haberl et al., this volume).

Forests versus Agriculture

In sub-Saharan Africa, between 1975 and 2000, land used for agriculture expanded largely at the expense of forests, which decreased by 16% over the same period (Brink and Eva 2009). Between 1980 and 2000, 83% of the agricultural expansion in the tropics came at the expense of intact or disturbed forests (Gibbs et al. 2010), mostly in the Amazon Basin, Southeast Asia, and to a lesser extent West and Central Africa. Forested areas have been highly affected by the recent wave of large-scale, cross-border land transactions carried out by transnational corporations: about 24% of the land deals are located in forested areas, representing 31% of the total surface of land acquisitions (Anseeuw et al. 2012b). Continuing the recent trends, the deforestation from 2000 to 2030 might represent 152–303 Mha (Lambin and Meyfroidt 2011). With more proactive policy interventions, trade-offs between conserving forests and feeding the world's population could be minimized given the low opportunity costs of avoided deforestation and the small contribution of deforested areas to the recent increases in food production (Angelsen 2010). In recent decades, only around 10% of the increase in agricultural output came from expansion of agricultural lands over forests, the rest stemmed from productivity increases. With appropriate policies, most of the future increases in food production might thus be achieved, in theory, without further forest encroachment (Meyfroidt and Lambin 2011). Note that agricultural intensification has externalities: it is often associated with increased use of energy, water, fertilizers, and pesticides; it has on-site and downstream impacts on terrestrial, freshwater, and marine ecosystems through various forms of pollution (Matson and Vitousek 2006); and, if it involves mechanization, it frees up labor that may migrate and convert more land to low-input agriculture.

Although global rates of tropical deforestation remain alarmingly high, they have decreased over the past decade. A handful of developing tropical countries (e.g., Vietnam, Costa Rica, China, India) have recently been through a forest transition to create a shift from shrinking to expanding forests (including tree plantations) on a national scale (Meyfroidt and Lambin 2011). Forests and plantations of mostly exotic species have expanded primarily on abandoned agricultural lands. These forest transitions occur through different pathways that are contingent upon the local socioeconomic and ecological contexts. The following generic processes of forest transition have been identified (Rudel et al. 2005; Meyfroidt and Lambin 2011):

- Agricultural intensification and industrialization drive labor scarcity in agriculture and concentration of production in the most suitable land, possibly influenced by global markets.
- Scarcity of forest products and services drives tree plantations, forestry intensification, and forest protection by private and public actors. This may be influenced by global environmental ideologies as well as national political factors external to the forest sector.
- Smallholder labor-intensive, tree-based land-use intensification represents forest transitions that are generally associated with a significant outsourcing of forest exploitation to neighboring countries via increased imports of wood and sometimes agricultural products.

The ecological value of this reforestation depends on the residual deforestation of old-growth forests, the proportions of natural regeneration of forests and tree plantations, as well as the location and spatial patterns of the different types of forests.

Current international negotiations on reducing emissions from deforestation and forest degradation (REDD+) and other emerging market-based approaches could provide a political momentum for forest transitions. In reality, this will depend on future productivity gains and on how much agriculture and other land uses expand on forested versus nonforested lands, as well as on land with marginal versus high potential for agriculture.

Urban versus Agriculture

Urbanization as land cover, in the form of built-up or paved-over areas, occupies less than 0.5–2% of Earth's land surface, but it is growing at a fast rate: urban land cover grows on average at more than double the rate of growth of the urban population (Angel et al. 2011). This rapid growth is likely to continue in parallel with urban population growth and income growth. Urbanization affects land change well beyond the limits of urban areas through its wide networks of influence, or urban land teleconnections (Seto et al. 2012b). Urban lifestyles raise consumption expectations. In the developed world, large-scale urban agglomerations and extended peri-urban settlements fragment the agricultural landscapes (Seto et al. 2011), thus decreasing the profitability of large-scale mechanized farming. In the less-developed world, urbanization outbids all other uses for land adjacent to a city, including prime croplands under intensive agriculture. Preliminary results by Angel et al. (2011) suggest that, on average, half of the projected urban expansion will take place on cultivated land. This is likely to take place mostly on prime agricultural land under intensive use, located in coastal plains and in river valleys. Not only do natural habitats deserve to be protected against pressures from other land uses, high-yielding farmlands should also be protected. One possible approach could be along the lines of the agricultural easement purchase program in the United States, which aims to maintain local food production capacity.

Tree Plantations versus Natural Forests

Some argue that forestry intensification could satisfy global timber needs from limited areas of high-yielding tree plantations, thus saving remaining forests from exploitation pressure (Sedjo and Botkin 1997). In reality, although forest plantations can expand on former agricultural land (Sedjo and Botkin 1997), they often compete for space with natural forests. For example, in Vietnam, expansion of tree plantations had a negative effect on the amount of natural forest regrowth at the district level, providing indirect evidence for a competition between the two land uses (Meyfroidt and Lambin 2011). In Chile, commercial pine and eucalyptus plantations expanded over native forests, and their exploitation did not relieve pressure on old-growth forests. New Zealand is often cited as a case where plantations relieved pressure on natural forests, but expansion of exotic plantation forestry proved to be the single most important driver of recent deforestation there. In India, the forest transition was due to increases in tree plantations, and the area of natural forest is still decreasing. In rural areas dominated by smallholder agriculture, silvopastoral and agroforestry systems combine grazing, crop production, and forestry on the same area, thus creating multifunctional landscapes which share rather than monopolize land.

Bioenergy and Feed Crops versus Food Crops

Policy mandates in the United States and the European Union on biofuel[2] use have promoted crops for the bioenergy[3] markets, curtailing potential expansion of food production globally (Searchinger et al. 2008). Estimates of the amount of bioenergy that could be technically supplied in the future tend to overestimate both the area available for bioenergy crops, due to insufficient consideration of constraints and competing land uses, and crop yields (Haberl et al. 2010). Bioenergy crops replace food crops in fields already under cultivation, expand into natural ecosystems, or divert crop production away from the

---

[2] A biofuel is a fuel that contains energy from geologically recent carbon fixation. These fuels are made by a biomass conversion.

[3] Bioenergy is renewable energy made available from materials derived from biological sources.

food market to the bioenergy market. In the latter case, the decrease in supply of food crops (e.g., for corn, sugarcane, potato, wheat used for ethanol, or palm or rapeseed oil used for biodiesel) leads to an increase in the market price for the replaced crop, thus causing more land to be allocated to grow that crop elsewhere. These indirect land-use changes trigger a cascade of crop-by-crop substitutions, which eventually causes land conversion at the margins and a loss in ecosystem services (e.g., carbon storage and sequestration potential) (Searchinger et al. 2008). When cultivation expands on abandoned croplands, there is still an ecological loss because natural vegetation regrowth on these areas is sacrificed (Searchinger et al. 2008). Similarly, the decoupling of livestock from the land is accompanied by a greater demand for feed, and thus cropland. In 2000, around 35% of the world's croplands were already used to produce animal feed. This shift from pastures to grain-feeding from intensive croplands, combined with a shift to monogastric animals and breeding improvements, have the potential to reduce the expansion of the land footprint of livestock per unit produced while increasing the competition on land suitable for crops (Steinfeld et al. 2006; Wirsenius et al. 2010).

Intensive Cropland versus Extensive Agriculture

Given the increasing constraints on agricultural expansion, intensification of land use, made possible by technological progress and better management practices, is likely to increase to satisfy the bulk of future increases in demand. Land-use intensification takes place along a gradient of input use. The most significant land-use competition is between (a) smallholder, labor-intensive systems, which produce mainly for subsistence needs and local markets, and (b) large-scale agri-business—capital-intensive systems that produce commodities for distant, often urban, markets. The world's pasturelands that are used extensively (one head of cattle per hectare, or less) have been increasingly identified for further expansion of highly mechanized, rainfed grain production, assuming appropriate investments and land-use governance. This is especially the case in Latin America. In the state of Mato Grosso, Brazil, between 2006 and 2010, 91% of the expansion of soybeans fields occurred on previously cleared land, mostly pastures. This did not result in a decrease in cattle production; pastures that were carrying very low cattle densities were used more efficiently (Macedo et al. 2012). In this case, increased agricultural production can occur simultaneously with reduced deforestation in tropical forest frontier regions. Expanding cropland at low environmental costs by converting some pastures to croplands, while intensifying cattle ranching and restoring degraded pastures, could spare a lot of land for forests: up to 10 Mha in the Brazilian Cerrado (Bustamante et al. 2012).

Global demands for commodities also contribute to the conversion of traditional shifting cultivation systems toward more intensive and permanent crops for commodity markets (van Vliet et al. 2012). These changes are particularly rapid in Southeast Asia, where land zoning and other policies restrict shifting cultivation and promote its conversion. It may also have important social impacts on livelihoods of specific groups, including uplands dwellers and ethnic minorities (Cramb et al. 2009). These changes may contribute to raising average incomes but it can also increase insecurity of livelihoods through exposure to the instability of global markets, to widening inequities, and to marginalization of already poor communities. In addition, conversion of shifting cultivation mosaics to permanent crops creates significant environmental impacts, including an increase in deforestation. If labor demand in expanding crops is small, shifting cultivators may move their fields elsewhere, possibly encroaching onto forests. However, in areas that are poorly connected to urban centers—especially in tropical highlands, Central Africa, and Madagascar, and for farmers that have low access to other opportunities—shifting cultivation systems remains a vital activity (van Vliet et al. 2012).

Globalization of Land Use

Land changes are increasingly caused by global-scale factors, with a growing separation between the locations of production and consumption of land-based commodities. For example, a large and growing fraction of forest conversion today is associated with commodities produced for global markets. Land-use decisions related to these commodities are largely driven by factors in distant markets, mostly associated with wealthy urban consumers, in addition to local-scale factors (DeFries et al. 2010). Consumers outsource their land use to other countries, and a virtual land trade develops through the land use embodied in the international trade in agricultural and wood products (Anderson 2010).

The distant factors which affect land use are not restricted to trade patterns. They also include remittances sent by migrants, the specific organization of global commodity value chains, channels of foreign investments in land, the transfer of market or technological information to producers via a diversity of networks (from farmer associations to Internet and cell phones), and the development and promotion of niche commodities that target narrow but wealthy market segments with high value commodities produced in limited quantities (Le Polain and Lambin 2013). To understand the influence of these complex and global social networks on land use requires new approaches to land-change studies (see Eakin et al., this volume).

Other trends are associated with a globalization of the direct or indirect policy interventions on land use. The final consumers of agricultural and wood commodities, the corporations involved in their transformation and retailing, and civil society show a growing concern for sustainability (Dauvergne and Lister 2012). These actors are beginning to express a preference for goods whose supply chain has been certified as meeting sustainability criteria. Simultaneously, large agri-business corporations are increasingly adopting sustainability standards and applying these to their suppliers. In parallel, several countries have pursued more traditional command and control policies, such as land-use zoning or harvest regulations, to protect and restore their forests and other valuable ecosystems; they are also demanding greater enforcement of, and compliance with, existing regulations. Payments for ecosystem services (PES) programs—either implemented at the national (as in Costa Rica) or international level (as REDD+)—combine aspects of demand-driven and command-and-control interventions on land use, as they are market-based but the governments or international organizations act as intermediaries between suppliers and consumers of ecosystem services as part of their public regulatory and policy efforts. Thus, the range of instruments that affect land use include: national land-use zoning policies; support to land-use intensification; multilateral trade agreements; eco-certification and other forms of corporate sustainable sourcing strategies; industry roundtables and working groups around specific commodities, moratoria, payments for ecosystem services; and NGO campaigns. Nonetheless, evidence about the actual effectiveness of these various instruments on land use "on the ground" remains insufficient.

Conclusion

With increasing competition between different land uses for accessing productive lands, greater attention must be given to the geographic distribution of land use. Improving the adjustment of actual land use to land potential may offer the possibility of sparing land. For example, some lands with a high potential for food production may have to be taken away from growing fuels and animal feed, and will have to be protected from urban sprawl and land degradation. Most of the additional future demands for agricultural and forest products are likely to come, however, from increasing efficiency of land use through productivity increases. This will require a better management of the externalities of land-use intensification on other natural resources and adjacent ecosystems.

Future research will need to address such questions as:

1. How should we manage trade-offs between food, fiber, and fuel production and nature conservation? How will economic, social, and ecological costs of land conversion increase with land-use expansion?
2. How likely is it that forest transitions, which have recently taken place in a handful of developing countries, will emerge in additional countries? Alternatively, will reforestation in some countries be achieved solely through a displacement of land use elsewhere, thus merely outsourcing deforestation to other places?
3. Will the rapid increase in global trade in wood and food commodities lead to a more efficient use of land globally, through a better adjustment

of actual use to land suitability and to potential environmental services provided by the land?
4. Will urbanization cause major losses of prime cropland under intensive use, thus leading to indirect land-use changes at the margins and causing more conversion of natural ecosystems in cascade (or spill over) effects?
5. Is the looming scarcity of productive land going to trigger market responses, causing changes in consumption patterns (e.g., adoption of more vegetarian diets) and rapid innovations in land-use efficiency in response to price increases?
6. Are policy mandates on biofuels going to be responsive to indirect land-use changes, food price volatility, and productive land scarcity to which they contribute, among other factors?
7. Are international concerns on climate change mitigation via forestry (e.g., via REDD+ policies) and biofuel production going to conflict with concerns on food security, access to land by indigenous communities, and biodiversity preservation?
8. What type or combination of policy interventions on land use—direct or indirect, regulatory or market-based—are more likely to spare land for nature while improving access to basic goods for all, thus avoiding a geographic displacement of land conversion and degradation to the periphery?

3

Food Production and Land Use

Peter J. Gregory and John S. I. Ingram

Abstract

This chapter examines the interactions between food production and land use in the context of how a future global population of another 2–3 billion over the next 50 years can all be fed to deliver food security for all. Increased crop production in the last 70 years has occurred as a result of both expansion of cropland (altering natural ecosystems to produce products) and intensification (producing more of the desired products per unit area of land already used for agriculture or forestry). For the future, it is widely recognized that, globally, only a small proportion of future increases in crop production will come from the cultivation of new land (about 20%); the majority will come from intensification via increased yield (67%) and higher cropping intensity (12%). Because the area of cropped land is likely to increase proportionately less than the future demand for food, reducing the gap between current yields and potential yields is a major goal for the future.

Urbanization affects the use of land to produce food, but it also has major effects on nutrient budgets with a major shift of nutrients from rural to urban areas. In addition, distinct and disparate views of urban communities have emerged in terms of the value of food associated with a decrease in the ratio of food producers to food consumers.

Finally, changing land use is only one of a number of global environmental changes affecting food production and food systems. Multiple incremental adaptations to agricultural systems are possible to cope with climate and other global changes, but transformational adaptation will be required in some regions.

Introduction

A key challenge facing humanity is how a future global population of another 2–3 billion over the next 50 years can all be fed nutritionally and healthily, thereby delivering food security for all. This chapter examines the interactions between the production of food from crops, crop yield, and the land area required for cropping if the projections for food are to be met. Urbanization affects the use of land to produce food, but it also has major effects on the distribution, accessibility, and utilization of food. These issues are explored in relation to influences on nutrient budgets and the disparate views of urban communities about the value of food. Finally, changing land use is only one of a number of global environmental changes affecting food production and food systems. The need to produce more food while simultaneously delivering other valuable services from land is briefly explored.

Competition for Land

While food production alone cannot guarantee food security, it is an essential component of food systems, and it is this activity that constitutes a major competitor for land. In assessing the factors contributing to competition for land, it is useful to distinguish between the drivers (underlying causes such as societal trends, institutional factors, and socioeconomic and technological factors) and the pressures (direct causes such as natural causes, land transition, and land degradation) that result in competition at different geographical scales (Smith et al. 2010). Competition is thus an emergent property of these drivers and pressures (Smith et al. 2010). During the last 60 years, the world's human population has increased from about 2.2 billion in 1950 to about 7 billion in 2011, and this has been sustained by substantial increases in crop and animal production (Godfray et al. 2010). Increased crop production has occurred as a result of both expansion of cropland (altering natural ecosystems to produce products) and intensification (producing more of the desired products per unit area of land already used for agriculture or forestry). Only about 3 billion hectares of the world's 13.4 billion hectare land surface is suitable for crop production, and about one half of this is already cultivated (1.4 billion ha in 2008 [Greenland et al. 1998; Smith et al. 2010]; as a comparison, 1 billion ha is equivalent to 250 countries the size of Switzerland or 1.2 times the size of Brazil). The remaining potentially cultivatable land lies currently beneath tropical forests and some grassland areas, but it would be undesirable to convert this to arable land because of the effects that this would have on conserving biodiversity, greenhouse gas emissions, regional climate and hydrological changes, and because of the high costs of providing the requisite infrastructure.

It is widely recognized that, globally, only a small proportion of future increases in crop production will come from the cultivation of new land (about 20%); the majority will come from intensification via increased yield (67%) and higher cropping intensity (12%; see Table 3.1). However, while intensification will dominate, cultivation of "new" land will contribute significantly to crop production in sub-Saharan Africa (27%) as well as in Latin America and the Caribbean (33%). Almost no land is available for expansion of agriculture in South and East Asia, in the Near East or in North Africa; indeed, these areas may actually lose agricultural land to urban development. Thus, intensification will be the primary means of increasing production in these regions (Gregory et al. 2002; Bruinsma 2003).

Table 3.1 Projected contributions (%) to increased crop production between 1997/1999 and 2030. Derived from Bruinsma (2003).

	Land area expansion	Increase in cropping intensity	Yield increase
All developing countries	21	12	67
Sub-Saharan Africa	27	12	61
Near East/North Africa	13	19	68
Latin America and Caribbean	33	21	46
South Asia	6	13	81
East Asia	5	14	81

A consequence of these changes is that per capita arable land area will continue to decrease (it decreased from 0.415 ha in 1961 to 0.214 ha in 2007) while average cereal yield will need to increase by about 25%: from 3.23 t ha^{-1} in 2005/2007 to 4.34 t ha^{-1} in 2030 (Bruinsma 2003). Put another way, had the increases in yield over the last 60–70 years not been achieved, almost three times more land would have been required to produce crops to sustain the present population; land that, as indicated above, does not exist except by using some that is unsuitable and/or undesirable for cropping. Without the increased yields that resulted from intensified practices, competition for land for different purposes would have been greater (Smith et al. 2010). As demands for land to satisfy the required multiple ecosystem services increase, competition for land will intensify, although there are major uncertainties in such projections (Smith et al. 2010).

Yield results from the interaction of three factors: genotype (G), environment (E), and management (M). Evans (1998) highlights how the synergistic effects of these interactions, linked to innovative technologies, have contributed to past increases in yield. Among the important contributors to these have been:

1. Improved germplasm able to grow vigorously (e.g., hybrids), resist pathogens, and respond to fertilizers without lodging (in particular, the use of dwarfing and semi-dwarfing genes in rice and wheat).
2. The application of fertilizers, particularly the availability of affordable nitrogen fertilizer.
3. The development of chemicals to control weeds, pests, and diseases.
4. Improved irrigation systems, especially in rice-producing countries and for some previously rainfed crops.

Together these technological innovations, combined with institutional and market reforms, have modified G, E, and M to increase yields greatly. As a result, over large areas the yields of many crops have increased year on year (Gregory and George 2011). Global yield increases for a number of crops have typically been linear, with values of 53 kg ha^{-1} yr^{-1} for rice, 41 kg ha^{-1} yr^{-1} for wheat, and 63 kg ha^{-1} yr^{-1} for maize over the period 1961 to 2004. Increases in yield have also been linear with time in many individual countries (e.g., wheat yields in several European countries), although in a few instances, technological innovations have produced more rapid, stepwise increases in yield (e.g., Australia). Whereas crop yields have increased globally and throughout North and South America, Europe, Australia and much of Asia, a notable exception has been that of Africa where, for example, per capita food production decreased by about 5–10% between 1980 and 1995. The reasons for the poor performance in Africa, relative to other countries, are many and include social unrest and war, poor institutions and governance, climatic variability which makes reliable irrigation difficult, and weathered soils that are deficient in nutrients. It has been argued (e.g., Bationo et al. 2012b; Buresh et al. 1997) that a major factor behind many of the observed decreases in yield in African countries was the decline of soil fertility accompanied by the lack of fertilizer application. Much has been written about the need to "re-capitalize" the soils of Africa, especially with regard to phosphorus status, but progress has been limited (Greenland et al. 1998).

Because the area of cropped land is likely to increase proportionately less than the future demand for food, reducing the gap between current yields and potential yields is a major goal for the future (Jaggard et al. 2010). Potential yield is a theoretical upper limit to yield imposed by solar radiation (affecting growth), temperature (affecting development and growth), and water supply (affecting mainly growth but also development). A review of data from crops of maize, rice, and wheat grown in a range of countries showed that the gap between potential and actual yields ranged from about 20–80% (Lobell et al. 2009). In many irrigated cereal systems, yield appeared to plateau at or about 80% of potential yield, whereas in rainfed systems, average yields were commonly 50% or less of potential (Lobell et al. 2009). While part of the yield gap is inevitable because of crop losses during harvest, storage, and transport, as well as in the way that land areas are reported (Jaggard et al. 2010), there are still large differences in performance between adjoining farms. A fundamental constraint in many irrigated systems is the uncertainty in growing season weather; this is also a factor in rainfed systems, where interactions between water and nutrient availability are complex. Raising yields above 80% of yield potential is possible, but only if technologies can be developed and adopted that reduce the uncertainties faced by farmers in assessing soil and climatic conditions or that respond dynamically to these conditions or both (e.g., installation of nutrient and water sensors; Lobell et al. 2009). Such technologies may have the added benefits of increasing the efficiency of use of inputs and reducing losses off-site as well as increasing yields.

While the immediate loss of biodiversity as a consequence of clearing land for crop production has been well documented, the higher-level consequences of increased crop yields on sparing land for biodiversity conservation is a matter of considerable debate. This is because:

1. On-farm losses of biodiversity due to practices giving high yields may outweigh the benefits of sparing biodiverse habitats.
2. High-yielding crops may have negative effects on offsite biodiversity.
3. Land sparing does not occur or is imperfect.

The complexity of the factors involved is evident in the findings of Ewers et al. (2009), who analyzed the changes in yields of 23 staple crops for 124 countries between 1979 and 1999. While per capita area of the 23 staple crops decreased in developing countries where large yield increases occurred, this was counteracted by a tendency for an increased area of nonstaple crops, which led to only a weak tendency for land sparing overall. In developed countries there was no evidence that higher yields reduced per capita cropped area, possibly because of the role of agricultural subsidies in promoting production, thereby overriding any land-sparing effects. Ewers et al. (2009) concluded that land sparing is a weak process, but that improved agricultural technology may have contributed to the maintenance of natural vegetation cover in the past, and that future conservation benefits, while debatable, are potentially available if land-use policies are also modified.

Another factor influencing the future use of land will be the changing climate. For example, warming of the climate in Scotland, as documented by Gregory and Marshall (2012), could have substantial beneficial effects on the land-use potential for agriculture in Scotland; potential expansion of areas of "prime" agricultural land could result, especially in the eastern and southern areas of the country (Brown et al. 2008). However, while warming is well established, a crucial and presently largely unknown factor influencing future land use is precipitation. In Scotland, wetter winters are already being experienced, leading to waterlogging of some autumn-sown crops (e.g., wheat and rape) and to delays in land preparation for spring-sown crops. However, should summer rainfall decline (no significant decrease has currently been identified using long-term data), then this could act in the future to reduce land capability (Brown et al. 2011). In parts of the world in which water already limits crop production, the combined effects of increased temperature and decreased rainfall could have severe detrimental effects on land capability and yields (Fischer et al. 2005).

Institutions and policies also have a profound effect on the competition for land. In many developed countries, trade tariffs and subsidies affect decisions about land use by influencing the crops that are grown or the type of production system employed. Subsidies tend to limit competition for land, distort markets on a global scale, and influence the use of land in other regions of the world (Smith et al. 2010). Some countries and regions (e.g., Europe) have moved away from production subsidies toward environmental protection payments. These, too, have the effect of supporting farmer incomes, preserving the current landscape, and reducing the evolution of new land uses. In addition, policies influencing land ownership have big effects on the practices used by farmers.

A final factor influencing competition for land is the demand for non-agricultural crops, such as forestry and energy. In a range of modeling studies, Smith et al. (2010) demonstrate that future policy decisions in the agriculture, forestry, energy, and conservation sectors could have profound effects, usually intensifying competition for land to supply multiple ecosystem services. An example of this is seen in the intense debate about the use of land to produce ethanol from maize in the United States. Although the actual amount of land converted from food production to biofuel production is slight, the effect has been for the increased production of maize bioethanol in the United States to reduce the amount of maize available for trade on international markets. About 25% of the U.S. maize crop was used for bioethanol in 2007, and this significantly reduced the 55–60 % of maize that the United States normally contributes to global trade (Naylor et al. 2007). Coupled with the other factors that drive up food commodity prices, this affected the food security of low-income people in the developing world.

Food and Urbanization

In 2008, the world's urban population exceeded its rural population for the first time. This process of urbanization has been going on for some time, but it accelerated significantly during the twentieth century. In 1900 there were, globally, an average of 6.7 rural dwellers for each urban dweller, but by 2025 it is projected that there will be three urban dwellers for every rural dweller (Satterthwaite et al. 2010). Along with this change has been a decrease in the ratio of food producers to food consumers, brought about by rapid growth in the world's economy and the shift from primary industries (agriculture, forestry, mining, and fishing) to manufacturing and services.

Expansion of towns and cities often occurs at the expense of agricultural land, not least because most cities have been there for some time and were located where they are because their hinterlands could supply food from fertile soils. Satterthwaite et al. (2010) suggest, however, that the loss of agricultural land is often overstated, with only Western Europe having > 1% of its land area as urban among all of the world's regions. Moreover, while land is clearly used for buildings and roads, etc., the remaining land may be used for more intensive production (e.g., vegetable and fruit crops) or for a variety of urban and peri-urban agricultural schemes. Many people living in urban areas rely on urban or peri-urban agriculture for part of their food. For example, in urban areas of East Africa during the 1990s, 17–36% of the population grew crops (often maize) or kept livestock.

In many urban centers, waste materials are efficiently recycled through livestock and composting. For example, in Nairobi, municipal solid waste is used for composting by a range of community-based organizations, and animal manure is imported into the city from the arid and semiarid livestock-producing areas from up to 300 km away (Njenga et al. 2007). These represent business opportunities for some urban dwellers. However, the quality of the solid wastes used to produce food is crucial, as heavy metals may accumulate causing toxicities in leafy vegetable crops. In Vietnam, Khai et al. (2007) measured the flows of nitrogen (N), phosphorus (P), potassium (K), copper (Cu), and zinc (Zn) for intensive small-scale aquatic and terrestrial vegetable systems in two peri-urban areas of Hanoi City. They found that high inputs on chemical fertilizers, chicken manure, and irrigation water were used, resulting in surpluses of all nutrients: 85–882 kg N ha^{-1} yr^{-1}, 109–96 kg P ha^{-1} yr^{-1}, 20–306 kg K ha^{-1} yr^{-1}, 0.2–2.7 kg Cu ha^{-1} yr^{-1}, and 0.6–7.7 kg Zn ha^{-1} yr^{-1}. They concluded that irrigation with wastewater contributed to the high inputs and that inputs at these levels constituted a major threat to the soil and water environment.

One major consequence of urbanization that is little reported is the major shift of nutrients from rural to urban areas. Nutrient budgets for different production systems have been conducted at a range of scales in many countries to demonstrate either gains or losses of nutrients from soils (Smaling et al. 1996; Vitousek et al. 2009). With the increasing movement of the nutrients present in crops from rural to urban areas, nutrients in the resulting wastes are likely to become concentrated in towns and cities unless effective methods of recycling can be found. Urbanization has also meant that less food can be consumed directly from the field which has implications for storage, processing, and distribution facilities. Food has to be transported from sites of production; thus the quantity and quality of fresh produce such as fruit and vegetables are often low among the urban poor, leading to poor nutrition and health problems such as rickets and mineral disorders.

Another consequence of urbanization is that different, distinct, and contradictory views can emerge on the value of food, based on the understandings that different communities have about food and the relations between food, society, and the environment. Eakin et al. (2010) explore three different perspectives on food that highlight these different modes of discourse and result in different policies and courses of action. One view perceives food as a global commodity in which a diverse range of products are bought and sold competitively: markets operate to reward efficiencies in production, distribution, and processing, and favor product differentiation. This "exchange value" of food is one in which there is demand without differentiation across a market. Maize is maize, food is a commodity to be traded, and ideas of "terroir" or soils contributing to product quality have little resonance. This perspective, in which the economic value of food is paramount, has contributed to policies relating to failures of entitlement as major drivers of food insecurity as well as to regulations to make globalized food commodity markets work for all. A second view was formally stated in 2005 in the Millennium Ecosystem Assessment (MEA 2005): it perceives food as one of several provisioning, regulating, and supporting services provided by natural ecosystems. This view has rapidly gained credence in scientific discourse, but is often in conflict with the commoditization and globalization perspective, which seeks to secure production by controlling the environmental conditions in which food is produced through technological advances. Resolving these conflicts through the development of systems of food production that can utilize ecological perspectives and achieve the yields required from food provisioning is a substantial challenge for the future. This challenge sits alongside that of reducing poverty in both rural and urban populations, as without this, food insecurity will continue to afflict those without the means to access the food that is available.

Finally, globalization of food as a commodity and increasing pressures brought about by global environmental change have given renewed life to a third view of food as a basic human right. Communities to whom this right is denied are likely to rise up and protest, as happened in many countries during the "food crisis" of 2008; indeed urbanization can facilitate the ease of political protest on the street about food insecurity. Associated with this basic right are also ideas of food sovereignty and the right to choose food of particular types, produced in specific ways. Again, this view of food is at odds with food as a commodity. Eakin et al. (2010) conclude that future food security requires governance systems that can reconcile these contrasting views and achieve, in addition to freedoms of availability, access and utilization as well as the freedom to make food choices that support individual, communal (and environmental) preferences.

Adapting Food Production to Global Environmental Changes

The intensification of crop and animal production systems to meet human demands for food has often been achieved at some cost to other ecological goods and services. For example, excessive nutrient inputs, especially of nitrogen and phosphorus, have resulted in coastal eutrophication and reduced the quality of water in reservoirs used for drinking water (Vitousek et al. 2009). Similarly, cultivating soils for crop production has often increased the frequency of substantial soil erosion by either water, tillage, or wind so that the current mobilization of soil globally is about 5 t yr^{-1} for every person on the planet (Quinton et al. 2010). Degradation of land intensifies competition for land, since it reduces the quantity of land suitable for a range of uses such as food production. ISRIC (1991) produced a world map of human-induced soil degradation based on the knowledge of 250 experts from six continents; this map shows that of the 11.5 billion ha of vegetated land, 15% was degraded. Erosion was the main process of degradation, and about 20% of the agricultural land worldwide was moderately degraded and 6% strongly degraded (Oldeman 1994). A more recent global assessment of land degradation (ISRIC 2008) identifies 24% of the land area as degrading, mainly in Africa (south of the equator), SE Asia and southern China, North and Central Australia, the Pampas, and parts of the boreal forest in Siberia and North America. Although cropland occupies only 12% of land area, almost 20% of the degrading land is cropland, with forests also overrepresented (28% of area but 42% of degrading land). Some 16% of the land area is improving including cropland, rangeland, and forests. Overall, the assessment shows the importance of natural catastrophic phenomena and human management in driving degradation, with the latter also instrumental in speeding up rehabilitation.

The degradation of soils, the pollution of water, and the increasing need to obtain multiple ecosystem services from defined areas has led many to challenge the technologies that have resulted in today's intensified agriculture and to call for the development of sustainable production practices to ensure that the multiple functions of land, and the many ecosystem goods and services provided by land, are conserved and sustained for future generations (Pretty 2008; Foresight 2011; Powlson et al. 2011). Although there is general agreement that agricultural sustainability includes elements of profitable production, environmental stewardship, and social responsibility, there is much less agreement as to how sustainability is to be achieved in practice beyond the need to integrate biological and ecological insights into the production process. Pretty (2008) suggested that sustainability encompasses four key principles:

1. Integrate biological and ecological processes such as nutrient cycling, nitrogen fixation, soil regeneration, allelopathy, competition, predation, and parasitism into food production processes.
2. Minimize the use of those nonrenewable inputs that cause harm to the environment or to the health of farmers and consumers.
3. Make productive use of the knowledge and skills of farmers, thus improving their self-reliance and substituting human capital for costly external inputs.
4. Make productive use of people's collective capacities to work together to solve common agricultural and natural resource problems (e.g., for pest, watershed, irrigation, forest and credit management).

Such principles go well beyond the need for the continued technological innovations, such as new germplasm underpinning increases in yield, and embrace the need to develop important capital assets for agricultural systems, including natural, social, human, physical, and financial capital. A corollary of this analysis is that many disciplines and ways of thinking will be required to develop sustainable systems and that there is unlikely to be a single solution appropriate to all soils and production systems.

Cassman et al. (2003) concluded that to avoid severe degradation of natural resources and to reduce greenhouse gas emissions, crop production systems would need to produce higher yields on existing cropland, limit expansion of the cultivated area, achieve a substantial increase in nitrogen fertilizer efficiency, and improve soil quality through increasing soil organic matter. There is little doubt that for most soils, sustainable production is inextricably linked to the maintenance of soil organic matter contents through appropriate additions to offset the losses caused by cultivation and nutrient depletion (Greenland et al. 1998; Bationo et al. 2012a).

Tilman et al. (2011) approached the problem of determining what might constitute sustainable intensification by first defining the global demand for crop production in 2050 and then examining the environmental impacts of different ways of achieving this. Their analysis, based on measured relations between per capita energy and protein demands with per capita gross domestic product, and allowing for different economic groups, indicates a 100–110% increase in global crop production. Using past nitrogen fertilizer rates as a surrogate for soil fertility enhancement (while recognizing that soil fertility can be enhanced in other ways), various pathways to increasing yields to the required levels were investigated. The options examined were:

1. Current technology in which each economic group retained its 2005 N-dependent yield function.
2. Technological improvement in which technological advances continue along existing temporal trends to 2050.
3. Technology transfer in which low-yielding countries adopt and adapt the existing high-yielding technologies of high-yielding countries.
4. Technology improvement and transfer in which all countries achieve soil- and climate-adjusted yields.

In summary, with present trends of intensification in rich nations and expansion of cropland in poor nations, by 2050 an additional 1 billion ha would be cleared with greenhouse gas emissions of 3 Gt yr^{-1} and N use of 250 Mt yr^{-1}. However, if intensification were concentrated on existing cropland and transfer and adoption of high-yielding technologies were successful, then only 0.2 billion ha would be cleared, greenhouse gas emissions would be reduced to one-third (1 Gt yr^{-1}), and global N use would be 225 Mt yr^{-1}. Although this analysis omits any effects of future climate change, it indicates what might be possible with investment in innovative technologies and infrastructure.

Concluding Remarks

While multiple incremental adaptations to systems of agricultural production are possible to cope with climate and other global changes, transformational adaptation is emerging as a topic in agriculture (Rickards and Howden 2012). Figure 3.1 shows that transformational changes in agricultural systems are at one end of a spectrum that begins with changes in crop husbandry (such as planting time and row spacing) via system-wide changes (such as new crops and new balances between crops and livestock) to wholesale changes in land use. As societies seek to develop the entire food system underpinning their

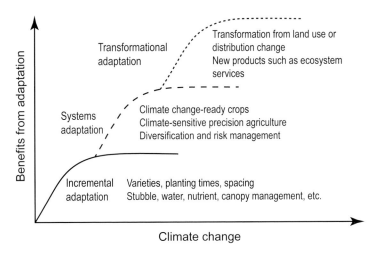

Figure 3.1 Levels of adaptation in relation to benefits from adaptation actions and degree of climate change (reprinted with permission from Rickards and Howden 2012; this figure was adapted from Howden et al. 2010).

food security, it is clear that the response they take will have profound effects on the ways in which land is used (Ericksen 2008).

4

Finite Land Resources and Competition

Helmut Haberl, Cheikh Mbow,
Xiangzheng Deng, Elena G. Irwin, Suzi Kerr,
Tobias Kuemmerle, Ole Mertz,
Patrick Meyfroidt, and B. L. Turner II

Abstract

Rising demand for land-based products (food, feed, fiber, and bioenergy) as well as conservation of forests and carbon sinks create increasing competition for land. Land-use competition has many drivers, takes different forms, and can have many significant implications for ecosystems as well as societal well-being. This chapter discusses several emerging issues, including the effect of increased demand for nonprovisioning ecosystem services (biodiversity conservation and carbon sequestration), urbanization, bioenergy, and teleconnections. Three major types of land-use competition are discerned: production versus production (e.g., food vs. fuel), production versus conservation (e.g., food production vs. conservation), and built-up environment versus production or conservation (e.g., food vs. urban). Sustainability impacts that result from land-use competition are analyzed and found to differ strongly between the different types of land-use competition. They are associated with important trade-offs and high uncertainty. Institutional aspects related to land-use competition are discussed using a conceptual model that distinguishes types of institutions (government, private, community) as well as their functions (objectives, distribution/equity, effectiveness/efficiency). Analysis of long-term trajectories suggests that land-use competition is likely to intensify in the medium- to long-term future, mainly in the face of expected scarcities in resource supply (e.g., in terms of limited resources such as fossil fuels), mitigation and adaptation policies related to climate change, as well as climate change impacts and demographic pressures. The chapter concludes with a discussion of major research gaps, and it outlines priority research topics, including the improved analysis of interdependencies of land and energy systems, "land architecture" (i.e., the significance of spatial configurations), and multiscale models to assess local-global connections and impacts.

Introduction

Competition for land is emerging as a globally pressing issue due to the sheer scale of global demand for land-based products and critical changes in processes of society-nature interaction that affect land use. The potential magnitude of the changes to the land surface of Earth that result from this increased competition is large, and land-use competition can have major implications for ecosystems and societal well-being (Coelho et al. 2012; Smith et al. 2010). In some cases, increased competition is due to new sources of demand (e.g., nascent markets for ecosystem services that have arisen from increased global demand for biodiversity conservation, climate change mitigation, and other services) whereas in others, long-standing forms of land-use competition reach threshold levels due to changes in environmental processes (e.g., climate change) that have intensified the biophysical and human impacts of land competition (Andersen et al. 2009). Indeed, Lambin and Meyfroidt (2011) suggest that society may face a looming global scarcity of productive lands over the coming decades. Land-use competition is a systemic phenomenon characterized by complex feedback processes between human and biophysical components in the land system (Figure 4.1).

Limits to land-based production, beyond those set by net primary production (Vitousek et al. 1986), and the losses in ecosystem services which a land system can withstand, are set by ambient environmental conditions of the land and the technomanagerial system employed on it, both of which vary considerably. For example, until recently the Chaco region of Argentina was not considered ideal for cultivation; however, drought-tolerant soybean strains have made it so (Zak et al. 2008). Perhaps even more difficult is the complexity of decision factors that determine which land-use system is employed. Consistent

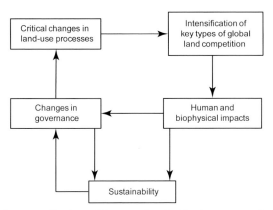

Figure 4.1 Conceptualization of the complex feedback processes involved in land-use competition. Competition for land both results from and affects a host of other factors such as governance, land-use change, sustainability as well as human and biophysical impacts.

with the proposition that intensive agriculture spares other lands from cultivation (Balmford et al. 2005; Phalan et al. 2011), mechanized commercial cultivation based on synthetic fertilizers and pesticides in the developed world has outpaced consumption, thereby allowing substantial contraction of agricultural land without substantially increasing net imports from abroad (Krausmann et al. 2012). In developing economies, by contrast, there is little evidence that agricultural intensification takes marginal lands out of cultivation or reduces the clearing of new lands (Rudel et al. 2009b). Indeed, intensification may result in a rebound effect which increases the total demand for land and production, as it makes farmers more competitive, or in a rise in consumption of products with elastic demand (Lambin and Meyfroidt 2011).

Whether or not land limits are being reached, the potential magnitude of the changes in Earth's land surface that result from the increased land competition is large and carries significant trade-offs for human-environment well-being. Understanding the trajectory of competition-driven changes requires a better understanding of the sociopolitical and economic dimensions of the problem (Coelho et al. 2012; Smith et al. 2010), while understanding the trade-offs and their implications demands improved data and new analyses to address the potential extent, pattern, and magnitude of the various forms of land competition (Turner et al. 2013).

We define land-use competition as a contest between different purposes, or functions, for which a given piece of land, or the resources flowing to and from such land, could be used. This competition involves the outcomes of land cover and services provided by land use (e.g., forest for carbon storage vs. opening land for agriculture) as well as the use of the products produced on the land (e.g., maize for food vs. maize for feed vs. maize for biofuel). This perspective goes beyond typical land-use change studies, as the main analytical entry point is the competition and inherent trade-offs between several (two or more) land uses. Land-use competition may involve the character of the land unit itself, as in the case of agriculture competing with protected areas. Alternatively, the competition can be disconnected from the land itself and occur further along the supply chain, as in the case of crops that can be sold for food, biofuels, or fodder. Some types of land-use competition involve different actors competing for the land, whereas others involve the same actor switching between several possible uses.

In this chapter, we identify critical forms of land-use competition that are currently underway, the ways in which competition takes place, the trade-offs it generates, and its implications for sustainability. We identify specific land-competition trends and link these to the types of land competition that are emerging or expected to emerge globally. We discuss impacts of major types of land-use competition across human and biophysical dimensions, in particular with respect to their implications for sustainability, and consider the role of institutions and examine how intended outcomes compare to the actual impacts and implications of unintended outcomes for potential changes to governance.

Finally, we discuss long-term biophysical and human trends and how they may potentially affect changes in global land-use competition on the medium to long term.

We conclude that land-use competition is a major phenomenon driving land-use change. Although there are few types of land-use competition that are entirely new, the sheer scale of ongoing changes is likely to result in substantial challenges on short-, medium-, and probably long-term timescales. Major categories include land demand for conservation (biodiversity, carbon storage), bioenergy, and livestock feed (grazing and fodder crops). We identify important research gaps that are needed to work toward more sustainable future land-use systems and cope with the intertwined challenges of biodiversity loss, climate change, as well as food and energy supply.

Trends in Land-Use Competition

There are few, if any, major types of land competition underway that are entirely new. Land set aside for conservation, urban impacts on hinterlands, wood-fuel forest land, and long-distant transport of food—versions of the categories of trends we discuss below—have a long history. Those which we single out below, however, have entered a new stage of significance and thus constitute trends in competition that warrant attention.

Ecosystems and Biodiversity Conservation As a Land Use

Historically, land change has primarily taken place to increase the production of goods or resources, provide ecosystem services (e.g., agriculture and forestry), and meet urban and infrastructure needs, including recreation and waste deposition (Dunlap and Catton 2002). Protected areas, in contrast, tend to be established on the remaining marginal lands, usually at a distance from "productive" lands (Joppa and Pfaff 2009). Major initiatives are currently underway to set aside land for nonprovisioning ecosystem services (regulatory, supporting, and cultural; MEA 2005) and biodiversity. This change involves:

- protected area networks (Batisse 1982; Soares-Filho et al. 2010) such as the Meso-American Biological Corridor,
- carbon offsets and payments for ecosystem services (PES) initiatives such as Reducing Emissions from Deforestation and Forest Degradation (REDD+) (Angelsen 2009), or
- land-use restrictions to protect watersheds in various ways (Pires 2005).

Increasingly, private NGO trust initiatives are acquiring land for conservation purposes, thus affecting future development which could otherwise take place on it (Kull et al. 2007; Langholz and Lassoie 2001). Conservation is increasingly being driven by the official recognition of indigenous lands and reserves (Ricketts et al. 2010), as is the "rewilding" of land—a practice that often occurs in communities which have lost agriculture or forestry activities in which the rewilding stimulates outdoor tourism.

Potential or actual agricultural land competes with demands for nonprovisioning ecosystem services and the conservation of natural ecosystems and biodiversity. Although rural land managers have long been aware of the importance of most ecosystem services, demand for nonprovisioning ecosystem services has become significant in spatial amplitude and intensity as well as in penetration into markets and cultural/institutional spheres (see Table 13.1 in Gentry et al., this volume). Land conservation and restoration must now be considered a new form of land use—one that actively competes with others—rather than a passive "victim" of land-use competition or residual category of land.

Urbanization

We are entering a world in which the large majority of humankind lives in urban areas. The most recent projections by the United Nations estimate a global population of 9.3 billion by 2050, with an estimated 6.3 billion living in urban areas, nearly one-third of which will reside in the cities of China and India. Most of this population growth will occur in small- and medium-sized cities, not in mega-cities (Seto et al. 2011). Globally, urban land area is expected to triple in the time period from 2000 to 2030 (Seto et al. 2012b). In the developing world, urbanization involves a significant increase in the expected material standard of living which, in turn, will increase demand of land-based products. In the developed world, urban living can trigger significant economies of scale in efficiency of resource use when it results in increased population density. In the developing world, however, rural to urban migration tends to increase expectation for consumption, resulting in an increased demand for resources. Increased "urbanity," defined by urban livelihoods and lifestyles (see Boone et al., this volume), occurs in peri-urban and rural areas as the result of ecosystem amenities that attract "footloose" (i.e., nonpermanent or transient) households and firms away from traditional urban areas to places with mountains, coastlines, and other high-valued natural amenities (Deller et al. 2001; McGranahan 2008; Carruthers and Vias 2005). Regardless of where it occurs, this massive urban growth creates demands on rural and often distal populations to provide land-based resources and is a strong force behind contemporary telecouplings (see Eakin et al., this volume). An immediate land competition that results from this growth is the capacity of urban land uses to outbid peri-urban and rural land uses, pushing agriculture elsewhere (Seto et al. 2002). Indirect competition follows from the restructuring of the urban hinterlands owing to the demands from the city (Seto et al. 2012b).

Nonfossil Land-Based Energy Provision

Bioenergy use is expected to rise substantially by 2050 and might require approximately as much biomass as is currently harvested for food, feed, and fiber (Chum et al. 2012; GEA 2012; Krausmann et al. 2008). Burning biomass is an ancient means of generating energy, but the sheer scale at which this resource might be used in the future is unprecedented. New, and with profound current or future impacts on land systems, is the use of land to grow energy crops such as sugarcane, maize, switch grass, and oil palm, although many of these crops are still mainly, or at least partly, used for nonfuel purposes. Future bioenergy demand will further contribute to oil palm transforming tree cover in parts of southeastern Asia; sugarcane is commanding large tracts of land in Brazil; and a significant portion of the corn crop in the United States has already been allocated to fuel so that food maize prices have risen substantially on the international market (Chum et al. 2012; McNew and Griffith 2005). Furthermore, lands are also increasingly allocated for water power as well as wind and solar farms, yet with much smaller impacts on the land surface than with bioenergy (Coelho et al. 2012). If global reliance on fossil fuels and nuclear power continues to decline, lands devoted to these types of energy sources can be expected to continue to increase globally.

Teleconnections

Global-scale land changes and teleconnections (i.e., an exogeneous driver acting on a distant system, such as long-distant transport of land-based products) have a long history (Crosby 1986; Turner and Butzer 1992). However, as the separation of land production from resource consumption grows larger (Erb et al. 2009) and the speed by which social processes operate increases (Foley et al. 2011), a case can be made that the scale of current socioeconomic teleconnections is unprecedented. This is exemplified by the Chaco soybean example discussed above and illustrates Harvey's space-time compression concept (Harvey 1990). For example, subsidized maize ethanol production in the U.S. Midwest has decreased soybean production there, which in turn has led to the expansion of soybean crops in the Chaco region to meet the demand of fodder for Asia's boom in meat consumption (Macedo et al. 2012; Morton et al. 2006). In yet another example, forest transitions documented for various regions across the world appear to be related to the transfer of production elsewhere (Meyfroidt et al. 2010; Meyfroidt and Lambin 2011). In addition, informational feedbacks between cities and their hinterlands are creating socioecological systems that are "teleconnected" over increasing distances (Seto et al. 2012b; Eakin et al., this volume). Also new are biophysical teleconnections driven by anthropogenic global environmental change, which affect the suitability of land for different uses (Pielke et al. 2002). As an example, tropical deforestation affects global climate, changing temperature and precipitation for rainfed and irrigation farming everywhere. Farming is expected to be pushed poleward (Olesen and Bindi 2002), while snowpack-fed waters for irrigation in the U.S. Southwest are expected to decline (MacDonald 2010).

Summary

The emergence of these trends in land competition is associated with general trends in the characteristics of land users and the land systems they produce (Figure 4.2). Worldwide the proportion of land used by large, capital-intensive agri-businesses producing for distal markets has risen relative to small, labor-intensive family farms responding to more proximate demands (McCullough et al. 2008). This shift is often accompanied by a transition from heterogeneous, multifunctional landscapes providing not only goods but also nonprovisioning ecosystem services, toward homogeneous, monofunctional landscapes where many nonprovisioning ecosystem services deteriorate (Clough et al. 2011; Perfecto and Vandermeer 2010; Tscharntke et al. 2012). Land-use transitions need not involve moving along all these axes at the same time, and trends may go in opposite directions depending on the places or actors involved. Yet, in areas where land-use competition is rife, the changes shown in Figure 4.2 often bundle together.

These and other changes in land use that result from land competition are largely related to the neoliberalization of the global economy and, increasingly, to the rise in importance of ecosystem and landscape preservation (see Gentry et al., this volume). Economic instruments are increasingly used to regulate land production and environmental governance: witness the spread of carbon markets (e.g., Angelsen 2008), water quantity (Grafton 2011) and quality markets (Shortle and Horan 2008), and PES schemes (Kinzig et al. 2011). These

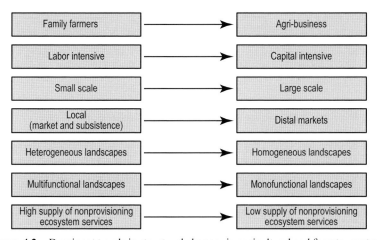

Figure 4.2 Dominant trends in structural changes in agricultural and forestry systems.

Global Land-Use Competition: Major Categories

market mechanisms in tandem with urbanization, in turn, affect local cultures, knowledge systems, and land-use practices.

Land use can be broadly characterized as involving conservation, production, and the built environment:

- *Conservation*: the maintenance of land for nonprovisioning ecosystem services. Important services are biodiversity, carbon sequestration and storage, watershed protection, erosion control, detoxification and purification in soils, pollination, pest control, etc. As detailed above, maintenance of cultural ecosystem services (among others as part of indigenous cultures) or of environmental amenities are also increasingly important forms of this land use.
- *Production*: the use of the land for production or most provisioning services, essentially agriculture and forestry. Important production types are timber extraction, nonfood tree crops, annual and perennial food crops, and nonfood annual crops for feed, biofuel, clothing, water yield, among others.
- *Built environment*: the construction of settlements (cities to villages), impervious surface infrastructure such as roads, waste deposits as well as large dams and surface mining. These changes in land cover are either irreversible or require large investments to be reversed.

Land-use competition involves (a) *production* versus *production*, as in the case of land being used for either food or biofuel production; (b) *production* versus *conservation* (e.g., when forests are protected to preserve carbon stocks and prevent agricultural expansion); and (c) *built environment* versus *conservation* or *production*. Within these categories, we identify forms of land competition that appear to be especially significant in regard to its global reach or significance for large regions, its escalating pace of change, and its large social and environmental significance (Tables 4.1 and 4.3; see also Figure 1.1 in Seto and Reenberg, this volume).

Local Food or Feed versus Food for Distant Markets

The conversion of food production for local consumption or local markets to food production for distant markets has been occurring for centuries, but current escalating competition in much of the Global South is affecting a wide range of production systems and crops. This competition has intensified as a result of international demand for commerical products and opportunities to engage the appropriate markets, often through state- or NGO-led agents or private middlemen. Two specific categories are the change to commercial production of feed and biofuel, which are discussed separately below. In some instances, such as in Brazil, cascading changes in land use are involved (e.g., sugarcane pushes soybeans to other lands which, in turn, pushes pasture into tropical forest lands; Lapola et al. 2010).

Traditional Multifunctional Agriculture versus Agri-business Agriculture

This competition typically involves change from highly diverse land-use systems, often integrated agriculture and forestry with high levels of nonprovisioning ecosystem services, to monoculture-like cropping with little, if any, integration of forestry and low levels of nonprovisioning ecosystem services (Carlson et al. 2013; Ziegler et al. 2009). Agri-businesses and foreign country agencies are important actors; some of these changes may be part of the "land grabbing" phenomenon. Reduced landscape diversity, a breakup of previously closed nutrient cycles, large-scale inputs of fertilizers, pesticides, and mechanization typically follows (Krausmann et al. 2003).

Food versus Feed

Increasing global demand for animal products—especially beef—has gradually caused over the past decades a shift from human food to livestock feed production. This trend is being amplified by the surging demand of animal products as a result of rising household incomes, especially in Asia and other parts of the Global South (FAO 2008). The production of feed is escalating in many parts of the world, incuding sub-Saharan Africa and Latin America (Havlík et al. 2013). The main actors involved are large agri-businesses and companies trading in commodity markets as well as farmers responding to markets.

Food or Feed versus Bioenergy

This land-use competition is driven by the global search for solutions to mitigate climate change, which is driven by expected increases in fossil fuel prices and explicit policies that mandate or encourage biofuel production and consumption. The term "bioenergy" refers to any kind of biogenic source of technical energy and includes solid, liquid, and gaseous energy carriers. The food/feed versus bioenergy competition can take many forms, ranging from competition for feedstocks (maize for ethanol vs. maize for food or feed) to competition for land and water. In the United States and Brazil, this competition is spreading rapidly. However, cultivation of energy crops does not always compete with food production; they may even help to restore degraded lands (e.g., in salinized Australian drylands; Harper et al. 2009). Various actors are involved, ranging from smallholders that produce for new lucrative markets to large-scale agri-businesses. Large transnational energy companies are also

Table 4.1 Major ongoing forms of land competition and their geographical location; arrows indicate the flow from the former land use/system toward the land use/system that is replacing it. Note: "geographic location" indicates areas where this type of land use competition is prominent.

Types of land-use competition	Shifts in land-use/system characteristics	Geographic location	Examples of competition
Production vs. production	Local food or feed ↓ Food for distant markets	Global South	In Senegal, groundnuts or sesame produced for local markets are now shifted toward the global market. In Vietnam, smallholders grow coffee for export instead of rice for local markets.
	Traditional multifunctional farm ↓ Agri-business	Global especially the Global South	In Laos, rubber plantations are taking over swidden mosaics. In Indonesia and Malaysia, oil palm estates replace smallholder swidden mosaics of fields, fallows, and agroforests. In Ghana and Cote d'Ivoire, cocoa and coffee are competing with food crops.
	Food ↓ Feed	Global	Newly cleared croplands in Brazil and Argentina are used for soybean for animal feed rather than food production. In Vietnam, smallholders formerly growing maize for human consumption now grow maize as animal feed for China.
	Food or feed ↓ Bioenergy	Global	Maize in the United States and Europe, sugarcane in Brazil is increasingly used for biofuels. In Europe, sunflower and canola for bioenergy replace other crops.
Production vs. conservation	Food or feed ↓ Conservation	Global South, especially the Tropics	Reconstruction of some wildlife corridors in Latin America, protected areas in Africa, REDD+ in many developing countries are now actively restricting cropland expansion, or reclaiming agricultural areas. Recultivation of abandoned lands in the former Soviet Union goes in the opposite direction.
	Conservation ↓ Timber extraction	Global, especially the Tropics and the boreal	Logging in remaining forest frontiers (e.g., in Papua New Guinea, Laos, Surinam, British Columbia, Siberia, or the Pacific Northwest) are leading to biodiversity loss and forest degradation.
	Conservation ↓ Tree plantations for timber	Global	Forest plantations in China, Vietnam, Brazil, Chile, or Ecuador are being promoted to meet timber demand, and expand partly on natural forests or grasslands.

Table 4.1 *continued*

Types of land-use competition	Shifts in land-use/system characteristics	Geographic location	Examples of competition
Built-up environment vs. production or conservation	Food or feed ↓ Urban areas	Global in reach, but localized impact	Cities, urban development, and increasing "urbanity" globally often occurs in prime agricultural locations (e.g., prime rice lands in the Pearl River Delta, China).
	Food or feed or conservation ↓ Mining	Global in reach, but localized impact	Mining is contributing to deforestation in Cameroon, Sierra Leone, and Ghana. Impacts of oil spills in the Niger Delta is compromising the quality and diversity of mangrove ecosystems.

important actors, as they are diverting a part (albeit small) of their business from fossil energy to bioenergy. Other actors include national governments and international environmental policy makers that favor the transition from fossil energy to bioenergy.

Food or Feed versus Conservation

Increasingly, land is being valued for its nonprovisioning ecosystem services and maintenance of biotic diversity, resulting in efforts to conserve "wild lands" from development. Strong policies for preserving natural habitats are increasing, from biopshere reserves and corridors to individual or community old-growth forest patches maintained as part of REDD+ (Kinzig et al. 2011). The primary actors involved are the international environmental policy makers, conservation-oriented NGOs, private actors, and national governments, the latter not only having conservation objectives, but also seeking financial compensation.

Conservation versus Timber Extraction

High-value timber is becoming increasingly scarce as many natural forests have been either replaced by secondary forests (in much of the Western world) or logged for valuable timber (in much of the Tropics, where peak timber may be in sight for some regions such as Southeast Asia; Shearman et al. 2012). Demand for this timber remains high, leading to pressures on logging conservation lands. The main actors are logging companies and urban-based industries and buyers who use the timber for a wide range of purposes.

Conservation versus Tree Plantations for Timber

Tree plantations are rapidly expanding, mainly in Asia, subtropical and temperate South America, and peri-urban regions of the Global South due to demand for paper pulp and other wood-based products. The main actors are national governments who see the dual objective of promoting business and increasing forest cover with tree plantations and forest-based industries. These actors generally claim that forest plantations preferentially expand onto former agricultural areas, which are more accessible, have better infrastructure and labor force, and more suitable biophysical conditions such as gentle slopes (Sedjo and Botkin 1997). Yet, the evidence for that claim is mixed, confirmed for Europe and North America, where tree plantations constitute an economically viable activity for abandoned agricultural lands. But in tropical regions, the available evidence suggests that tree plantations often compete for space with natural forests and grasslands (Gerber 2011).

Food or Feed versus Urbanization

While relatively small in spatial scale, the replacement of prime agricultural land through urbanization is significant in many places (Seto et al. 2011, 2012b). Moreover, the indirect land effects of urbanization are considerable, as it alters land competition elsewhere to meet market demand. Typically, suburbanization uses up the largest tracts of former agricultural (and often highly productive) land. Urban-commercial land rents are too high for agriculture to compete. This urban deconcentration has led to a blurring between urban versus rural in many parts of North America and Europe; increasingly, places that appear "rural" based on their location and landscape form are nonetheless "urban" in their higher-order economic functioning and composition (Irwin et al. 2009). Key actors include footloose households and firms who can choose their location and policy makers who intentionally or unintentionally seek to concentrate or disperse urban growth.

Food or Feed Production or Conservation versus Surface Mining

Like urbanization, the competition of land for surface mining has a small spatial extent when viewed globally, but it constitutes an important land competition in several regions: gold mining in Western Africa or Western Amazonia, copper mining in Indonesia (Potapov et al. 2012). Surface mining for metals or rare earth minerals typically replaces forests or areas of highly productive agriculture and engender substantial spillover effects on neighboring areas, such as water and soil pollution (Hilson 2002; Schueler et al. 2011). The main actors are mining companies, industries using the minerals, and the growing urban population with its growing wealth, whose demand for new technologies has fueled increased mining activity.

Institutional change at various scales affects these types of land-use competition and change. International institutions adopt new objectives, forms, and means of actions. Since the end of the Cold War, trade liberalization, the increasing recognition of ecological conservation, and perhaps even the new forms of international terrorism have altered the nature of conflict over land use and the institutions involved in them (Campbell et al. 2000; McLaughlin Mitchell and Hensel 2007). In land and environmental governance, economic instruments are increasing in prominence, as seen in the spread of carbon markets (e.g., Angelsen 2008), water quantity (Grafton 2011), quality markets (Shortle and Horan 2008), and PES schemes (Kinzig et al. 2011). Local cultures, knowledge systems, and land-use practices are increasingly influenced by environmental changes, urbanization, and the land-use competition processes described above.

Mapping Land-Use Competition

Mapping the spatial patterns of various types of land-use competition provides substantial opportunities:

1. to further our understanding of the processes involved in land-use competition (e.g., to uncover the actors involved in competing land uses, to identify regions affected by or prone to intense competition, or to detect strong or surprising telecouplings in land systems),
2. to improve the assessment of the impacts of land-use competition and the associated social and environmental trade-offs,
3. to unearth the spatial dynamics involved in competition (e.g., leakage effects, displacement),
4. to enhance our ability to foresee future types of competition and the locations where they could occur, and
5. to target policies that will mitigate or resolve land-use competition.

Mapping land-use competition globally, however, is challenging because some types of land-use competition take place on the plot level (e.g., a palm oil plantation replacing a natural forest), whereas others only become apparent further down the supply chain (e.g., maize to be used as feed, food, or feedstock for biofuel). Moreover, at the plot-level not all land-use changes are necessarily a result of land-use competition. Conversely, land-use competition can be prevalent even in regions where land use is stable. Table 4.2 outlines the opportunities and challenges for mapping the nine key land-use competitions.

Generally, mapping land-use competition requires two types of data at relatively fine spatial scales (grid level or small administrative units). First, information on land use is needed, both regarding the extent of land use (e.g., extent of cropping or forestry) and its intensity (e.g., labor, fertilizer). In this context it is important to note that information on all land uses is required,

Table 4.2 Options and data needs for mapping the different types of land-use competition. Although the directionality of the competition may theoretically go both ways, the land-use competions listed below are suggested to play out primarily by the land use in *italics* replacing the other land use.

Land-use competition	Datasets needed	Scale
Local food or feed vs. *food for distant markets*	Crop type maps Population density Market data (or market accessibility maps as a proxy)	Grid level or administrative units, global coverage
Traditional multifunctional vs. *agri-business*	Land systems map (e.g., Ellis 2011; van Asselen and Verburg 2012) Market data or market accessibility Population density	Grid level
Food vs. *feed*	Information on the actual use of crops that can be used for feed or food	Administrative units, global coverage (food-producing regions)
Food or feed vs. *bioenergy*	Information on the actual use of crops that can be used for bioenergy or other uses (feed/food)	Administrative units, global coverage (food-producing regions)
Food or feed vs. *conservation*	Cropland extent and potential yields Extent of protected lands (state-owned and private reserves, areas where REDD projects are implemented, other set-aside land)	Grid level, global coverage
Conservation vs. *timber extraction*	Logging in natural forests (either actual harvests or a map of logging concessions) Extent of protected lands	Grid level
Conservation vs. *tree plantations (for timber)*	Extent of tree plantations Extent of protected lands	Grid level
Food or feed vs. *urban*	Urban extent Cropland adjacent to cities Map of cropland to urban conversions	Grid level
Food or feed or conservation vs. *mining*	Map of mining concessions and pre-mining/current land use	Grid level

including "nonproductive" land use such as protected or set-aside areas (such as REDD+ areas), indigenous territories, or private game reserves. Information on land tenure is also needed to provide information about how the land is used. Second, information on the characteristics of the land system is required, both in terms of land resources (e.g., soil fertility, water availability), which may spur the competition, and in terms of the socioeconomic attributes of the local area and broader region (e.g., tenure, market access) that influence the nature and intensity of the competition. In addition, if impacts or trade-offs of land-use competition are to be assessed, spatial information on the relevant environmental and human outcomes (e.g., biodiversity, carbon density, income inequality) is needed. Table 4.2 summarizes the data that would be needed for mapping each type of land competition.

Implications of Land-Use Competition for Sustainability

Identifying the implications of these various types of land-use competition for sustainability requires an articulation of the environmental and social trade-offs that arise from the impacts of intensifying land-use competition. These impacts can be both direct (i.e., the result of the land-use competition at that location) or indirect (or cumulative, e.g., multimarket effects, spatial spillovers or cumulative effects across space and time). In addition, gains and losses of services may accrue to private actors: as additional revenues or costs to companies, farmers or individual land owners, or to public interests. The latter impacts accrue to society or a community as a whole rather than only to individual agents and include, for example, environmental damages, improved or degraded ecosystem services, and changing cultural landscapes. A full accounting of trade-offs considers all types of impacts: environmental and socioeconomic, direct and indirect, private and public. Table 4.3 reports observed and hypothesized impacts of the major types of land-use competition outlined in the previous discussion.

In many cases, land-use competition is brought on by private market forces and benefits primarily private interests while the costs are mostly borne by the public. For example, competition for agricultural land by urban use generates economic gains for individual land owners as they sell rural land for development, and for peri-urban farmers who improve their access to urban consumers and markets. However, urbanization also generates multiple environmental costs, including increased impervious surfaces and urban run-off, and social effects in terms of rural cohesion and livelihoods. Land sparing is a potential benefit: in China there is evidence that migration of rural populations to urban areas reduces rural residential land use and makes land available for agriculture (Huang et al. 2007b). In other cases, land is not preserved because urbanization is driven by suburban decentralization rather than rural to urban migration.

Table 4.3 Impacts of various types of land competition on sustainability (ENV: environmental; SE: socioeconomic). Although the directionality of the competition may theoretically go both ways, the land-use competitions listed below are suggested to play out primarily when the land use in *italics* replaces the other land use.

Types of Competition		Types of Impacts	
		Direct	Indirect (or Cumulative)
Local food or feed vs. *food for distant markets*	ENV	Landscape heterogeneity; loss of nonprovisioning ecosystem services; environmental effects of industrialized farming	Changes in spatial structure of land use (land architecture)
	SE	Loss of livelihoods of smallholders; higher yields; higher volumes of products	Rebound effects of increased efficiency and availability of resources
Traditional multifunctional vs. *agri-business*	ENV	Landscape heterogeneity; ecosystem services	Pesticides and nutrient effluents; greenhouse gas emissions related to fossil fuels; potentially reduced land demand
	SE	Increased volume of products; loss of livelihoods of smallholders; increased inequity	Economic growth; opportunities to raise taxes
Food vs. *feed*	ENV	Environmental impacts from intensification of agricultural run-off including nutrient run-off, pesticides, soil degradation, etc.	Expansion of farming to new marginal land
	SE	Increased land productivity	Increased land prices; changes in grain prices
Food or feed vs. *bioenergy*	ENV	Reduction in greenhouse gas emissions from fossil fuels; environmental impacts from intensification of agricultural production including nutrient run-off, pesticides, soil degradation, etc.	Possible acceleration of deforestation; may fail to reach stated goals in terms of greenhouse gas emissions; spillovers push other land uses outward
	SE	Increased food-security risks; loss of livelihoods of smallholders; higher food prices; increased land productivity	May induce innovation and technology development
Food or feed vs. *conservation*	ENV	Preservation of valuable ecosystems; ecological recovery of degraded land	Intensification of nonconserved pasture and cropland; displacement of agricultural to marginal lands; change in hydrology
	SE	Increased access to ecosystem services; change in cultural landscapes; new income sources for local residents	Higher agricultural land rents

Table 4.3 *continued*

Types of Competition		Types of Impacts	
		Direct	Indirect (or Cumulative)
Conservation vs. *timber extraction*	ENV	Decline in natural habitats; decline in nonprovisioning ecosystem services; ecological succession; soil acidification; decreased landscape diversity	Potentially large-scale cumulative effects on regional hydrology; carbon emissions
	SE	Increased risks for food security; new income source for local workers; more nonfood products from forestry	Economic growth; opportunity to raise taxes
Conservation vs. *tree plantations (for timber)*	ENV	Decline in natural habitats; decline in nonprovisioning services; ecological succession; soil acidification; decreased landscape diversity	Potentially large-scale cumulative effects on regional hydrology; carbon emissions
	SE	Increased labor demand and thus rural wages; deterioration of livelihoods for forest-dependent people	Economic growth; opportunities to raise taxes; potential sparing of natural forests
Food or feed vs. *urban*	ENV	Increased impervious surfaces and urban run-off; reduced agricultural run-off; potential land sparing if rural residents abandon land when moving to city; less reversibility in land use; loss of ecosystem services	Increased temperatures in microclimate
	SE	Loss of rural lifestyle and culture; increased incomes for rural landowners selling land; greater opportunities for (peri)urban agriculture	Increased concerns about food security
Food or feed or conservation vs. *mining*	ENV	Decline in natural habitats; decline in nonprovisioning ecosystem services	
	SE	Increased labor demand and wages; possible effects on local food security	

Other types of land-use competition generate a more complex set of trade-offs across human and biophysical systems and private and public interests. In the case of agricultural land competition for bioenergy production, one motivation is to reduce greenhouse gas emissions, which is in the public interest. However, the resulting land-use competition may also result in environmental costs if food crop production moves somewhere else, perhaps driving deforestation ("indirect land-use change"), thereby undermining the stated goals, and sometimes creating detrimental effects for other public interests (e.g., biodiversity). Expanded bioenergy production may also imply economic trade-offs, by inducing increases in agricultural productivity, which raises the incomes of farmers but reduces food or feed crop availability and may ultimately drive up food prices. The latter is a pecuniary externality[1] caused by the reduction in food supply, which generates a public cost—one that is especially large for poor households that have a low elasticity of demand for food.

These examples illustrate the potential complexity of the trade-offs related to land-use competition. Any assessment in terms of sustainability thus requires an understanding of the dynamic socioeconomic and environmental processes and their interactions that generate both the direct and indirect impacts listed in Table 4.3. Doing so requires not only data and mapping, but also the development of models to permit a better understanding of these system dynamics and assessment of these impacts and their trade-offs (Figure 4.1; see also the discussion on institutions by Eakin et al., this volume). However, our understanding of the environmental and socioeconomic processes and their interactions is limited both by a lack of data and scientific understanding as well as the inherent uncertainty of dynamic systems. Thus accounting for uncertainty in assessing trade-offs and sustainability is critically important.

Managing Land-Use Competition

Governance, Land Management, and Institutions

Institutions evolve from specific contexts to address particular challenges and are partly determined by culture and history. Institutions can have a large influence on the way land-use competition occurs as well as on its outcomes. In this section we propose a framework to identify mismatches between institutions and the land-use competition they mediate. By "institutions" we mean a wide range of organizations and formal or informal rules that influence decisions.[2]

A large scholarly literature has emerged on how institutions affect land management, for example related to the governance of common lands (Ostrom 1990), urban land and sustainable cities (Bai et al. 2010; Evans et al. 2004),

---

[1] Pecuniary externalities operate within the market as opposed to technological externalities (e.g., environmental degradation) which generate effects external to the market. Pecuniary externalities arise when markets are related (e.g., through input and output linkages) and a change in one market generates price effects in a related market. Because these price effects are external to individuals, they are considered externalities and can be counted as public costs (e.g., for consumers in the case of a price increase) or benefits (e.g., for producers in the case of a price increase).

[2] Turner (1997:6) defines an institution as "a complex of positions, roles, norms, and values lodged in particular types of social structures and organizing relatively stable patterns of human activity with respect to fundamental problems in producing life-sustaining resources, in reproducing individuals, and in sustaining viable societal structures within a given environment."

agricultural land and production systems (Binswanger et al. 1995; Deininger and Feder 2001; Palmer et al. 2009; Stavins et al. 1998), management of conservation and protected areas (Joppa and Pfaff 2011), PES (Jack et al. 2008; Wunder et al. 2008; Robalino et al. 2008), or REDD+ (Angelsen 2010; Kerr 2013; Lubowski and Rose 2013; Matthews and Dyer 2011). Since Hardin's (1968) claim—land used for production that is neither private nor under government control inevitably degrades due to the "tragedy of the commons"— it has been repeatedly demonstrated that resources held by local commons (i.e., that is communally owned and used) are often managed efficiently and sustainably (Ostrom 1990, 1999). Moreover, it has been shown that private owners can degrade their land (Kirby and Blyth 1987) and that the same can happen with command economy institutions, as in the case of the Aral Sea (Micklin 1988).

There are examples of both successful (e.g., the U.S. Conservation Reserve Program, which initially had problems but gradually evolved; Roberts and Lubowski 2007) and unsuccessful institutions (e.g., the initial form of the PES system in Costa Rica; Sánchez-Azofeifa et al. 2007) created to mediate specific cases of land-use competition. Many cases exist in which institutions fail to resolve competition in ways that are successful, and no single institutional form (or set of institutions) has emerged to solve all problems. Within environmental economics, there is considerable literature on "instrument choice" within developed countries (Stavins et al. 1998) that discusses which instrument best addresses environmental issues under which circumstances.

Types of Institutions and Their Functions: Conceptual Considerations

Here we focus on what the types of land-use competition identified above require in terms of the structure and functioning of institutions to regulate competition among land uses (see also Gentry et al., this volume). Potentially new directions of research are discussed using two examples: competition between (a) food and feed and (b) food and urban areas.

Institutions can have a combination of organizational "functions." Institutions may be able to set society's agenda and define objectives, which could be a combination of environmental, economic, and social outcomes or impacts. Institutions can act to protect interests, share resources, and aim at a more equitable distribution of resources. Laws (and the institutions that enforce them) mostly protect existing defined property rights. In cases where property rights or sharing rules are poorly defined, "community" institutions can play a role in finding a consensus. Alternatively, private actors could claim these poorly defined resources and cause costs that were not agreed to by society more widely. Institutions will affect how efficiently or effectively objectives are achieved once they are set.

Based on these distinctions, we offer an analytical framework (Table 4.4), in the form of a matrix, to help explore how institutions do and could affect

Table 4.4 Conceptual framework for analyzing institutions and their functions in mediating land-use competition, with an example for water quality trading in Lake Taupo, New Zealand. For more information on the system, see Duhon et al. (2012).

Types of Institutions	Functions		
	Define objectives	Distribution/equity	Effectiveness/efficiency
Government:			
Scientists	Provided evidence on historic leaching and lake quality and projections without control		Provided consistent monitoring tool that enabled use of economic instrument
Regional council	Jointly determined goal of no deterioration in quality	Negotiated free allocation to farmers and Maori and paid for some reductions	Continue to develop mitigation options
			Designed and implemented first non-point-source water quality trading system
Central government		Provided share of funds to buy back leaching allowances	Provided overall legal structure; Resource Management Act
Private:			
Farmers	(Ultimately) accepted the need for control		Changed land use and mitigated leaching on farms; engaged in trade
Agricultural consultants		Benchmarked farms leaching consistently	Assisted in compliance and disseminated mitigation options
Community:			
Landowners' groups[1]		Engaged in debates on different structures for free allocation	Advocated for flexible instruments
Iwi (local Maori tribe)	Advocated for protection of lake	Negotiated for protection of their interests—current low emitters	
Lake Taupo Protection Trust		Ensured that all traders are well informed	Identified and negotiated buy-back of allowances

[1] Mike Barton (2005), a local beef farmer, played a key role in coordinating the local voice.

situations where there is "complex" land-use competition. In particular, we are looking for new or exacerbated situations of land-use competition where existing institutions are leading to outcomes that are considered unfavorable. "Complex" could involve multiple jurisdictions (local to international), multiple services (production, biodiversity), and/or multiple actors in unusual connections (e.g., local landowners and households in Africa being affected by dietary choices in Europe in a contest between growing food to eat directly and feed for animals). By "institutions" we mean a wide range of organizations and rules (formal or informal) that aim to influence outcomes. We have grouped institutions as follows:

- those who have some formal power to coerce and some responsibility to represent a wider group,
- those primarily motivated by the interests only of their group, and
- those with an interest beyond their own but without strong coercive power.

We recognize that these are not clear distinctions and that some organizations will be hard to classify and may be differently classified in different applications. This framework is not intended to provide a taxonomy. Some examples are:

- "Government": national, local, and regional elected government bodies, government departments, laws, regulations, international agreements, and corrupt politicians.
- "Private sector": Companies, multinationals, smallholders, and private market institutions (e.g, banks, insurance companies).
- "Community": NGOs, media, universities, cultural norms, church, and family.

One way to use the matrix is to take one of the important land-use competition types identified above and its sustainability implications, identify the institutions currently involved and how their objectives have affected the overall objectives sought in this competition (that become "drivers" and desired outcomes), and the distributional and efficiency outcomes. This may help identify the source of any undesired outcomes, either through dominant power in specific institutions or weakness or absence of another institution.

To illustrate, Table 4.4 shows the set of institutions involved in the creation of the first non-point-source cap and trade water quality market in New Zealand. Together the institutions successfully defined the objectives and reached agreement on distributional issues while creating a potentially efficient mechanism. Looking forward we can compare critical institutional needs with the available institutions to diagnose likely challenges. One problem that may arise is that the compliance mechanisms available to the regional council are uncertain and weak (each noncomplier must be taken to court). This is likely to create problems when the cap becomes strongly binding. Solving this requires changing the Resource Management Act, which can be done only by the central government.

Example 1: Governance Challenges Related to Food versus Feed Competition

At present, there is a global trend toward greater consumption of animal products per capita, and this has substantial, growing impacts on greenhouse gas emissions, land use, and other environmental effects (Pelletier and Tyedmers 2010; Steinfeld et al. 2006; Wirsenius et al. 2010). As identified above, if diet change continues to involve increased animal products, more land will be needed for animal production for grazing or feed crops (Haberl et al. 2011a; Erb et al. 2012).

When seen from a large (regional or global) level, this is a significant environmental issue due to the increasing land demand that it creates as well as the thermodynamic inefficiency of converting feed to animal products. It will raise the price of agricultural land and increase the price of other agricultural products, especially food. Land owners are likely to benefit, but those who need to buy food, especially the very poor who eat little or no meat, are likely to lose. Indirectly, at a global scale, pressure to use land for agriculture and to intensify production will increase. This will tend to decrease conservation land, thus putting more pressure on natural ecosystems. If poorly managed, it will also reduce ecosystem services that are essential for ongoing production of food and feed, and thus will be unsustainable even in the narrow sense. Locally, in some cases, if local markets are not integrated globally or with other regions, a shift from food to feed production might affect local food prices. Mostly, however, the effects on food and land prices and environmental outcomes are likely to be indirect and global rather than direct and local. When, for example, one crop switches from one to the other use, or the same field is planted with another crop, the impacts at a local level are likely to be limited. For actors that have environmental interests on the local scale, the use of the crop is probably not a priority, and the effects on local food prices are small relative to the economic benefit from responding to the demand for feed. The cumulative global effects, however, can be large and potentially important (Steinfeld et al. 2006). Thus, the governance challenge is to manage this global-scale competition in such a way that it is acceptable and can be implemented at the local scale.

The increased demand for feed is driven by consumption patterns and affected by institutions within markets that coordinate the production and distribution of food. Consumption drivers are largely at an international scale and, other than population and general rises in material well-being, might include cultural influences such as the media. While there are some examples of institutions acting to change or enforce some diets (e.g., WHO diet recommendations), few institutions deliberately address the environmental implications of increased meat demand. Local responses are more likely to be driven by social concerns (e.g., the landless movement protesting growing food insecurity), while in distant consumer places environmental issues might be more important. Important knowledge gaps exist regarding how changes in information, media, or incentives (such as a tax on the consumption of animal products) would affect food production and consumption decisions at local and global scales.

Example 2: Governance Challenges Related to the Competition between Food/Feed Production versus Urban Land Expansion

As described above, urban encroachment on cropland pushes food production for city dwellers further away. Given the historical location of cities, encroachment onto "prime agricultural land" often occurs and can be very costly to reverse. The direct effect of this land transition is a different set of environmental impacts that are generated by the city relative to the previous impacts from food production. On a larger scale, urbanization implies that people move away from rural areas, which can reduce environmental pressure in these areas. The net impacts are, however, unclear: food and environmental impacts may simply have moved or, if urban dwellers have a lower "land footprint," then overall pressure can be reduced. The main actors are private land owners, land developers, local authorities who define zoning constraints and make decisions on infrastructure, and farmland conservation NGOs. Many of the issues related to these institutions have been well studied (Cheshire and Sheppard 2002; Fischel 1985; Glaeser and Ward 2009), especially in the developed world. New institutional responses to urban land demand include, for example, the "food mile" critique, which originally applied a simplistic approach to assess the environmental costs associated with long-distance transportation. Later assessments pointed out, however, that reducing long-distance transportation may, in some cases, even increase emissions if it results in larger production emissions, as the latter are usually much larger than the emissions related to transport (Weber and Matthews 2008; Blanke and Burdick 2005). This may be the case, for example, if local production results in higher greenhouse gas emissions due to unfavorable soil or climate conditions. This demonstrates why policies need to be based on accurate and comprehensive indicators so as to avoid creating unintended detrimental effects.

Incremental versus Transformative Changes in Land-Use Governance

The above examples show that (a) what looks like a competition between food and feed at local scale may also involve competition between production and conservation at the global level and (b) by solely focusing on some social or environmental impact, new institutions to manage competition might produce unintended counterproductive consequences (e.g., "food miles"). Thus policies which seek to manage the impacts of various types of land competition must account for the variety of social and environmental trade-offs that exist across multiple scales. This complexity requires new institutions to manage these trade-offs as well as to encourage synergies among land uses whenever possible. One question that arises is the extent to which we need some form of global governance to manage land competition. Obviously, many aspects of the governance of land use and land-use competition will remain at the local or national level. For example, many issues of food security involve not only overall production or efficiency but distribution, access, and other issues that concern national governments. The question is: Which aspects, issues, and decisions might be addressed more appropriately on the global level, and how might global land governance emerge? Who could benefit from some form of global land governance?

Some forms of global land governance might emerge from biophysical and environmental concerns (e.g., in terms of land set aside for habitats) whereas other forms might emerge from social issues (e.g., related to global food systems) (see Margulis as well as Gentry et al., this volume). Other innovative ways to address land-use competitions might be explored, such as forms of land allocation or land resource quotas per individual or household. However, this raises serious issues, including question of scale, social acceptability of such approaches, and the technical difficulties involved in accounting for the resources necessary for each product. Other approaches might directly constrain production or provide incentives to producers by assessing, for example, a sustainability user cost for resource extraction.

Land Competition in the Context of Long-Term Transitional Dynamics

Global population growth, shifting consumption patterns and dietary habits, and growing economic activity (GDP) are major drivers of long-term expansion of urban areas, croplands, pastures and secondary vegetation at the expense of primary (natural) vegetation (Figure 4.3a; Hurtt et al. 2011; see also van Vuuren et al. 2012). The underlying causes for the changes in the volume and pattern of resource demand are ongoing transitions from agrarian to industrial societies (Haberl et al. 2011b). While this transition may seem to be more or less completed in industrialized countries, it is taking off rapidly in developing regions (e.g., in China and India; Fischer-Kowalski and Haberl 2007), where it currently affects more than two-thirds of the world population. The agrarian-industrial transition results in surging global resource use (Figure 4.3b; Krausmann et al. 2009) and greenhouse gas emissions (Canadell et al. 2007), hence triggering growing concerns about resource scarcity (GEA 2012) and changing global biogeochemical cycles and the global climate (IPCC 2007).

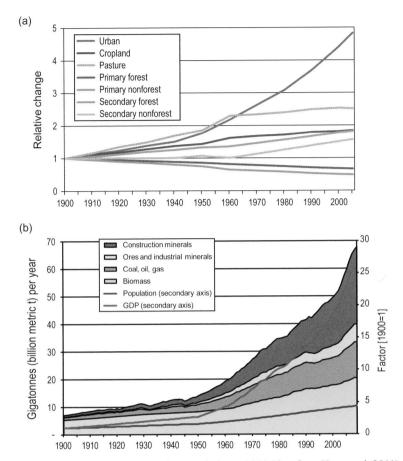

Figure 4.3 (a) Global land-cover change relative to 1900 (data from Hurtt et al. 2011). (b) Global use of materials compared to increases in the gross domestic product (GDP) and population (data from Krausmann et al. 2009, updated with data from http://www.uni-klu.ac.at/socec/inhalt/3133.htm). Left axis: global resource use (10^9 metric tons per year). Right axis: growth in population and GDP during the same interval (1900 = 1); GDP is measured in constant 1990 Geary-Khamis dollars.

These global sustainability challenges have not only motivated a new growth critique and the search for new welfare models (Jackson 2009)—issues outside the scope of this chapter—they are also bound to affect land systems in many ways. Furthermore, major global assessment reports (MEA 2005; GEA 2012; IPCC 2007; IAASTD 2009) suggest that the ongoing trajectory cannot be followed long term without causing massive biodiversity loss, soil degradation, resource shortage, and climate change. Thus, humanity faces a key challenge in how it will feed healthily and sustainably a future global population of 9 billion (Smith et al. 2010).

Competition for land is not in itself a driver that affects food and farming in the future, but is an emergent property of other drivers and pressures. Thus land-use competition arises from the above described transition from agrarian to industrialized societies. Some of the new types of land-use competition outlined above are a result of the sheer scale of these processes (e.g., those related to urbanization, the demand of land from mining, or increases in food production). Others are a result of policies aimed at addressing some of these challenges (e.g., the emergence of conservation and a major type of land use with the power to compete with other types of land use, such as food production or the competition between food/feed and bioenergy).

Over the last century, the rate of expansion of farmland (cropland and grazing land) was substantially lower than that of population, GDP, or food provision. This "decoupling" was achieved through a combination of increases in yields, on the one hand, and increases in conversion efficiencies from primary biomass harvest to final product supply, most notably those of feeding efficiencies (i.e., the amount of feed required per unit of animal product) on the other (Erb et al. 2012; Haberl et al. 2011a; Krausmann et al. 2009). Taken together, these increases in yield growth and biomass conversion efficiency were large enough to allow for increased per capita food supply, in terms of both calories and animal protein, as well as a reduction in malnutrition and hunger, at least until the food price surges in 2007 (Godfray et al. 2010). These efficiency increases have contributed to reduce deforestation and greenhouse gas emissions (Burney et al. 2010), although the actual effect is difficult to determine since population and per capita food consumption respond to increasing supply (Lambin and Meyfroidt 2011; see also Lambin and Meyfroidt, this volume). Increases in yield and conversion efficiencies are generally expected to continue over the next decades. Most global projections of the current generation of global assessment models project that (a) global demand for agricultural products will rise by 70–100% until 2050 and (b) most of that increase will come from increases in yields and conversion efficiency; cropland areas, by contrast, are expected to grow only moderately, perhaps by some 6–19% until 2050 (Coelho et al. 2012; FAO 2006; MEA 2005; Tilman et al. 2011). Future projections thus indicate only a moderate growth in farmland. This picture may well change, however, if the expected yield growth does not materialize, if bioenergy were to play a major role in the future, or if expected future trajectories of demand were to shift, as discussed below.

Limits to Continued Agricultural Intensification

While many scenarios and official forecasts more or less project past yield gains into the future, as shown in the examples cited above, several developments could counteract a continuation of past yield growth. First, yields could approach physiological limits in some regions (Cassman 1999; Peng et al. 2000; Tilman et al. 2002). Likewise, the most suitable agricultural lands in some regions are already in use, and thus further agricultural expansion would occur on land less suited; soil erosion and depletion of nutrient stocks could also hamper yield growth. Together, this suggests that yield increases could slow down in the future. While improved management might help to sustain yield growth (IAASTD 2009; Coelho et al. 2012), substantial investments would be necessary to maintain yield increases (Kahn et al. 2009). In addition, past improvements in land-use efficiency depended on the availability of abundant energy to run agricultural machinery as well as for the production of fertilizers and pesticides (Pimentel et al. 1990; Krausmann et al. 2003). Energy scarcities could reduce the future growth of yields. Finally, organic agriculture is arguably associated with important environmental benefits compared to conventional, industrialized farming (e.g., lower impacts on biodiversity, lower nutrient runoff or leaching, increased soil quality; IAASTD 2009). However, the yields of organic agriculture are substantially below those of intensive, industrialized farming, especially when the whole crop rotation cycle is taken into account (Seufert et al. 2012). Thus, an expansion of organic agriculture would result in a larger demand for farmland, unless this is accompanied by a simultaneous change in diet toward a more vegetarian diet (Erb et al. 2012).

Changes in Demand for Land-Based Services and Products, in Particular Bioenergy

So far, income growth has been almost universally associated with dietary changes toward a higher share of animal-based products (Haberl et al. 2011a, 2012). Analyses of regional time series of dietary change suggest that the consumption of animal products tends to stagnate in affluent regions such as in the United States and Europe, whereas it continues to grow in developing regions (Coelho et al. 2012). Changes in diets toward fewer animal products could help to reduce emissions significantly (Tukker et al. 2011; Stehfest et al. 2009; Wirsenius et al. 2010), as well as farmland demand, and thus lead to a relaxation of land-use competition (Erb et al. 2012; Popp et al. 2010; Smith et al. 2013; Stehfest et al. 2009).

Rising fossil energy prices, growing concerns about resource scarcity and nuclear risks, and increasing attemps to reduce CO_2 emissions are motivating policies to develop renewable energy sources, including hydro and wind power, geothermal energy, and different types of bioenergy (solid, liquid, or gaseous). While all renewable energy technologies require land (and may hence result in land-use competition), area demand per unit of energy is by far largest in the case of bioenergy (Coelho et al. 2012). Some studies even suggest that up to 36% of all land (except Greenland and Antarctica) could be used for bioenergy production (Smeets et al. 2007)—an area similar to that currently farmed for food and fiber. Less ambitious bioenergy supply targets, including those put forward by major global assessments, would still require the cultivation of energy crops on some 2–10% of all land (Beringer et al. 2011; Haberl et al. 2010; van Vuuren et al. 2009; Chum et al. 2012; GEA 2012).

Embarking on such policies would change the global land system fundamentally because global energy demand is practically unlimited compared to the capacity of terrestrial ecosystems to supply biomass (i.e., their net primary production, NPP[3]). Once biomass becomes competitive on the energy market, the very structure of agricultural commodity markets will likely change substantially (FAO 2009a). Moreover, surging biofuel production would entail major land-use and land-cover changes, mainly in the developing world (Danielsen et al. 2009; Warren 2011) and at the expense of forests and pastures (Mbow 2010). This would significantly reduce the mitigation effect of biofuel; forest loss could even alter the greenhouse gas balance (replacing fossil fuels with bioenergy) from savings to an increase (Lapola et al. 2010; Searchinger 2010).

Land required for sequestration and storage of carbon represents another type of demand. The challenge is to know how much land-based greenhouse gas mitigation can be achieved without compromising food security and environmental goals (Smith et al. 2013). These issues point to the need of scrutinizing the social and ecological impact of pursuing global climate change goals without nesting it to local land needs and development priorities. Land-use policies based on environmental zoning approaches can be beneficial in this context (Coelho et al. 2012; Macedo et al. 2012). At present it is unknown whether, where, and under which circumstances use of land for bioenergy or carbon sequestration provides greater benefits in terms of greenhouse gas reduction (Smith et al. 2013).

Climate Change Effects and Adaption to Climate Change

Land use is strongly influenced by climate change (Giannini et al. 2008; van Vuuren et al. 2011). Changes in rainfall and temperature modify growing conditions and primary productivity (Pettorelli et al. 2005; Fensholt et al. 2009; Hiernaux et al. 2009; Zhao and Running 2010) and may affect yields negatively or positively. Unfortunately, there are very large knowledge gaps and uncertainties with respect to many important potential impacts, in particular the magnitude of a possible CO_2 fertilization effect as well as its interaction with land management, nutrient availability, and yields (Haberl et al. 2011a). Moreover, the need to adapt to climate change will require modifications in land-management practices toward improved resilience to climate fluctuations and other climate changes. This may include the diversification of cultivars, changes in pasture management, adoption of new production systems, and changes in sowing dates (Waha et al. 2012). In developing countries, there is

---

[3] Aboveground terrestrial NPP is approximately double humanity's total energy needs, including food and feed (Haberl et al. 2011a).

growing agreement that agroforestry (i.e., the integration of trees with crops and livestock systems) offers a good strategy for improving soil properties and increasing yields, while at the same time improving nonprovisioning ecosystem services and bioenergy production and limiting the human impact on remaining forests (Smith and Wollenberg 2012). Agroforestry may also be one particularly rewarding option to help sequester carbon and reduce greenhouse gas emissions that result from land-use change (Zomer et al. 2009). Adapatation of land use to current and future climate change is an emerging research area that goes beyond carbon-related issues.

A Long-Term Perspective on Land Demand for Conservation

Several considerations suggest that land demand for conservation may increase also in the medium to long term. Climate change mitigation has motivated initiatives for substantially increasing global forest cover (e.g., REDD+, the clean development mechanism, and national afforestation programs) (Mertz 2009; Alig et al. 2010). Indeed, ambitious climate change mitigation scenarios mostly require "negative" carbon emissions in the second half of the twenty-first century (van Vuuren et al. 2011), and increasing carbon stocks in forests is an option to achieve negative emissions as is the combination of bioenergy with carbon capture and storage (Popp et al. 2011).

Another reason why land demand for conservation may be assumed to rise over the long term is related to ecosystems and biodiversity. Long-term environmental change includes the loss of biodiversity, which in turn affects ecosystems services and functions (Toit et al. 2004; Mbow et al. 2010; Gonzalez et al. 2012). Although the interactions between biodiversity and ecosystem services are widely, and controversially, debated, there are concerns that their possible links with critical "tipping points"[4] could lead to large, rapid, and potentially irreversible changes in land cover or land productivity. The Global Biodiversity Outlook (GBO3) has adopted the concept of tipping points to better understand trends in biodiversity. Climate change is thought to be a strong driver of biodiversity loss, which in turn would likely have irreversible consequences for ecosystem functions and services (Mooney et al. 2009; Pimm 2009). It has therefore been argued that long-term strategies need to be implemented to conserve biodiversity and ecosystems and avoid long-term detrimental effects of biodiversity loss (MEA 2005).

Both climate change and biodiversity conservation are likely to remain on the agenda for the near foreseeable future, thereby creating increasing funds, economic opportunities, and institutions to devote land for conservation activities worldwide, but particularly in developing countries (Rosendal and Andresen 2011). These concerns may well be exacerbated by a growing

---

[4] "Tipping point" commonly refers to a critical threshold at which a tiny perturbation can qualitatively alter the state or development of a system (Lenton et al. 2008).

recognition of water scarcity and limits to water resources, for example due to the water demand of bioenergy crops or the overexploitation of groundwater resources in many large aquifers, especially in Asia and North America (Gleeson et al. 2012). Competition for water and land are linked processes; they may reinforce one another and contribute to further growth of ecological injustice between rich and poor regions (Coelho et al. 2012).

Knowledge Gaps

The novel challenges of increased land-use competition in a rapidly urbanizing and more teleconnected world poses a large array of important research questions and knowledge gaps. In this section we highlight some that emerge from the above discussion.

Future Interaction of Land and Energy Systems

The many types of competition between land uses and future trajectories of urbanization are strongly related to changes in energy supply. Hence one emerging scientific challenge is to underpin strategies that better manage agriculture and land use in an increasingly energy-scarce world. For example, there are complex trade-offs between the socioecological costs and benefits of land-use intensification and increased adoption of organic agriculture. While there are obvious trade-offs between food production, bioenergy production, and the use of land for carbon sequestration and biodiversity conservation (Erb et al. 2012), there are also possible synergies from an integrated optimization of food and bioenergy supply chains, for example, through use of agricultural residues or waste flows (Haberl and Geissler 2000; WBGU 2009). All these complex feedback processes will likely be strongly affected by climate change impacts (e.g., the impacts of climate change on food, feed, and energy crop yields; Haberl et al. 2011a), which at present are poorly understood, in particular due to the difficulties in modeling the responses of farmers to changes in temperature, water availability, or extreme events (e.g., Waha et al. 2012). Although a move toward less energy consumption and environmentally less demanding and more healthy diets poses few, if any, risks of detrimental environmental feedbacks, it has been notoriously difficult to achieve such changes through political action or institutional change (Haberl et al. 2011b). Some views argue for a world where energy efficiency would increase with agricultural improvements, hence leading to an agriculture system where production would be centralized (and concentrated in a very few places); this would result in high efficiency and therefore reduce the competition for land or conversion to other uses.

Finite Land Resources and Tipping Points

Overall, care must be given to the underlying assumption that society is about to reach the productive land limits of Earth. History is marked by many ill-fated pronouncements that such limits were about to be breached. While land limits and productive limits are real, what constitutes prime agricultural land has long been contingent on the management strategies and technologies employed, which in turn are contingent on political and economic conditions (e.g., advances made in tropical agriculture relative to past views about the paucity of productivity of many tropical soils). "Finite," perhaps, is better understood in terms of the negative human-environment trade-offs among competing land uses (Lambin 2012) and the possibility that the totality of land changes may reach thresholds or tipping points that affect the functioning of ecosystems or the Earth system as a whole. In particular, priority should be given to improving our understanding of the systemic linkages between biodiversity, ecosystem services, land-use change, and climate change, given rapidly growing greenhouse gas emissions, strong increases of demand, and the lack of success in both biodiversity and climate change mitigation policies.

Land Architecture

The future of land dynamics will increasingly be affected by complex trade-offs (and synergies) among multiple land-use/cover units. Mosaics of these units constitute a land system, in which the number, kinds, size, shape, distribution, and connectivity of the units constitute the architecture. This architecture and the associated trade-offs affect human and environmental well-being. Understanding the human and environmental outcomes of land system architectures—from the urban center to the distant wildlands—is necessary to evaluate which alterations of the architecture will provide societal preferences in a more sustainable manner (Chan et al. 2006; Goldstein et al. 2012; Polasky et al. 2005). To improve understanding requires attention to the interactions among land units within the land system, including all of the dimensions noted above as well as interactions among different land systems. To date, assessments of full land system architecture are unavailable, but partial assessments (either of a few land units or a few trade-offs) provide clues that the architecture matters. Recent work, for example, shows that the design of green spaces in Beijing affects its urban heat island (Li et al. 2012), whereas the design of mesic neighborhoods and commercial areas in Phoenix affects land-surface temperature as does the urban design in Baltimore (Connors et al. 2013; Zhou et al. 2011). In terms of rural wildland, local-to-regional precipitation is affected by the architecture of deforestation, the size and distributions by which forest patches are cut (Malhi et al. 2007). The landscape level might often be appropriate to balance the trade-offs between competing land uses and maximize the synergies and multiple functions of land (Fischer et al. 2008; Koh et al. 2009).

Need for Multiscale Models to Assess Local-Global Connections and Impacts

Improved governance requires an assessment of trade-offs, beyond the general summary of impacts presented in Table 4.3, and must include a quantitative assessment using careful empirical analysis with the necessary data. Many case studies of specific land-use changes have been done (Lambin and Geist 2006). While the development of local case studies is critical for understanding the place-specific set of constraints and attributes that influence land-use competition and its impacts, understanding the sustainability impacts of changing land-use systems on a global scale requires much more. Global integrated assessment models are useful for understanding the direct and indirect impacts of price changes on input and output markets, including changes in land used in the production of agriculture or forest commodities. However, space is highly aggregated in these models and thus it is not possible to articulate how local impacts, which may be quite heterogeneous across space, aggregate up to regional or global scales. To answer the question of global impacts and feedbacks, an understanding of how changes at the global scale (e.g., due to consumer demand or technological adoption) influence local land-system changes is necessary. For example, Verburg et al. (2008) developed a multiscale, multimodel approach for analyzing future land-use changes in Europe. In this approach, flows of economic inputs and outputs within Europe are, on one side, nested within global flows, and, on the other side, they condition a spatial allocation model of land use at a grid level. If local changes generate spillover effects across space that aggregate up to influence regional or even global outcomes, then an understanding of the reverse linkage (from local to global) is also necessary. This is particularly important for assessing the global implications of land-use competition, which may occur in many places locally, but in aggregate have impacts that accrue globally. Spatially explicit models that account for local conditions and spatial heterogeneity at microscales are needed to represent these local processes (Irwin and Wrenn 2013) and can be used to assess their cumulative effects on regional and global outcomes (Partridge and Rickman 2013). In addition to multiscale models, a range of new analytical approaches can contribute to improve the understanding of teleconnections and how they influence land-use competitions (see Eakin et al., this volume).

Institutions for Managing Complexity under Uncertainty

Institutions have always needed to address competition, but as competition increases, the stakes rise and institutional forms become more important. Equity is likely to be a growing issue on all scales, as competition for basic needs increases. Greater uncertainty from, for example, climate change also adds pressure on institutions. Fine tuning or, in some cases, redesigning institutions to be more efficient, more stable, and better able to respond to shocks will be increasingly important. Accelerated teleconnections could raise the efficiency of systems, but also make our systems more vulnerable, possibly requiring new institutional forms.

To respond to the needs of policy makers for advice on how to improve, replace, or supplement institutions, empirical analysis of existing institutional performance is needed. We already face competition and uncertainty, and thus can explore the responses of existing institutions to them; to a certain extent, what is changing is the intensity of the problem. However, simulation models are also needed; models that are simpler in geographic and other details than those discussed above, but which capture the essence of institutional forms and the pressures on them, so that we can simulate the likely behavior of potential future institutions. These are likely to be agent-based models that allow careful experimentation so as to lead to a new theoretical understanding of the robustness of institutions.

Future Land-Use Competition

Many other forms of land-use competition than those highlighted here are occurring, some of which are likely to become more important in the future, and new types are likely to emerge. For example, if demand for biofuels continues to grow and traditional food crops like maize, soybean, and oil palm are increasingly sold for energy, biofuel rather than food production could become the dominant competitor for conservation areas. Likewise, if an increasingly urban population in the developing world adopts diets with a higher share of organic food (similar to the Western world), land-use competition between organic (usually more extensive) agriculture and conventional agriculture may arise. Moreover, urban sprawl and vacation home construction could in some regions compete with conservation goals on a massive scale. Likewise, carbon sequestration and storage in forests via afforestation to mitigate climate change could compete with food or feed production. More attention needs to be given to the identification of emerging competitions and understanding the processes behind them.

Conclusions

In our increasingly urbanized and teleconnected world, land-system changes are characterized by different processes of land-use competition. Major drivers include rapid and massive increases in resource demand that result from economic and population growth, changes in diet toward more animal products, attempts to replace fossil fuels with land-based renewable energy, bioenergy, the growing demand for the conservation of nonprovisioning ecosystems, urbanization, and teleconnections. This competition is being played out in a global arena.

Three broad categories of land competition are (a) production versus production, (b) production versus provisioning services, and (c) built-up environment versus production or conservation. We have provided several examples of concrete competition processes (e.g., food vs. feed and food vs. bioenergy). Analysis of such concrete land-competition processes allows a better understanding of the trade-offs involved as well as the geographic location and socioeconomic/political contexts in which these competition processes occur. Various institutions at different scales are involved in defining the objectives for which land and resources are used, and they contribute to manage the competition between land uses. An analysis of long-term drivers and trajectories reveals that changes in energy systems are likely to have a strong influence on how these conflicts will shape the future development of the land system.

Knowledge gaps include the need for an improved analysis of future interdependencies of land and energy systems, the need to understand the importance of patterns in land systems (i.e., land architecture), and the need for multiscale models to assess local-global connections and impacts.

Many dimensions of land competition and its emerging dynamics and outcomes have yet to be adequately addressed by the land-change science community. A range of important research topics have been identified in this chapter. Some of these topics, however, appear to hold special significance, either in terms of their relatively recent emergence, and thus paucity of research attention given to them, or in regard to their looming implications for sustainability:

- Land for various forms of nonprovisioning ecosystem services and biodiversity maintenance, including carbon sequestration and environmental amenities, has become a major form of land use that is actively competing worldwide with other land uses; it is not a passive or residual category of land. These lands range from biosphere reserves and land acquired by NGOs or large corporations to smallholder forests preserved by PES or by local indigenous institutions.
- The growth of land required for bioenergy (solid, liquid, or gaseous) potentially looms large. Currently, land taken for this use may have a cascading effect on other agricultural lands and may directly or indirectly affect forests or other ecologically valuable land.
- Multifunction and heterogeneous land systems are being increasingly transformed to monofunctional and homogenous systems. The human-environment dimensions of this transformation require system-wide assessments. Analyses and planning at the landscape level might be increasingly important to minimize and balance the trade-offs that arise from land-use competitions as well as maximize the synergies across land uses.

- The critical land uses and systems generating "high-cost-to-advert" damages need to be identified, as do the institutional failures associated with them.
- Land system assessments require methodological advances to consider trade-offs among different land uses and land covers, as well as to account for multiple, interacting human-environment dimensions and the spatial interactions among these dimensions.

Making research progress on these pressing issues of land competition and sustainability requires better data, data integration, and modeling to assess the trade-offs across the many, but interdependent, social and environmental scales that both influence and are impacted by land competition.

5

Land-Use Competition between Food Production and Urban Expansion in China

Xiangzheng Deng, Yingzhi Lin, and Karen C. Seto

Abstract

Land-use competition for urbanization and food production has created a significant challenge in China for which there is no quick solution. Maintaining cultivated land is essential to food security, while the trend of urban expansion along with socioeconomic development is irreversible. This chapter analyzes the total effects of urbanization on cultivated land and food production. The mode of urbanization is crucially important, as cultivated land area and food production can be promoted under certain conditions. The problem of food security is not rooted in the occupation of land by urbanization and will not be solved by suppressing urbanization. Instead, developing a rational method of urbanization offers the most effective solution.

Introduction: Context of Land-Use Competition in China

For over thirty years, China has experienced a period of rapid economic growth and is gradually transitioning to a stage of sustained, stable, and integrated economic and social development. In this stage, land use for food production and urbanization often conflicts with each other, causing new phenomena to occur, such as massive production factor flow and reiterative land-use change (Deng et al. 2006). Systematic and integrated research is needed to explain these new phenomena and to provide points of reference for decision makers.

Urbanization and cultivated land protection are two important and interrelated issues in China, and coordinating their relationship constitutes the key problem. It is evident that urban expansion has led to reductions in the area of cultivated land in China. In recent years, more than 70% of the increase in urban land area took place in what was previously cultivated land (Hao et al. 2011). Although the central government has tried to halt the loss of cultivated land, the expansion of urban areas and cultivated land protection are likely to be in conflict for the foreseeable future.

Limited Cultivated Land

Cultivated land protection is a key issue of sustainable socioeconomic development in China. "Rationally utilizing and steadily protecting each inch of farmland" is a basic national policy. Until now, the world's strictest cultivated land-protection systems have been implemented in China. These include the basic farmland-protection system, a total dynamic balance of cultivated land, and a land-use regulation system (Li et al. 2009; He et al. 2011). However, a significant gap exists between policies and measures and their actual implementation.

The total area of cultivated land in China accounts for 9.5% of the world total, ranking the fourth largest after the United States, India, and Russia. However, when one considers its huge population, China has inadequate cultivated land. According to an investigation by the Chinese Ministry of Land and Resources, China presently has a total of 1,826 million mu (about 122 million hectares) of cultivated land, which is equivalent to 0.093 hectares per capita. This represents only one-third of the average cultivated land area per capita in the world, yet China has the largest population in the world: 1.37 billion inhabitants in 2010, accounting for 19.6% of the world's total. In other words, if China did not import food and relied solely on domestic supplies, it would need to feed nearly 20% of the world's population with less than 10% of the world's total cultivated land. The combined effects of a large and growing population, continued economic development, concerns about national food supplies, and the scarcity of land resources have increased the demand for cultivated land (Yang 2004). However, the actual supply of cultivated land has been declining: the area of cultivated land has steadily diminished from 130.2 million hectares (Mha) in 1996 to 121.7 Mha in 2008 (Figure 5.1). This implies that more than 6.5% of the cultivated land area disappeared during a 12-year period. This has created a major challenge for China to guarantee a stable domestic food supply for its population.

In addition to urbanization, some ecological and environmental protection policies and projects have been the primary causes for the reduction in cultivated land area (Wang et al. 2012a). These eco-environmental protection policies and projects include:

- *Grain for Green Program*: This program is one of the six great ecological forest programs in China, whose aim is to convert tracts of cultivated land with serious soil erosion, desertification, or salinization to grassland or forest in line with local climatic and geological conditions, and to restore natural vegetation in a planned, gradual manner. To date,

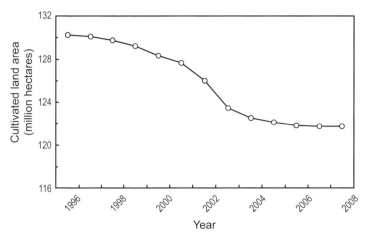

Figure 5.1 Change in area of cultivated land from 1996 to 2008.

the program has converted 14.67 Mha of cultivated land to forest and grassland.

- *Reclaimed Farmland to Lake Program*: This program is an engineering project designed to return reclaimed cultivated land to lake ecosystems, with the aim of restoring the regulatory functions of river flow and eubiosis.
- *Fast-Growing and High-Yielding Plantation Program*: This program aims to fill the gap between timber supply and demand, against the background of forest resource preservation, by planting short rotation and high-yield forests. The higher profits associated with such forest plantations promote the illegal conversion of cultivated land to forest, leading to significant reduction in cultivated land area.

The eco-environmental protection policies and projects led to large-scale conversions of cultivated land. Moreover, these policies and projects are expected to remain in place in response to increasing public concern about environmental problems. Currently, it is estimated that no more than 70% of the existing cultivated land is used for growing grain. According to the cropping index, the area sown for all crops in China is 153 Mha, of which 107 Mha are used for grain. A significant proportion of crop demand (e.g., oilseeds, cotton, and sugar) in China are met by imports, adding the equivalent of up to about 47 Mha of cultivated land imported every year.

Most researchers believe that the area of cultivated land in China will decline steadily (Figure 5.2). According to recent estimates, by 2020 the area of cultivated land will decline to 109.4–120 Mha which is below the threshold of sustainable cultivated land area of 180 million *mu* (about 120 Mha) (Cai et al. 2007; Ma and Niu 2009; Yue et al. 2010). A more long-range pessimistic

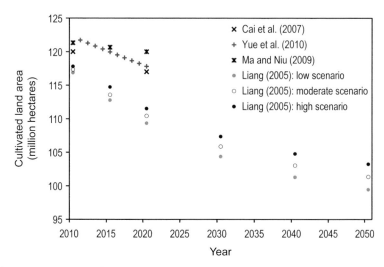

Figure 5.2 Forecast for cultivated land area.

forecast suggests that the area of cultivated land will decline to 100–105 Mha in 2050 (Liang 2005).

Accelerated Urbanization

Rapid urbanization in China began in the late 1970s and has been the dominant social and economic phenomenon ever since (Hao et al. 2011). The urban population in China increased from 172.45 million in 1978 to 665.58 million in 2010, and the urban share of the total population increased from 17.9% to 49.7% during the same period. Although this is still far below the average urbanization levels of developed countries (Liu et al. 2012a), the increase in urban population, 493.13 million, is larger than the combined total population of the United States and Japan (Wang et al. 2012b). In particular, during 2000–2010, the urban population in China increased by 207.14 million and its share of the total population increased by 13.5% (Li 2011; Ma et al. 2012).

Since the implementation of economic reforms and associated policies, urbanization development in China can be divided into three phases (Liu et al. 2012b). During the first phase, from 1978 to 1986, employment in the secondary and tertiary industries increased steadily while employment in the primary industry declined.[1] The development of secondary and tertiary industries

---

[1] In China, the primary industry includes agriculture, forestry, animal husbandry, and fishery; the secondary industry includes mining, manufacturing, production, and supply of electricity, gas and water, and construction; other industries such as transport, storage, post, wholesale and retail trades, hotels and catering services, belong to the tertiary industry.

absorbed large amounts of surplus rural labor force, which further stimulated the urbanization process. During this period, the transition from a largely agricultural economy to a manufacturing economy resulted in the construction of factories and worker housing, which in turn lead to large-scale urban expansion and land-use change.

In the second phase, from 1987 to 1998, many regions of the country experienced accelerated urbanization. Per capita income increased and urban expansion was no longer primarily due to the construction of factories and worker housing. Rather, luxury housing, commercial properties, and recreational spaces began to be developed to accommodate the rising middle and managerial class. For example, this period witnessed the development of large single-family homes, shopping malls, and golf courses throughout the country. The construction of luxury housing and golf courses on agricultural land was so prevalent that by 2006, the central government issued restrictions of constructing luxury villas and golf courses on abandoned land.

During the third phase, from 1999 to the present, the rate of urbanization rate has increased by as much as 1.5–1.7%. In addition, the urban-rural income gap has widened. The secondary and tertiary industries continue to develop and mature, while the employment ratio of the primary industry is still declining at remarkable rates, especially as of 2001. This urbanization process has resulted in the widespread conversion of agricultural land to nonagricultural uses, and in raising the tension between land use for food production or for urbanization.

The competition between these two land uses is further exacerbated by the disparate goals that these land uses achieve. Urbanization (and associated urban land change) is considered the engine of economic growth. It has long been observed that countries with higher percentages of their populations in agriculture have a low per capita gross domestic product (GDP). One target of the central government is to increase the level of urban population to 70% by 2050, approaching the average urbanization level of developed countries. Such rapid increase in urban population will inevitably parallel high rates of urban land change. During the period between 1981 and 2005, the urban area in China expanded from 7,438 km^2 to 32,521 km^2, an increase of 337% (Song and Zenou 2012; Xu et al. 2011). On the other hand, maintaining the country's cultivated land area has long been viewed as the key to food security as well as national security. Since 1978, the country has pursued a policy of agricultural self-sufficiency. However, this goal was abandoned in early 2013 when the central government announced that the country would stop pursuing the goal of agricultural self-sufficiency. This announcement has considerable implications for cultivated land area. Even when the government pursued agricultural self-sufficiency and cultivated land was protected by law, urban expansion claimed a lot of prime agricultural land.

To illustrate the scope of Chinese urban expansion, we constructed maps from Landsat images for five years (1988, 1995, 2000, 2005, and 2008) from three selected cities (Figure 5.3). Shanghai was included as an example of a large metropolitan region in the rapidly developing coastal area; Kunming represents a large city in China's inland region; and Yibin was included to illustrate typical changes in a small, prefectural level city. Although the scales of the maps are different, examining their changes over time allows us to observe the rapid expansion that occurred in all of them (Deng et al. 2008).

Accelerated urbanization in China is likely to continue in the future for two reasons. First, the wage differential between urban and rural areas is increasing, making urban areas increasingly attractive to rural laborers. Statistics indicate that the market price of nonagricultural labor has increased to up to five times as much as that of rural labor. In 2010, the per capita net income of rural households was 5919 yuan, less than one-third of those of migrant workers in urban areas. This wide rural-urban income gap prompts an ever-increasing number of rural workers, especially youth, to move to urban areas. Second, the abandonment of the agricultural self-sufficiency policy is likely to lead to the modernization of agricultural production, as farmers have more incentives to increase efficiency and become more competitive. In turn, mechanization, a prerequisite for increased efficiency and modernization, will make many agricultural workers obsolete, freeing up more laborers to seek urban work.

A recent study suggests that China's urban area is highly likely to expand by 219,700 km^2 between 2000 and 2030, nearly tripling the country's current urban footprint (Seto et al. 2012a). To illustrate the future of urban expansion for specific cities, we illustrate urban planning examples from Shanghai, Kunming, and Yibin:

- According to the municipal government planning document, *Urban Planning of Shanghai*, the population of Shanghai will increase to 16 million by 2020 and the urbanization level will reach 85%. The constructed area (in vertical and horizontal dimensions) in the urban core will expand to 1,500 km^2 in 2020, up from 174 km^2 in 2010.
- According to the *Urban Planning of Kunming*, the urbanization level of Kunming in 2020 will increase to 73%. The constructed area in the urban core will expand to 430 km^2 in 2020, up from 258 km^2 in 2010.
- According to the *Urban Planning of Yibin*, the population of Yibin in 2020 will increase to 5.7 million and the urbanization level will reach 54%. The constructed area in the urban core will expand to 102.6 km^2 in 2020, up from 49.6 km^2 in 2008.

These urban planning publications for Shanghai, Kunming, and Yibin indicate that urbanization will bring about extensive urban land expansion over the next ten years. This applies particularly to small cities, where much of the new urban development will take place.

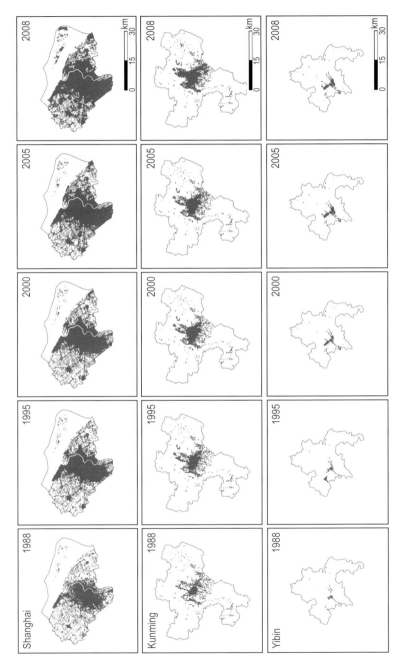

Figure 5.3 Maps illustrating the expansion of the urban cores from 1988 to 2008 in Shanghai, Kunming, and Yibin.

Cultivated Land-Area Reduction Due to Urbanization

There is no doubt that the dramatic economic development in China has brought about massive urban land conversion all over the country, leading to the loss of cultivated land at an alarming rate. Rapid urbanization contributes significantly to the loss of cultivated land (Lu et al. 2011; Su et al. 2011) through several ways. One of the most direct ways is that urban expansion occupies cultivated land. According to an investigation of 145 cities in China conducted by the Chinese Academy of Sciences, conversion of cultivated land to urban area accounts for 70% of the urban expansion. For certain cities in western China, this proportion is as high as 81%.

To illustrate the extent to which cultivated land has been taken over by urban expansion, we analyzed Landsat images of Shanghai, Kunming, and Yibin to generate maps of land change (Figure 5.4). Within an eight-year timeframe, 2000–2008 (Deng et al. 2008), these maps show that:

- Shanghai's urban land area expanded by 35.9%, and that 86.4% of the newly expanded built-up area occurred through conversion of cultivated land.
- Kunming expanded its urban area by about 6,440 hectares, accounting for 27.8% of its urban area in 2000 and occupying 4,610 hectares of cultivated land.
- Yibin, a typical small city, expanded its urban area by 1,180 hectares, accounting for 43.9% of its urban area in 2000. In addition, 942.7 hectares of newly developed urban areas were obtained through conversion from cultivated land; this accounted for 79.9% of the newly built-up area. In general, especially on the outskirts of cities, areas of cultivated land have been largely built over in the past few decades.

The direct conversion of cultivated land by urban expansion almost always takes place on high-yielding cultivated land with good agricultural infrastructure and high soil fertility (Wu et al. 2011; Wu and Zhang 2012). During the period from 2000 to 2008, approximately 1.0 million hectares of cultivated land (0.72% of total cultivated area) were converted into urban built-up area in China (Figure 5.5).

A second way in which urbanization promotes cultivated land reduction is through the loss of agricultural labor. The rural-urban income gap has encouraged a large number of agricultural workers to migrate to urban areas, which in turn drives further urbanization (Feng et al. 2010). Statistics show that 250 million migrant workers from rural areas were employed in urban areas at the end of 2011. This means that about half of Chinese agricultural workers are partially or completely separated from agricultural production. Since most of the migrant workers are middle-aged and possess technical expertise, this has left the very young and the elderly in rural areas, causing a structural scarcity of agricultural labor.

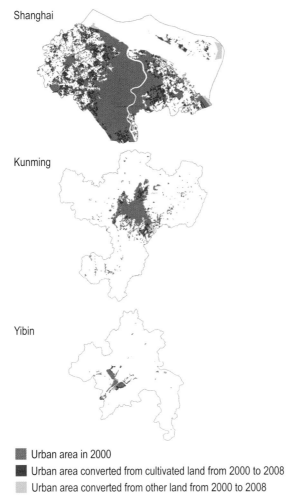

Figure 5.4 Maps indicating the occupation of cultivated land by urban expansion from 1988 to 2008 in Shanghai, Kunming, and Yibin.

The idea that agricultural labor loss can be balanced by mechanization of farming is unlikely because at present, the popularization of agricultural mechanization in China is insufficient. This holds particularly true for many hilly areas, where the rate of agricultural mechanization is high, but considerable labor and expertise are needed to operate the machinery because most of the machines are small in size and labor intensive (Siciliano 2012). Whether cultivated land abandoned by migrants may be cultivated by the remaining farmers in rural areas is an issue under debate. In reality, migration of the agricultural population always occurs in large magnitudes and, generally, all of the rural workers from a

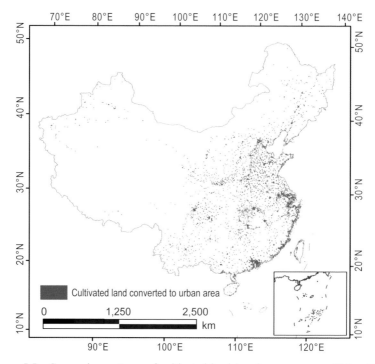

Figure 5.5 Conversion patterns of cultivated land to urban area across China, 2000 to 2008.

particular village migrate to the same or neighboring cities. Even if some famers remain, the contractual use of cultivated land belonging to former farmers is not common because the labor demand per unit area of cultivated land does not meet basic requirements. In addition, the cultivated land market has not been developed and the contractual rights transaction of cultivated land is scarce in China. According to an investigation implemented by the Ministry of Land and Resources, about 2,000,000 hectares of cultivated land are abandoned every year due to the migration of agricultural labor.

Loss of Agricultural Production Due to Urbanization

Scarcity of food threatens the sustainability of the socioeconomic system and forces price increases, which can eventually lead to famine and social unrest. Currently, China is confronted with a threat to its food security, second only to Latin American and African countries (Xiang et al. 2011). In 1985, China shifted from being an importer of food grains to an exporter for the first time since 1960. Now, however, after thirty years of industrialization and urbanization,

China has become a major importer of food grains (Deng et al. 2006). In the future, the disparities between supply and demand of food in China will grow to be even more severe.

Analyzing the net changes by land type, we assessed the extent to which the conversion of cultivated land to different uses has affected total production potential (Figure 5.6). In total, the conversion of cultivated land to other uses led to a net loss of 347.56 million tons, or 0.1% of total potential productivity in 1988–2008. Of this total amount, a decrease of 499.28 million tons (about 65.4%) of the total decreased production potential was due to the conversion of cultivated land to built-up areas.

It is important to note that not all of the built-up area converted from cultivated land is being used as urban land. In addition, it is important to consider the abandonment of cultivated land, including conversions of cultivated land to grassland or unused land, due to urbanization. Thus, we can still conclude that urbanization is a major cause of loss of production potential.

The reduction in cultivated land area in China has been blamed on irrational urban planning but not necessarily urbanization. In the views of some scholars, urbanization with rational urban planning will economize and realize the intensiveness of cultivated land use. Others argue that the duality in land tenure is the reason for the rapid reduction in cultivated land area. The differences in forms of land ownership between state-owned land and collective land leads to the occupation of cultivated land or other rural lands that are too cheap for other land uses. Compensation fees for requisitioned land include land compensation fees, resettlement fees, and compensation for attachments to or

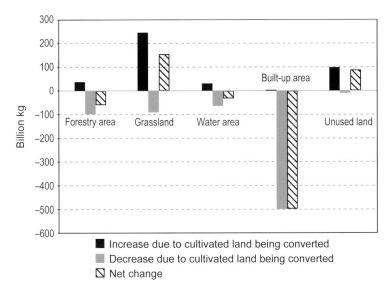

Figure 5.6 Changes in total agricultural production potential associated with changes in cultivated area in China, 1988–2008.

green crops on the land. According to the Land Management Law of China, the total fees of land compensation and resettlement should not exceed thirty times the average output value of the three years preceding the requisition of the cultivated land. This implies that the average price of land will soar after changing the form of land ownership from collective land to state-owned land. Subsequently, economic interests motivate the local government to occupy the cultivated land.

Some measures have been implemented with urban expansion to counter the negative impacts on agricultural production potential. In 2001, the cultivated land dynamic balance policy was implemented to compensate for the loss of cultivated land to built-up areas throughout the country (Guo 2004). This policy compels the same amounts of grassland, forestry, or unused land areas to be converted to cultivated land and at least ensures the stability of cultivated land area. Although the quality of cultivated land converted to built-up area is always higher than lands converted to cultivated land, the cultivated land dynamic balance policy mitigates the decline of agricultural production potential to a great extent.

According to forecasts, China's population will reach its maximum, 1.62 billion, by 2030 and will inevitably be paralleled by increases in food demand (Ma and Niu 2009). The per capita annual food demand may grow by 70 kg to 460 kg as per capita incomes rise and dietary patterns change. This means that by 2030, the total demand for food will be 7.43 billion tons and that a 30% increase in food production will be required to maintain the balance between food demand and supply over the next twenty years.

As early as 1994, the question of how to feed China was raised by Lester R. Brown (1995). In tandem with the sharp increases in international grain market prices, greater attention is now being paid to food security. Accordingly, some new agricultural policies—removal of agricultural tax, use of agricultural subsidies, and new rural construction—have been implemented to guarantee national grain production and to improve the stability of the national grain market (You et al. 2011). However, the increasingly upward trend of grain prices has not been fundamentally reversed in China because of the contradiction between food security and urbanization.

A low agricultural income, which determines the low opportunity cost of agricultural labor movement to cities, is the guarantee and premise of cheap labor supply for industrialization and urbanization. The relative surplus in grain supply and the inferior status of farmers in wealth redistribution supports the prolonged existence of this guarantee and premise. Migrating away from cultivated land becomes the most popular option for farmers when the incomes offered by urban jobs greatly exceed that of farming. One of the serious consequences of this phenomenon is that regional and seasonal labor shortages emerge in rural areas, vast areas of cultivated land are abandoned, and grain supply drops.

Over the past several years, in the process of urbanization in China, a large number of agricultural workers migrated to urban areas; this led not only to a labor shortage in rural areas and a reduction in grain supply but also to a housing shortage and overcrowding. The improvement in housing demand typically creates a rapidly developing and extremely profitable real estate industry. The presence of fertile farmland around cities is an attractive target for real estate development; therefore, large-scale infrastructure and industrial park construction leads to the decrease in grain supply.

Since 1990 food production in China has been volatile. For example, during the period 1999–2003, Chinese food production declined rapidly from 51.33 million tons to 43.07 million tons, with an annual rate of decrease of 3.2% (Zhang et al. 2008). The combination of food production fluctuations and population growth results in a decline of food output per capita. The food output per capita across the country varied from 375 kg to 415 kg during 1990–1998 and rapidly declined to 333.6 kg in 2003. After 2003, there were limited, transient increases in the food output per capita, but the overall trend was that of decline.

While the cultivated land dynamic balance policy ensures a stable amount of cultivated land area, it cannot do so for productivity. The cultivated land occupied by urban expansion is generally both high quality and high yielding, while the compensatory cultivated land is nearly always barren. Thus, more inputs such as fertilizers, pesticides, and water are required to bring these barren lands up to a suitable level of productivity. Considering the cultivated land balance in 2004, 72% of the cultivated land occupied by urban areas was well-appointed with irrigation facilities, and no more than 34% of compensatory cultivated land possessed irrigation infrastructure. According to the simulation results, potential productivity declined by an average of 2.2% annually for the cultivated land involved in the cultivated land dynamic balance policy (Deng et al. 2006). The degradation in the quality of cultivated land due to urbanization is one of the major causes of food production losses in China.

For many countries, the transition process from low to moderate urbanization levels is accompanied by food security challenges, which reach their peak during the phase of rapid and maturing urbanization. Thereafter food security challenges begin to ease up during the phase of high urbanization rates. This is because during the transitional process (from low to moderate urbanization rate), rapid urban population and economic growth, and decline of the rural population cause the agricultural sector to lag behind the industrial and service sectors. Once the urbanization rate reaches a high level, increased food demand due to urbanization serves to promote agricultural production and relieve food crises (Guo 2004; Zhang and Jia 2005; Zhang et al. 2007). Ironically, the stress of food security is related to urbanization in an inverse U pattern (Figure 5.7).

China is currently in the stage of urbanization with the most rapid increases, and the development of industrial and agricultural sectors is uneven. The

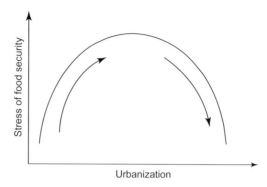

Figure 5.7 Stress changes of food security in the process of urbanization.

productivity gap between industrial and agricultural sectors leads to a lack of capital investment in the agricultural sector, thus inhibiting Chinese agricultural development. This indicates that the expected substitution of capital for cultivated land loss due to urbanization cannot be realized.

Concluding Comments on the Debate on the Impacts of Urbanization on Food Production

Urbanization has clearly led to the loss of cultivated land area and food production in China. Although some researchers argue that there are positive or at least no significant negative effects of urbanization on cultivated land and food production, statistical data and satellite data suggest otherwise. Since 1978, the Chinese rural population has decreased by 144 million but the rural constructed area has increased by 1.31 million hectares. This implies a loss of about 1.10 million tons of crops. During the period from 1978 to 2010, the urban population rapidly increased by 488 million (National Bureau of Statistics of China 1988–2011a); the total urban area expanded by only 0.32 million hectares (National Bureau of Statistics of China 1988–2011b), which implies a reduction of 0.28 million tons of crop production (Guo 2011). These statistics show that urbanization may play a positive role in maintaining the total area of cultivated land and food production.

The total effects of urbanization on cultivated land and food production are contingent upon the mode of urbanization. The relatively intensive utilization of land in urban areas is evidence of the positive effects of urbanization on maintaining cultivated land area and food production. The migration of rural population to urban areas heightens the efficiency of land use per capita (Zhao 2011). Meanwhile, population reduction in rural areas favors the merging of villages/towns and land consolidation, which is one of the main ways by which the area of cultivated land can be increased. Therefore, the process of urbanization may promote the maintenance of cultivated land area and food production, especially if urbanization leads to industrialization and villages and towns merging appropriately.

It is evident that the intensity of utilization of built-up areas in urban areas is much higher than in rural areas. Population migration from rural to urban areas may lead to the reduction of total built-up area and ease the pressure on cultivated land loss. To heighten these positive effects of urbanization, several recommendations should be followed:

- Irrational expansion of urban areas should be prohibited. One of the most prevalent ideas about urbanization in China is that an increase in urban area is always better for economic and social development. This idea ignores the fundamental issues and challenges associated with the growing urban-rural gap.
- In terms of intensive land utilization, three points should be stressed: First, intensive land utilization in urban areas could reduce the cultivated land lost to urban expansion. Second, intensive noncultivated land utilization in rural areas could reduce the occupation of cultivated land by rural residential areas. In addition, abandoned residential real estate projects should be reclaimed and farmed again. Third, intensive cultivated land utilization through modern and diverse methods of improving productivity can increase the value of cultivated land and help reduce its loss.
- A legal framework of urbanization and cultivated land protection must be put into place to enable a more comprehensive perspective on urban expansion and conservation of cultivated land.

In conclusion, the competition of land use for urbanization and food production poses a significant challenge in China, for which no quick solution is available. The food security problem is not rooted in the land occupation of urbanization. Thus, food security will not be achieved simply by suppressing urbanization. Instead, developing a rational method of urbanization appears to be the most effective solution.

Distal Land Connections

6

Globalization, Economic Flows, and Land-Use Transitions

Peter J. Marcotullio

Abstract

This review examines contemporary globalization and economic flows and the impact of these forces on land-use transitions. It finds that the contemporary period of globalization is defined by an integrated global system and massive flows of trade, finance, and investments. The global conditions, including these flows, are both indirect and direct drivers of land-use and land-cover change. The resultant global land-use system dynamics have shifted in well-defined ways, including the speed, direction, location, and timing of land-use transitions. Studying the contemporary global system of land use and land cover requires new models capable of tracing the sources and effects of globalization and economic flows.

Introduction

This chapter has several goals. First, it reviews trends in globalization or the constellation of events, processes, and linkages that have created "transplanetary and supra-territorial connections between people" (Scholte 2005:8) resulting in a "widening, deepening and speeding up of cross-border interconnectedness in all aspects of contemporary social life" (Held et al. 1998:2). The outcome of this review suggests that while globalization has been growing over the past decades, if not centuries, the contemporary period is new and has resulted in a globally integrated system that operates differently than anything previously.

Second, selected elements of globalization are highlighted to demonstrate the quantitative differences between the contemporary speeds, intensities, and extents of connectivities compared to previous times. In this case, a representative sample of economic trade, investments, and financial connections or "flows" (Ritzer 2009) delineate fundamental global linkages. Economic flows of trade, investments, finances, and remittances have never been higher (whether measured in absolute terms or with respect to global gross domestic product, GDP), never reached more locations around the world, and never have been transmitted faster than those of today. These changes are indicative of a functionally altered global economic system. Contemporary economic flows are helping to create an integrated transnational economy, where local decisions and outcomes are embedded in regional and global processes.

Third, an abstract framework is provided to understand how globalization and economic flows have influenced land-use and land-cover dynamics, both in terms of general mechanisms and outcomes. The focus includes connecting the bundle of processes that define globalization and the individual types of economic flows to dimensions of local decisions about land. Land use is characterized by the human arrangements, activities, and inputs for a certain land cover, or observed biophysical cover on Earth's surface, to produce or maintain it (Choudhury and Jansen 1998). While numerous land-use and land-cover dynamics exist (Geist et al. 2006), the focus here is on changes in land-use transitions. Transitions are inflections in long-term trends, including both the quantities and rates of change in quantities of interest (National Research Academy 1999). In the case of land-use and land-cover dynamics, transitions identify changes in the amount or intensity of a particular land use or land cover over time (see, e.g., Mather 2007; Mather and Needle 1998).

The extent and intensity of globalization has increased over the past decades such that it has become a major primary (indirect) driver in global biophysical system change (MEA 2005). At the same time, over the past decades land-use and land-cover change has been unprecedented (for review, see Lambin and Meyfroidt, this volume) and subsequently has also significantly affected Earth system functioning (Foley et al. 2005). Scholars have identified the importance of globalization generally as a cause of land-use, land-cover change (Lambin and Meyfroidt 2011; DeFries et al. 2010), but more research is needed to detail the actors, mechanisms, and distal linkages involved (Ramankutty et al. 2006).

This chapter posits that the influence of globalization and economic flows for land-use, land-cover change dynamics cannot be underestimated. A review of land-use transition case studies suggests that local land-use decisions are increasingly being influenced by events and activities in distal locations. While globalization acts to change the context of land-use dynamics, economic flows have become both direct and indirect drivers of change. These influences have altered previously well-defined land-use transitions in some general ways. The result of these changes suggests that the global land-use, land-cover system has been systemically altered.

Further deciphering the details of how globalization and economic flows affect land-use change dynamics requires new models and tools. The concept of urban land teleconnections promises to aid our understanding of these linkages and therefore enlighten land-use policy (Seto et al. 2012b). Urban land teleconnections is an increasingly relevant concept and tool as both the character of twenty-first century globalization and the sources of the economic flows are bound up with the urbanizing population (for an explanation of land teleconnections and the significance of this concept for land-use change see Eakin et al. as well as Liu et al., this volume).

Here, I present the elements of contemporary globalization, arguing that it is both the result and perpetrator of a new and unique global social system. Phases of globalization are outlined that identify the unique aspects of the contemporary period (1973 to the present). Thereafter I review the trends in selective economic flows, focusing largely on trade, finances, and investments. Discussion follows to provide a number of potential mechanisms by which these flows can be linked to land-use, land-cover change, with some generalizations offered on how land-use transitions have been altered over the recent past. I conclude with some thoughts on the importance of new models to elucidate these linkages between economic flows and land-use, land-cover change.

Is Contemporary Globalization Different from That of the Past?

Global capitalist economic development has not been smooth. There exist distinct phases of growth and recession throughout modern history, which have impacted multiple economies. While the dating of these phases is subject to historical debate (see, e.g., Berry 1991; Maddison 1991; Reijnders 1990), economists have long believed in distinct temporal fluctuations in economic activity at multiple scales. For example, analysts have identified a multiplicity of economic cycles, including 3.5- to 4-yr Kitchin cycles, 9- to 12-yr Juglar cycles, 15- to 25-yr Kuznets cycles, and 50- to 60-yr Kondratieff cycles (for a review, see Korotayev and Tsirei 2010).

The ebb and flow of international trade, investment, and finance has been associated with long-wave periods of capitalist development. Long-wave cycles were first identified by van Gelderen (1913; cited in Berry 1991), but it was Kondratieff (1926) who developed the full theory of their existence. Subsequently, while many economic historians have based their research into rhythmic movements in economic development on changes in output (GDP), interest rates, population, savings accumulation, and (wholesale) prices, studies have demonstrated that periods of growth and recession have also had significant impact on, and have been impacted by, international flows in trade, investments, and capital flows (Eichengreen 2008; Kindleberger and Aliber 2011; Smith et al. 2012). Indeed, scholars argue that shifts in international movements of goods, finance, and people can be matched with long waves of development.

While it may be true that globalization or "transplanetary relations between people" (Scholte 2005:8) and accompanying economic global flows have been under development for centuries, it seems that during modern times each long-wave period represents a distinct phase in its development (Held et al. 1998). In analyzing the sequence of historical phases, I will use Maddison's (2001) well-defended periodicity in global long-wave economic development: pre-1870,

1870–1913, 1913–1950, 1950–1973, and 1973 to the present.[1] During many of these different periods or eras, globalization and aspects of global economic flows advanced, but they also sometimes retreated.

Changing international economic relationships during these phases are underpinned by fundamental spatial distinctions in markets. Originally, these differences were thought of as differences in production and terms of trade. Countries exported or imported based on the comparative advantage for each set of goods and services. More recently, the concept of the spatial division of labor became a valuable, if not crucial, addition to explain the patterns of transnational flows. The division of labor refers to the separation of tasks in the production process, including both social and technical divisions (Painter 2000). Massey (1984) suggests that the spatial division of labor now refers to distinct geographic transnational locations associated with specific production stages, nodes, or sectors of any given production chain. Certainly, one of the most important trends in the movement of economic goods, services, and capital has been the rise in the "New International Division of Labor" (NIDL), associated with the internationalization of production (Freobel et al. 1980). The NIDL includes the differential expansion of the labor pool across different economies. Whereas there is a concentration of skilled labor in some economies, there are large pools of unskilled labor in others. As firms have become increasingly footloose, they have taken advantage of the different labor pools across borders. As markets for goods, services, and capital expand, so the division of labor increasingly stretches across geographic space, both incorporating and differentiating economic activities.

Whereas distinctions across the transnational economy are inherent in economic flows, distinct phases are defined by overall global trends that seemingly cut across cultures. These include, among other things, the adoption of transportation and communications technologies, the adoption of goods processing and production standardization processes, the emergence of transnational corporations (TNCs), and increasingly, a set of institutions and ideology upon which global flows are maintained and promoted (Murray 2006; Robertson 1992).

Importantly, these globally shared sets of changes have shifted over time, helping to define different periods of global economic development. Thus, for example, phases of economic development and international economic flows prior to 1870 were limited by a "cluster of technologies" focused on animate, wind, and biomass fuel energy. What helped change the dynamics of economic flows and further integrate the global economy were improvements in transportation, including the adoption of coal and steam locomotion in both trains and boats, and the adoption of telegraph technologies (Grubler 2003). Changes

---

[1] Others have distinguished phases of globalization including Robertson (1992) who identified five phases between the early fifteenth and later twentieth century; Held et al. (1998), who discerned four epochs of globalization; and Scholte (2005), who identified three phases. All, however, point to the contemporary phase as starting during the postwar period and intensifying after the mid-1970s.

in the fundamental structure of the global economy mean that trends in economic flows are more than simply the changes in quantities of goods, capital, finance, and movement of investments; they also represent advances across technologies, institutions, governance, and economic activities, all of which contribute to the development of a more *integrated* global economic system. Over time, not only do the intensity (deepening), extensity (widening), and speed of flows increase, the context within which the flows operate also changes. For example, the opening of markets through the lowering of trade barriers, the spread of management practices for different land uses, and ideologies that support consumption practices can either help to facilitate or attenuate the flows themselves.

Analysts have modeled the international movement of economic flows in the global economy across time. Figure 6.1 is a schematic representation of shifts in intensity of capital flows from the mid-nineteenth century to the present day. Obstfeld and Taylor (2003) show a clear picture emerging for the pattern of economic flows. During the second half of the nineteenth century (1870–1914) there was a remarkable rise in capital market global integration. A trough followed and globalization receded. The tendency for global economic disintegration lasted approximately forty years. A strong new upsurge started after World War II and has had two upward phases. This fluctuating trend is also evidenced for trade (Kuznets 1967) and foreign direct investment (FDI) (Dunning 1988).

Given the high levels of economic flows prior to 1913, some researchers argue that the global economy was more integrated during that period (Hirst and Thompson 1996; Hopkins 2002; Sachs and Warner 1995; Zevin 1992; Rodrik, unpublished). The perceived reemergence of the global, capitalist market economy since 1950, and more specifically since the mid-1980s, mirrors that which occurred from 1870 to 1913, at the height of the so-called "Gold Standard"

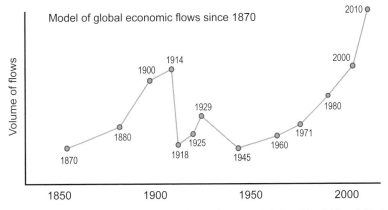

Figure 6.1 Model of global economic flows since 1870 (after Obstfeld and Taylor 2003, used with permission of University of Chicago Press).

period. Among the evidence provided by these analysts is that up until recently, the production of international tradable merchandise constituted a larger share of overall economic activity a century ago when compared to this current era. Even so, as Bordo et al. (1999) have demonstrated, commercial and financial integration before World War I was more limited than today. For example, trade played a much smaller role in pre-World War I production than today due to higher relative transportation costs, government trade barriers, and information asymmetries which served as obstacles to deeper integration during the previous era. Trade and direct investments have now opened up to international competition in what a century ago were "nontraded" sectors, such as services, retail trade, and public utilities. Moreover, whereas the number of TNCs was limited in the past, constituting only a small part of domestic economies, the reverse is true today. More importantly, however, globalization and economic flows have advanced dramatically over the past decade (2000–2010). Current levels of cross-border flows have now overwhelmed those of the past.

Global integration is being promoted by institutional harmonization with regard to trade policy, legal codes, tax systems, ownership patterns, and other regulatory arrangements (Sachs and Warner 1995). For example, between 1990 and 2002, average world tariff rates declined from 10.5% to 6.0% and the average ratio of global imports plus exports in GDP increased from 75.2% to 86.8%. The World Trade Organization, an important global institution promoting these changes, now includes 157 members.

The resultant economic growth for nations integrated into the global system has been spectacular over the past few decades. Some developing nations, such as China and India, have experienced annual growth rates exceeding 9% and 6%, respectively, something unheard of in prior eras. At the same time, however, the distribution of global income is increasingly polarized (Wade 2001). Research suggests that high-income OECD nations (representing approximately 14.8% of the world's population) hold 83.3 % of world household wealth and 76.9% of world GDP (based on official exchange rates) (Davies et al. 2006). Moreover, individual organizations, such as TNCs, increasingly reap enormous revenues making them more economically powerful. For example, between 1990 and 2003, the asset values of foreign affiliates of the world's TNCs increased fivefold and TNC sales threefold while world GDP in current prices increased by 160% (UNCTAD 2007). Moreover, the global economy is increasingly less stable. According to Kindleberger and Aliber (2011), the years since the early 1970s are unprecedented in terms of the volatility in the prices of commodities, currencies, real estate, and stocks, and there have been four waves of financial crises, where a large number of banks in three, four, or more countries collapsed simultaneously. The global economy continues to endure a roller coaster of growth and recession, as witnessed by the disruptions of the early 1980s and the early 1990s, the Mexican peso crisis (1994), followed by the Asian financial crisis of 1997–1998, the post-9/11 crash in 2002, and the economic slowdown that began in 2008, the latter being the most severe and most global since the Great Depression of the 1930s.

Economic Flows

Global economic integration has meant massive systemic change over the past sixty years. The era from the 1950s to the present is typically subdivided into two subperiods: the "Golden Age," from 1950 to 1973, and the present era, starting from 1973. The dividing point between these two sub-eras is marked by, among other things, the breakdown of the Bretton Woods agreements, the oil shocks and its repercussions, the global drop in primary commodity prices, and an episode of stagflation in the U.S. economy. Due to space limitations, this section neither separates the trends during these subperiods nor does it discuss the changes to the international economic structure that facilitate differential patterns (for reviews, see Dicken 2003; Held et al. 1998; Scholte 2005). Instead, this section presents overall postwar trends in trade, FDI, selected financial flows, and remittances.

Trade

During the postwar period, trade grew rapidly. World figures are crude, but they tell a persuasive story. Maddison (2001) argued that from the end of World War II to the early 1970s, world trade grew at over 7.0% per annum. Krugman (1995) calculates that in 1950, world exports amounted to 7% of world output, but rose gradually in the 1960s and then more sharply in the early 1970s to reach 11.7% of global GDP. By the mid-1970s, levels of global trade were similar to pre-World War I conditions.

The figures for the developed countries are indicative of the "Golden Age" of capitalism. The total value of merchandise exports from noncommunist countries rose from USD 53.3 billion in 1948 to USD 112.3 billion in 1960, or at an average annual growth rate of over 6%. From 1960 to 1973, growth rates were even higher at about 8% per annum (Kenwood and Lougheed 1999). During this period, Japan experienced the highest trade growth rates: over 15% per year.

Since 1973, trade has become an increasingly important component in national economic development. The 1970s, however, began with a change in the growth of exports, with global annual increases dropping to an average of 4.5% from 1973 to 1979. After this initial slowdown, however, world trade began to increase again at rates both higher than GDP growth and higher than those of the past. From 1980 to 2000, world trade increased by over 7% annually (Table 6.1) and much of this growth occurred in a spurt at the end of the century. The first decade of the twenty-first century has been extremely impressive, with world trade increasing by more than 10% annually.

Table 6.1 Value growth rates of merchandise exports: annual, 1980–2000, and 2000–2010. Data from UNCTAD and UNCTADstat (2012).

Economy	1980–2000 Growth Rate	2000–2010 Growth rate
World	7.10	10.89
Developing economies	7.66	14.37
Africa	1.90	16.31
Latin America	6.40	11.27
Asia	9.04	14.84
Oceania	5.56	7.71
Transition economies	1.36	18.25
Developed economies	7.17	8.51
North America	7.50	6.17
Asia	7.14	6.38
Europe	7.09	9.47
Oceania	6.56	13.31
Least developed countries	5.01	19.44
Africa and Haiti	3.04	22.70
Asia	10.96	13.03
Islands	1.82	11.26

As a result, global trade has grown from USD 2.03 trillion in 1980 to USD 15.25 trillion in 2010. According to the World Bank, global exports were approximately 30 % of global GDP in 2010, up from 24 % in 2000. Moreover, trade is less concentrated than in the past. The top ten exporters accounted for 58 % of all trade in 1980, but by 2010, the top ten exporters accounted for only 50.5 % of global trade. Furthermore, total primary and manufactured merchandise traded has not only increased in value, it has also increased in weight (Table 6.2).[2] Consumer goods are transported primarily in ocean containers enabling direct transportation from exporter to importer without the need for intermediate cargo transfer. Containerization for shipping was developed in the 1950s and international standardization followed in the 1960s and 1970s. By 1970, the volume of moved containerized goods totaled 2.5 billion tons. By 2000, 5.9 billion tons of container goods were shipped, and by 2010 the figure was 8.4 billion tons.

---

[2] Although ocean freight costs have risen in recent years, they are still among the lowest compared to other modes of transportation. The trade-off is the long voyage time. A typical trip from Hong Kong to Los Angeles can take up to ten days, and a trip from Hong Kong to New York through the Panama Canal can take up to 31 days. Seaborne trade does not cover all trade in merchandise. Air transport typically handles high value goods. For example, in 2000, the JFK airport in New York processed approximately 1.8 million tons of freight, everything from Old Masters, to diamonds, to racehorses (Ascher 2005).

Table 6.2 Overview of seaborne trade from 1970–2010; goods loaded (millions of tons). Data from UNCTAD (2010), Review of Maritime Transport (various years).

Economy	1970			1990			2010		
	Oil	Dry cargo	Total	Oil	Dry cargo	Total	Oil	Dry cargo	Total
Developed	118.5	686.7	816.2	325.4	1,345.5	1,670.9	544.2	2,288.2	2,832.5
North America	6.0	308.0	314.0	27.2	515.1	542.3	152.4	523.1	675.5
Europe	110.9	244.8	366.7	286.3	482.2	768.5	346.4	720.3	1,066.6
Japan*	0.3	41.6	41.9	1.2	81.9	83.1	24.5	151.2	175.7
Australia, New Zealand	1.3	92.3	93.6	10.7	266.3	277.0	21.0	893.6	914.6
Transition economies	64.5	80.8	145.3	115.3	92.6	207.9	196.1	319.7	515.7
Developing	1,158.4	394.9	1,553.3	1,313.9	814.9	2,128.8	2,012.1	3,047.9	5,060.1
Africa	289.7	119.1	408.8	345.3	179.0	524.3	425.1	308.2	733.3
Latin America	155.4	160.4	315.8	226.6	298.3	524.9	304.2	825.4	1,129.6
Asia	713.1	105.9	819.0	741.7	329.6	1,071.3	1,281.1	1,909.5	3,190.7
Oceania	0.2	9.5	9.7	0.3	8.0	8.3	1.7	4.8	6.5
World Total	1,341.4	1,162.4	2,514.8	1,754.6	2,253.0	4,007.6	2,752.4	5,655.8	8,408.3

* in 2010 combined with Israel

Despite globalization trends, countries still tend to trade with territories that are located closest to them, so much of global trade remains intraregional. Western Europe is the largest trader, although two-thirds of that trade takes place within the region. Canada and Mexico are two of the top three trading partners with United States.

Another important trend is the increasing trade of developing countries. In 1980, the developing world accounted for 29.5% of all trade; by 2010 this share was up to 42%. This rapid increase in trade from developing countries could be due to the global economic downturn experienced since 2008, which hit the developed world particularly hard. At the same time, however, the long-term trend for increasing shares of trade in the developing world is undeniable. For example, there are now three developing economies that typically appear in the top exporter or importer categories: China, Hong Kong, and South Korea. China is now the world's largest exporter of merchandise.

While trade in manufactured products has increased, so has that of primary commodities. For example, trends in the quantity of 15 primary commodities trade categories from 1961 to 2010 demonstrate annual average increases of approximately 3.7%. Trade in agricultural oil crops grew the fastest at approximately 5.5% (for details on oil palm production and trade, see Wicke, this volume), but trade in pulses, vegetables, beverages, natural rubber, meat, dairy, and eggs also increased by more than 4% annually. During this same period, trade in forest products increased by between 3.6 and 5.1% annually; in addition, the quantity of forest products trade increased from 12 million to 165 billion tons and from 44 million cubic meters of wood to 274 billion cubic meters of wood (FAO 2012a). According to UN trade data (UN 2012a), the monetary value of trade in crude minerals and metals increased from approximately USD 200 billion in 1990 to USD 1.2 trillion in 2010.

Services have become an important part of global trade, having grown in proportional economic importance over recent decades. Between 1989 and 2000, service exports grew at around 11% per annum while merchandise exports grew at 10%. Trade in services is concentrated, however. Approximately 65% of service exports come from 15 countries. The balance of service imports and exports is also different geographically to that of merchandise exports, with the United States being a major surplus exporter and Japan being a major service importer. When trade in goods and services are taken together, a clear regional concentration emerges. In 2002, Europe accounted for 42.2% of total goods and service exports and 40.3% of total imports. The Asia Pacific countries constitute the second most dynamic exporting region, accounting for 26.5% of global exports and 24.5% of imports—although these numbers are reduced significantly when Japan and China are removed from the calculation. The surplus in service exports from the United States is not great enough to offset the deficit in goods trade, with North America accounting for 15.9% and 21% of total global exports and imports, respectively (Murray 2006). Trade in services increased from 3.6% of global GDP in 1980 to almost 6% of global GDP in 2010.

Foreign Direct Investment

FDI includes ownership of, or investment in, overseas enterprises in which the investor plays a direct role in managing activities. FDI represents the activities of TNCs, which are firms that have the power to coordinate and control operations in several countries at one time and typically involve a web of collaborative relationships with other legally independent firms around the world (Dicken 2003).

By 1960, the total world stock of FDI was lower than during the "Gold Standard" period: USD 63 billion compared to USD 14 trillion in 1914. At the same time, however, it has been increasing steadily. By 1973, total world FDI stock more than tripled, reaching over USD 204 billion, and it has continued to grow faster than global output (Held et al. 1998).

Since 1973, several general trends in FDI have emerged. First, the total amount of FDI inflows has increased rapidly, albeit in a jerky uneven fashion (Figure 6.2). A quick glance at investment movements reveals large dips in the 1980s and 1990s experienced over a short period of time, with plunging retreats in FDI flows in 2002–2003, and most recently after 2008, followed by recovery. In terms of levels, inward FDI flows reached USD 12.3 billion

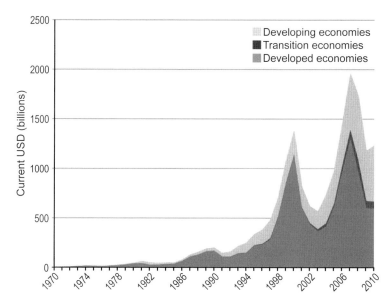

Figure 6.2 Global inward flows of foreign direct investment, 1970–2010. Data from UNCTADstat (2012).

in 1970. Following a relative slowdown in FDI during the early 1980s, FDI boomed in the late 1980s and reached record growth rates in the late 1990s. By 2000, FDI inflows exceeded USD 1.4 trillion, and in 2007, at its height, FDI inflows stood at approximately USD 1.97 trillion. By 2010, inflows were again climbing and had already exceeded USD 1.2 trillion.

The number of parent company TNCs and affiliates has grown with increasing levels of FDI. Currently there are almost 70,000 parent companies and over 690,000 foreign affiliates. TNCs account for one-quarter of world GDP, with 14% accounted for by activities of parent companies and 10% from foreign affiliates (UNCTAD 2010, 2011).TNCs are based increasingly in developing countries. The share of total TNCs from developing countries increased from 8% in 1992, to 21% in 2000, and approximately 28% in 2008. Moreover, 7% of the 100 largest TNCs are from developing countries (UNCTAD 2010). Within the developing world, Asia dominates the emergence of TNCs. For example, of the 100 largest TNCs from the developing world, only 18 are outside of Asia.

In general terms, the range of recipient countries has broadened, and while the developed world is the main destination for FDI, investment in developing countries in Latin America as well as East and Southeast Asia is also growing. In 1970, more than 70% of FDI inflows were to developed world countries, with North America (23%) and Europe (39%) taking the lion's share. This trend remained stable, and even increased, over time until around 2000. By 2005, however, developing world FDI inflows accounted for more than one-third of total. In 2010, the developing world received more than 46% of all FDI inflows. Moreover, in 2010, FDI inflows into the developing world and economies in transition together exceeded 50% of total.

This shift is clear from the changing status of the top FDI recipients. By 2010, four of the top ten recipients included China, Hong Kong, Brazil, and Singapore, and inflows into these countries accounted for more than one-third of the inflows into the top ten nations and more than 21% of total global FDI inflows. Furthermore, half of the top twenty host economies for FDI in 2010 were developing or transition economies (UNCTAD 2011).

The second new trend in FDI investment is that the level of outward FDI flows from developing countries has increased (Table 6.3). Prior to 1990, more than 95% of all FDI was from developed countries. By 2000, FDI from developing countries reached just under 11% of total. By 2010, developing countries accounted for almost a quarter of all FDI outflows, with economies from Asia accounting for almost 6%. Moreover, transition economies, from which FDI was negligible prior to 2000, now account for almost 5% of global FDI outflows.

Third, FDI stocks have not only increased, the location of these stocks is also shifting. Following the global increase of flows during the 1990s, FDI stocks have accumulated to reach USD 19 trillion in 2010, up from USD 700 billion in 1980. Since the postwar era, most inflows have been in the developed world: the "Triad" grouping of the European Union, United States, and Japan.

Table 6.3 Outward flows of foreign direct investment: annual, 1970–2000. US dollars (USD) are given at current prices and current exchange rates in millions and percent of global total. Data from UNCTADstat (2012).

Economy	1970 USD	1970 %	1990 USD	1990 %	2010 USD	2010 %
World	14,151		241,498		1,323,337	
Developing economies	51	0.4	11,914	4.9	327,564	24.8
Africa	19	0.1	659	0.3	6,636	0.5
Latin America	31	0.2	301	0.1	76,273	5.8
Asia	1	0.0	10,943	4.5	244,585	18.5
Oceania	0	0.0	11	0.0	71	0.0
Transition economies	0	0.0	0	0.0	60,584	4.6
Developed economies	14,100	99.6	229,584	95.1	935,190	70.7
North America	8,521	60.2	36,219	15.0	368,183	27.8
Asia	364	2.6	51,036	21.1	64,224	4.9
Europe	5,095	36.0	139,342	57.7	475,763	36.0
Oceania	120	0.9	2,988	1.2	27,020	2.0
Least developed countries	0	0.0	–3	0.0	1,819	0.1
Africa and Haiti	0	0.0	–5	0.0	1,702	0.1
Asia	0	0.0	1	0.0	108	0.0
Islands	0	0.0	1	0.0	8	0.0

As such, the accumulation of FDI stocks increased in share over time. In 1914, approximately 63% of world FDI stocks were located in developing countries. By 1980 the developing world share fell to 43%. By 2000, about 75% of world FDI stocks were located in the developed world. In the early twenty-first century, however, FDI stock shares in the developing world started to climb again. These increases occurred mainly in the developing economies of Asia. By 2010, Asian developing economies accumulated 19% of world FDI stocks and developing countries' share exceeded 31% of total (UNCTADstat 2012).

Fourth, the speed of FDI growth rates has increased. Importantly, FDI has increased more quickly than trade. Since the early 1990s the majority of countries liberalized their foreign investment regulations and actively encouraged inward investment. In the period 1991–1996, 95% of the 599 changes in national FDI regulations across the globe were in the direction of further liberalization (UNCTAD 1997:10). This trend has been particularly important with respect to investment in services, notably financial services, where foreign companies had previously faced various national restrictions. In response, during the mid- to late 1990s, FDI growth rates exploded (Table 6.4) with the result that by the early twentieth century, FDI flows reached up to 17% of trade.

Table 6.4 Selected indicators of foreign direct investment and international production, 1990–2010. Data from UNCTAD (2011:24).

Indicator	Value at current prices (billions of USD)				Annual growth rate			
	1990	2005–2007	2008	2010	1991–1995	1996–2000	2001–2005	2010
FDI inflows	207	1,472	1,744	1,244	22.5	40.1	5.3	4.9
FDI outflows	241	1,487	1,911	1,323	16.9	36.3	9.1	13.1
FDI inward stock	2,081	14,407	15,295	19,141	9.4	18.8	13.4	6.6
FDI outward stock	1,094	15,705	15,988	20,408	11.9	18.3	14.7	6.3
Sales of foreign affiliates	5,105	21,293	33,300	32,960	8.2	7.1	14.9	9.1
Exports of foreign afilates	1,498	5,003	6,599	6,239	8.6	3.6	14.7	18.6
Employment by foreign affiliates (thousands)	21,470	55,001	64,484	68,218	2.9	11.8	4.1	2.3
GDP	22,206	50,338	61,147	62,909	6.0	1.4	9.9	8.6
Exports of goods and services	4,382	15,008	19,794	18,713	8.1	3.7	14.7	18.6

Much of this growth was due to inflows into Asia and increasingly into the Least Developed Countries (LDCs) (although increases in Africa and Island LDCs start from lower levels). FDI is increasingly the vehicle for cross-border production. By 2010, the sales of foreign affiliates were 76% higher than the value of export trade in goods and services. This figure is up from 16% in 1990. While trade grew at impressive annual rates of over 8% during the early 1990s, FDI grew by over 22%. In the late 1990s, FDI grew by over 40% annually, compared to trade growth at almost 4%. Since 2000, however, trade has grown faster than FDI flows.

Finally, the dispersion of TNC activity, or the ability to create global value chains, has continued to increase. This has resulted in an increase of intra-firm trade, greater penetration of local markets, and, in general, advances in globalization of industries. There are now few countries outside the reach of TNCs.

Financial Flows

Changes in the structure of global finances combined with the deregulation of financial markets and technological advances have encouraged a dramatic expansion of the extensity and intensity of global financial flows and networks during the postwar period. Financial flows have increased faster than trade and FDI in growth and volume, and are the largest and potentially the most powerful economic flows of contemporary globalization.

Financial flows have increased since 1950, with the greatest increases occurring since 1990. Levels of international bank deposits, for example, expanded during the 1960s and accelerated after the 1970s. Global bank network savings by nonresidents rose to USD 7.9 trillion in 1995 and to over USD 19 trillion in 2011 (BIS 1996, 2012). In addition, over USD 4 trillion worth of bank deposits are now located in the 60 jurisdictional locations of offshore finance centers. By 1995, total international bank loans amounted to USD 7.1 trillion. By 2010, this figure increased to USD 22.3 trillion.

The annual volume of new Eurobonds grew to USD 5 billion in 1972, USD 43 billion in 1982, and USD 371 billion in 1995 (Kerr 1984; OECD 1996). Net issuance of all cross-border bonds and notes rose from USD 263.6 billion in 1995, to USD 1.0 trillion in 2000, and to USD 1.6 trillion in 2005. The total amount outstanding in claims for these bonds by 2011 amounted to over USD 27.5 trillion.

International equities markets have also dramatically increased in value. Since 1988, global market capitalization of listed companies increased from USD 9.7 trillion to USD 56.2 trillion in 2010 (World Bank 2012: Market capitalization of listed companies). Total stocks traded annually exceeded USD 97 trillion in 2007 before falling to USD 63 trillion in 2010. Most of this growth in market capitalization occurred in developed countries; however, during the first decade of the twenty-first century, markets in the developing world gained share. By 2010, the developing world accounted for 20% of total market capitalization (11% in Asia). The skyrocketing value of trading activity can be seen both in terms of the value of the stock traded with respect to global GDP and the value of the stocks traded in terms of percent GDP of the region (Figure 6.3). Cross-border investments have risen through investment portfolios. Today, numerous investors (especially institutions such as pension funds, insurance companies, unit trusts, and hedge funds) operate global portfolios. Many of these investment companies have further deepened their supraterritorial character by registering offshore, particularly in Luxembourg, the Bahamas, Dublin, and the Channel Islands. Global net portfolio equity inflows into different parts of the world were approximately USD 162 million in 1960 (Table 6.5). Cross-border portfolio investments increased to USD 1.4 billion by 1970 and USD 15.5 billion in 1980. By 2000, international portfolio flows exceeded USD 650 billion, and by 2010 they were USD 780 billion. At their height, in 2007, annual international portfolio flows exceeded USD 920 billion. These high levels of turnover reflect the fact that institutional investors are actively trading equities across borders.

Global markets for derivatives developed rapidly during and after the late 1980s. The notional amount of outstanding over-the-counter financial derivatives contracts reached USD 88 trillion at the end of 1999. In 2000, the value of these instruments exceeded USD 95 trillion, and by 2005 they exceeded USD 299 trillion. By the close of 2011, total contracts in over-the-counter derivatives reached USD 647.8 trillion (BIS 2012). Daily turnover rates have increased from USD 265 billion in 1998 to USD 2.06 trillion in 2010. Trading in derivatives has, in many instances, exceeded that of the underlying economic collateral. Total trades in derivatives in 2010 exceeded global GDP by a factor of ten.

In contrast to other financial flows, aid to developing countries in the form of development assistance has increased slowly over the past forty years, from an average of USD 47 billion during the 1970s to over USD 100 billion during the past decade (constant 2009 USD). During the 1970s and the most recent decade, official development assistance (ODA) into developing countries has been increasingly overwhelmed by private financial flows. For example, during the 1970s, ODA made up approximately 33% of all flows into developing countries. This ratio increased during the 1980s to reach 45% of all flows, but has since been decreasing. By the first decade of the twenty-first century, ODA made up approximately 38% of total flows. At the same time, however, a year-to-year picture demonstrates the variability of private flows into developing countries and the stability of ODA.

Foreign exchange markets, on the other hand, experienced enormous growth during this period. In 1998, USD 1.5 trillion of foreign exchange instruments were traded daily. By 2010, this figure reached USD 3.9 trillion (BIS 2012). These foreign exchanges included every national denomination in innumerable transactions that approach their "home" soil. Moreover, the aggregate volume of official foreign exchange reserves, or the foreign currency deposits and bonds held by central banks and monetary authorities of individual countries, in

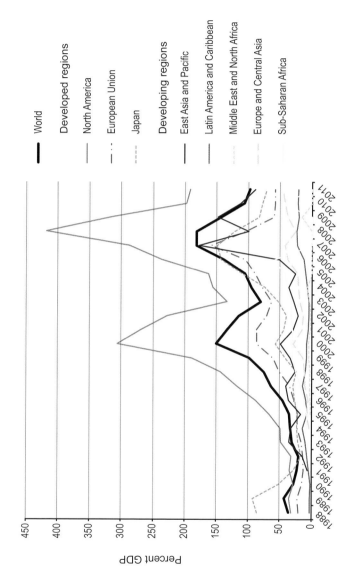

Figure 6.3 Stocks traded, total value (% of GDP). Data from World Bank (2012: Market capitalization of listed companies).

Table 6.5 Net inflows of portfolio equity, from 1960–2010 (balance of payments given in current USD millions). Data from the International Monetary Fund, Balance of Payments database.

Region	1960	1970	1980	1990	2000	2010
World	162	1,363	15,513	−11,586	657,427	779,547
Developed countries	162	1,363	15,452	−14,975	645,949	689,334
Europe	92	738	3,098	14,093	290,697	220,666
North America	70	624	5,523	−17,462	217,839	190,392
Japan			6,550	−13,357	−1,286	40,328
Other developed		1	282	1,750	138,700	237,948
Developing countries			61	3,390	11,477	90,213
Middle East & North Africa			17	5	265	1,973
East Asia & Pacific			51	440	6,589	39,715
Latin America & Caribbean			−11	2,464	−561	41,302
Europe & Central Asia				89	986	−840
Sub-Saharan Africa			4	393	4,198	8.063
World GDP				19,042,944	31,337,733	63,074,926

the world rose from USD 100 billion in 1970 to USD 1.39 trillion in 1995, USD 4.3 trillion in 2005, and USD 10.2 trillion in 2011 (Held et al. 1998; IMF 2012).

Several new forms of international money also emerged during this period including international credit cards from 1951, chip or "smart" cards from 1981, and debit cards in the 1990s. In "mature" economies, such as the United States and Western Europe, approximately 50% of personal consumption expenditure is in nonpaper form. As of March 2012, Visa's global network (known as VisaNet), incorporated over two billion credit cards in over 200 countries with a total volume of USD 6.3 trillion. In 2011, according to their corporate overview, VisaNet processed approximately 109 billion transactions. Visa, a name adopted by Bank of America in 1976, expounds a commitment to "doing business wherever our global clients need us to be to serve their needs."[3] Meanwhile, according to its annual report, rival MasterCard processed 23 billion transactions from over 1.6 billion cards for approximately USD 2.7 trillion in 2010. Both VisaNet and Mastercard operate ATMs around the world.

---

[3] Bank of America, Annual Report 2011. http://media.corporate-ir.net/Media_Files/IROL/71/71595/AR2011.pdf

Remittances

Remittances from a growing global migrant population have become significant. Total world remittances grew from USD 1.9 billion in 1970 to USD 448 billion in 2010, or over three times the ODA (Ratha et al. 2010; World Bank 2012: Workers' remittances and complensation of employees).The true size of remittances including unrecorded flow through formal and informal channels is likely to be even higher. Early on during the postwar period, the lion's share of remittances flowed into developed countries (81.3%). By the late 1980s, however, flows became equal, splitting 50% of world remittances to each of the developed and developing regions. In 2010, approximately 31% of all remittances were flowing into developed countries and 69% were destined for developing countries. About 37% of global remittances in 2010 went to five countries listed in the order of total money received: India, China, Mexico, the Philippines, and France. Given the size of these flows, many developing countries are implementing policies to nurture and improve these flows. With increasing migration, remittances have recently become an important policy issue. For the country sending migrants abroad, remittances sent home lead to increased incomes and poverty reduction, improved health and educational outcomes, and generally promotion of economic development (Ratha et al. 2011).

Summary

Table 6.6 presents economic flows during 2010 so that they can be put into relative perspective. Several things are worth noting from these figures. First, the absolute size of values for economic flows is daunting. Compared to global GDP, one can see how these flows are increasingly important in economic development. A second apparent trend is the relative size of the financial flows. International bank loans are now at approximately 38% of world GDP, international bond worth is 46% of global GDP, and the sales of foreign affiliates of TNCs exceed 57% of global GDP. Moreover, some of the values, particularly derivatives and forex transactions are over 10 and 18 times, respectively, the size of global GDP. This suggests that financial flows should not be overlooked when we consider environmental impact. The size and character of these flows has helped to make them both indirect and direct drivers of change for other systems.

Influence on Land-Use Transitions

Some analysts suggest that human transformation of land has had global effects as early as 8,000 years ago (Ruddiman 2003). Certainly, there are cases of region-wide changes in environmental conditions due to anthropogenic land-use shifts, starting about 6,000 years ago (Redman 1999). Some of these

Table 6.6 Relative strength of various global economic flows, 2010 (USD billions unless specified).

Globalization indicator	2010
GDP	63,075
Trade	
Exports (USD)	15,256
Weight (million tons)	8,408
FDI	
Inflows	1,244
Stocks	19,141
Sales of foreign affiliates	32,960
International bank loans	
Positions of banks	22,004
Syndicated credit	1,724
Debt securities	
Net issuance	1,514
End of year claims	26,784
Equities	
Market capitalization	56,173
Portfolio equities inflows	780
Derivatives	
Total contracts (annual)	601,046
Daily turnover	2,057
Foreign exchange	
Estimated annual	1,034,983
Daily turnover	3,981
Remittances	
Workers remittances and compensation	449
Development assistance	
Official development assistance (2009)	124
Net grants from NGOs and other official sources (2009)	31

examples include nonlocal change dynamics. For example, the deforestation of the Levant region during the pre-Greek era demonstrates how human activities at a distance affected land cover. By the seventh century BC, Levantine iron was being exported to Babylon. Because smelting requires high temperatures, deforestation occurred (Goldewijk and Ramankutty 2001). In classical Rome, forests were the source of fuel, building, and war material and were cleared for agriculture. Deforestation started in the area around Roman city-states and spread throughout the Roman provinces. When Rome's population was approximately 800,000, the city's leaders were ordering raids on granaries and agricultural areas throughout the entire Mediterranean basin, as well as from Iran to Scotland (Lowry 1990).

Much more dramatic changes have occurred over the past 300 years, and particularly over the past several decades (Ramankutty et al. 2006). Land-use change is now a global issue in terms of both share of the total terrestrial ecosystem that has changed and the impact of these changes on global systems (Houghton 1994; DeFries et al. 2004).

Until recently, however, the causes of land-use and land-cover change were seen largely as a local issue of cumulative importance. Studies suggested that population increases and poverty were the major underlying causes of land-use change transitions, such as the forest transition. The forest transition (Mather 1992; Mather and Needle 1998), defined as the long-term decrease in forest cover followed by a slow increase, was considered a good example of these dynamics; shifts in farmer population and behavior as they either let less productive lands go fallow or they left farms for urban jobs, resulted in forest regrowth in formerly worked fields (Mather et al. 1999).

Recent studies, however, suggest that while local issues are important to environmental change, they are enhanced or attenuated by nonlocal influences (Wilbanks and Kates 1999). These global forces include, among other things, trade, FDI, and financial and remittance flows. Analysts now suggest that global forces have become important, if not the main, determinants of land-use change (Geist and Lambin 2002; Lambin and Meyfroidt 2010, 2011; Lambin et al. 2001; Rudel et al. 2009a) and land-use transitions (Foley et al. 2005).

How do globalization and economic flows influence land-use and land-cover change? This discussion attempts to place the previous descriptions of globalization and economic flows in relation to land dynamics and specifically land-use transitions. Potential generalized ways in which the global economic system influences land-use dynamics are outlined, as are the results of these influences on land-use transitions.

Globalization and Economic Flow Influence on Land-Use Dynamics

Analysts suggest that there are two ways in which globalization and economic flows influence land-use dynamics: indirectly and directly. Together these modes of influence have created a new global land-use/land-cover system dynamic. That is, globalization processes and the resultant economic flows have created a new global system that is more integrated than previous global economic systems. The basis of global economic integration is a number of harmonized institutional arrangements that increase privatization efforts, expand markets, lower tariffs, diffuse information and cultural practices, etc., and therefore affect land-use dynamics. For example, the expansion of markets and the privatization of formerly public lands have helped to increase trade in forest and primary commodities, as evidenced above. The demand for goods across borders is facilitated by national and local factors, such as trade and investment opportunities and policies, and subsequently impacts agricultural and forestry development programs (Lambin and Meyfroidt 2011). The privatization of land has emerged as an important factor in deforestation (Geist and Lambin 2002). The diffusion of land-management systems from the developed to the developing world directly contributes to both social and spatial marginalization of particular urban groups in developing world urban systems (UN-Habitat 2009). Malfunctioning common property regulations and new land tenure and zoning measures are associated with land degradation in drylands (Geist et al. 2006). There is also a cultural component to globalization that spreads consumption to landscapes, particularly in urban and peri-urban areas. This can, for example, promote U.S.-style land-use patterns with golf courses, gated communities, and large lot single-family homes in unlikely places in the rapidly developing world (Leichenko and Solecki 2005).

These system changes have occurred with the growth of globalization and economic flows. The processes that create the flows, and some of the flows themselves, act as primary (or indirect) drivers which subsequently act upon proximate factors affecting local ecosystem change (MEA 2005). For example, the ability of large amounts of capital to flow across borders has affected the ability of local ranchers and agricultural producers to intensify land use more rapidly than during previous periods, when capital and technologies were largely domestic. Eco-labeling, organized food networks, and the diffusion of technologies which speed up communications and transportation all work to create incentives or constraints for land management (Geist et al. 2006). Trade expansion demands result in changes in land use along coasts for port development and road and train infrastructure. Road infrastructure subsequently opens up further opportunities for land-use change. The general mechanism by which globalization and economic flows act as indirect drivers affecting land-use dynamics is either to attenuate or amplify other driving forces of land-use and land-cover change by removing regional barriers, weakening connections within nations and increasing the interdependency among people and between nations (Lambin et al. 2001; Geist et al. 2006).

Global economic flows, however, can also act as direct drivers of land-use change dynamics. This occurs when a direct linkage is established between nonlocal actors and a land-use or land-cover decision-making process. One typical linkage, of great importance to urban and peri-urban areas, is through FDI flows. As mentioned earlier, FDI flows are capital investments made by TNCs. These investments have increasingly been flowing into developing countries, and most notably in Asia, in and around major metropolitan centers, affecting urban growth patterns in these locations (Lo and Marcotullio 2000). One typical pattern is a ring of industrial firms located directly outside large cities in Asia (Marcotullio 2003). While these types of flows have been facilitated by institutional actions, such as national and local privatization and liberalization policies, they represent a direct linkage of nonlocal factors in local land-use change. Direct influences, however, are also found in nonurban areas.

In the past, nonlocal actors of great wealth have occasionally purchased or leased large tracts of lands in foreign countries for their idiosyncratic desires. Examples include Henry Ford, who during the 1920s developed an industrial town in the Amazon along the Tapajos river, as a secure source of cultivated rubber for his automobile manufacturing operation (Galey 1979), and Daniel Ludwig, American billionaire, who began in 1967 a paper milling and pulp project in the Amazon rainforest along the Jari River, on a piece of land larger than the State of Connecticut (Jostock 2008). Today, however, there are multiplicities of different entities with large amounts of capital that can use them for investment, purchase, or lease of land. In addition to large TNCs, there are a number of new institutional actors that can invest. For example, as of 2010, the global total of assets under institutional management were estimated at close to USD 80 trillion, including USD 30 trillion in pension fund assets, about USD 25 trillion in mutual fund assets and another USD 25 trillion in insurance funds. Moreover, there are approximately 10,000 hedge funds with capital under management of nearly USD 1.9 trillion (Smith et al. 2012). Another powerful new organization is the sovereign wealth funds. A sovereign wealth fund is a state-owned investment fund or entity that is commonly established from balance of payments surpluses, official foreign currency operations, the proceeds of privatizations, governmental transfer payments, fiscal surpluses, and/ or receipts resulting from resource exports. In 2012, the world's top 36 sovereign wealth funds totaled nearly USD 5 trillion. These entities can buy or lease large tracts of land in foreign locations, and many of these types of organizations have been implicated in recent "land grabs" (GRAIN 2012). The general mechanisms for direct influence on land-use dynamics is through global flows to create ownership or management opportunities of land by distal sources.

Changes in Land-Use Transition Processes

What are the effects of the nonlocal influence on land-use and land-cover dynamics? Here I provide a generalized outline of changes. In drawing from the literature, several aspects of land-use transitions that are influenced by global economic flows are identified and compared to those in the past. That is, if we take, for example, the experiences of Europe and North America as that represented by the forest transition (Foley et al. 2005), contemporary changes in the global economic system have forced changes in these previous patterns, including the interrelated changes in speed, direction, location, and timing of decisions.

Speed

The speed of change refers to the amount of time it takes to undergo a transition or change to the land-use system. Contemporary speeds of transition are different from those in the past. Typically, the speeds of transition are faster, but they could also be slower. Conceptually the change in speed relates to the increase in the ability of capital and finance to flow into any one location. As countries are absorbed into the global economic flow network, the levels of capital and finance invested in land-use decisions increases faster than in previous eras.

A comparison of urbanization transition rates demonstrates the differences in speed between the development paths of the developed and rapidly developing world. For example, while the United Kingdom has urbanized at a slower rate than the United States, Japan and Korea are the fastest urbanizers in the world. Japan's urban residents shot up from 18% of the total population in 1920 to 71% in 1970; the same increase occurred over 130 years in the United States. South Korea experienced a similar urbanization transition over a period of 45 years, from around 1950 to 1995. China is also expected to increase its urban share from around 18% to 71% in 45 years (UN 2010).

Changes in speed are also reflected in urban land-use change. According to Angel et al. (2011) medium projections for the future suggest that urban land cover in developing countries will more than double over a thirty-year period, from 300,000 km^2 in 2000 to 770,000 km^2 in 2030, and then increase by another 64% by 2050 to reach 1.2 million km^2. Seto et al. (2011), however, predict that by 2030, total global urban land area will be more than 1.5 million km^2.

Lambin and Meyfroidt (2010) suggest that globalization has also accelerated other land-use changes, such as land-use intensification in some places and crop abandonment in others. An examination of the change in soybean production over the past thirty years demonstrates an acceleration of cultivation of this crop, which is related to, among other things, the globalization commodity chains (feed production process) and demand in distant places (Reenberg and Fenger 2011). A study of forest transitions, as experienced by a number of different nations, suggests that those in Europe (Denmark, France, Ireland, and the United Kingdom) occurred over longer periods of time than currently developing countries (Gambia, Costa Rica, China, and South Korea). Furthermore, the turning points in the transition during the twentieth century occurred with larger shares of remaining forest (Rudel et al. 2005).

Direction

The direction of land-use change refers to the level (share amount) and intensity of any given land use. For example, there is evidence that global trade flows in agricultural products have increased land under agricultural production as well as increased the intensification of agricultural production at the global level. As shown above, trade in agricultural products has increased over the past several decades and, at the global scale, agricultural trade has increased faster than agricultural production. Agricultural production has decoupled from cropland expansion during the past decades. At the same time, however, as trade has increased faster than cropland expansion, the amount of land irrigated and the annual rate of nitrogen and phosphorus fertilizer use has increased (intensity of land use) dramatically (Tilman 1999; MEA 2005). Forest product trade has also increased over the past few decades and with that increase amount of land with forests has declined. For example, between 1990 and 2009 the forested areas around the world declined by approximately 130 million hectares (FAO 2012a).

The direction of land-use change includes changes in expected patterns of development. In terms of land-use transitions, although there is evidence that some developing nations have undergone forest transitions and increased their share of forest land over the past years (Bae et al. 2012; Lambin and Meyfroidt 2011; Rudel et al. 2005), there is also evidence that some large forest-producing nations have not. In places such as Brazil, Cameroon, and Indonesia, continued deforestation is associated with expanding overseas markets for forest products as well as agricultural products. According to one group of researchers, if future global market conditions favor a boom in agricultural expansion in the future, it is questionable whether government and industry-led policies, in places such as Brazil, will be able to contain further deforestation (Macedo et al. 2012). Thus globalization may increase further and continue to influence deforestation in these places. It can be argued that in the absence of global demand signals, forest transition would play out differently.

Location

While global demand and increased transportation and communication technologies have facilitated the increase in land uses for various types of functions, there is a limit to what extent this trend can continue (for details on current and future land-use predictions, see Lambin and Meyfroidt, this volume). As the amount of land available has been increasingly under one or another type of use, trade-offs between land uses will be more frequent, as the remaining proportion of land is not economically and/or physically amenable for use (DeFries et al. 2004). In some cases, the increasing demand for a product (agricultural, forest, or urban) will help dictate the location of other specific types of uses. Indeed, this is already occurring and has been influenced by economic flows and globalization processes. As a result of displacement or leakage, agricultural intensification, for example, in one country triggers compensatory changes in trade flows and thus, indirectly affects land use in other countries, as these accommodate the displaced activities (Lambin and Meyfroidt 2011).

The change in location of land use or the displacement effect can occur in two different ways. First, with increasing global economic flows, more countries displace their demands for agricultural land abroad. Compared with patterns in the 1960s, a much larger proportion of the post-millennium displaced demand for timber and dairy products is now being absorbed by Brazil, Indonesia, Chile, and India (Meyfroidt et al. 2010). These locations have thus increasingly become the source of products that supply transnational demand.

Second, land-use restrictions to protect or conserve areas cause a displacement effect. Protected areas may displace land-use activities by forcing deforestation, settlement, or extraction activities to areas outside the reserve (Lambin and Meyfroidt 2011).

The change in location of land uses is not limited to primary commodity production. The displacement of industrial functions has been an ongoing process, associated with the growth of, among other things, the international division of labor, flexible production processes, and increases in FDI flows. The deindustrialization of the United States and Japan accompanied the growth of production in East and Southeast Asia. Typically, industrial firms locate factories on the outskirts or peri-urban areas of major metropolitan centers. This change in land use sets off a number of processes associated with urban growth (Lo and Marcotullio 2001).

While the displacement of industrial production is similar in outcome to that of agricultural and forest production, it represents a second and different type of displacement. In this second type, decisions for land-use change are generated from abroad. That is, in the displacement of agricultural and forest production, for example, local farmers, foresters, and government agencies respond to global trade and economic flow signals. In the industrial case, overseas firms look for favorable local signals to invest (e.g., local policies, business opportunities, labor and capital costs). While localities attempt to draw in investment from abroad, transnational firms also drive the process.

At the same time, a new form of agricultural and forest product displacement is emerging. This new form follows the patterns of industrial displacement, as entities from abroad invest in land to secure certain types of products. "Land grabbing" or "the land rush" occurs where large TNCs, institutional investors, sovereign entities, or other high net wealth entities acquire land. The process occurs in a large number of target countries and is slowly growing. By 2009, there were 1,217 agricultural land deals amounting to 83.2 million ha for land in developing countries, with Africa as the most targeted region (Anseeuw et al. 2012b; Friis and Reenberg 2010). The recent wave of investment differs from past trends, as it involves new types of investors, and focuses mainly on countries that did not appear to be attractive targets earlier, and which have very weak land governance (Deininger et al. 2011).

Timing

The timing of land-use change refers to the sequence of change as well as the point at which change occurs during the developmental process. Two aspects of the timing of land-use changes have emerged. First, current transitions occur at different times in the development process, as measured by GDP per capita (Table 6.7). In Asia, forest transitions seem to take place lower down the income scale. For example, the (unweighted) averages for Asian GDP per capita levels suggest that transition occurs at approximately USD 1,220 per

Table 6.7 Forest transition turning points and GDP per capita.

Economy	Approximate date of transition	GDP per capita
Europe		
Austria	1880	2.079
Belgium	1850	1.847
Denmark	1810	1.274
Finland	1900	1.668
France	1830	1.191
Germany	1890	2.428
Hungry	1930	2.404
Ireland	1930	2.897
Italy	1925	2.921
Poland	1890	1.536
Portugal	1870	975
Sweden	1900	2.209
Switzerland	1860	1.745
United Kingdom	1925	5.144
Average		2.166
European Offshoots		
USA	1920	5.552
New Zealand	1900	4.298
Average		4.925
Asia		
China	1970–1980	886
India	1950–1980	785
Philippines	1988	2.105
South Korea	1955	1.169
Vietnam	1991–1993	1.133
Average		1.216
Latin America		
Chile	1950	3.670
Costa Rica	1990	4.747
Cuba	1960–1970	2.045
El Salvador	1980–1990	2.128
Uruguay	1990	6.465
Average		3.811

Sources: Rudel (1998); Bae et al. (2012); Meyfroidt and Lambin (2011); Mather (1992); Rudel et al. (2009a); Kauppi et al. (2006); Houghton and Hackler (2000); Maddison (2001)

Notes:
GDP per capita in 1990 International Geary-Khamis dollars.
GDP per capita date for Denmark is 1820 and for Poland is 1900.
GDP per capita for multiple years was identified by the yearly mean for the selected period.

capita compared to the average for European nations of USD 2,170.[4] In Latin America, forest transitions occur at higher GDP levels, approximately USD 3,810 per capita; this is higher than in Europe, but lower than its European offshoots (USD 4,930).

Second, land-use transitions are occurring more simultaneously than in the past. The land-use transition model, as described by Foley et al. (2005), suggests that while the developed world experienced transitions in a sequential manner—moving from "presettlement," to "frontier," to "subsistence," to "intensifying," to "intensive" stages—those in the developing world may not follow the same pattern. Now, intensification can be found alongside frontier and subsistence land-use stages (e.g., Brazil and Indonesia). Nations may remain at one stage or potentially skip stages. The blurring of these sequential stages suggests that while the staged land-use transition adequately described the pattern of development to date, it no longer adequately does so for all nations.

Jumps and pauses between stages, as well as telescoping of stages, could be related to the flow of technology, capital, and knowledge to areas undergoing rapid land-use changes (for a description in urban areas, see Marcotullio 2005). As technologies previously unavailable at a given stage flow into developing nations, the ability of farmers and ranchers to effect land-use change and intensification shifts. The larger sums of capital available through loans, bonds, or foreign investment help to transform patterns of land development. The volatility of the commodities market, for example, is also important as previously used land can, during the following year, be left fallow or put to other use as agents respond to rapidly changing global market conditions.

Conclusions

Globalization as a concept helps explain the contemporary human condition defined by daily interactions with people and consumer products from foreign lands, and contact with intangible ideas, information, and images from around the world. Contemporary globalization—as both a set of institutions and massive flows of information, finance, people, and technologies—has changed not only economic dynamics but other systems as well. We are living in a new era, where there is "something new under the sun" (McNeill 2000). That "something new" includes changes in socioecological systems. Underpinning much of the new system dynamics are globalization institutions, processes, and flows—flows that are larger, move faster, farther and between a larger number of political entities than at any other time in history.

One system that has been altered by globalization and economic flows is the global land-use, land-cover system. The changes identified in this chapter

---

[4] The weighted average for Asia would be lower because both China and India, the largest populated countries, have GDP per capita levels lower than the unweighted average.

have been systemic; they have changed the way the global land-use, land-cover socioecological system operates. The implications of changes in the speed, direction, location, and timing of land-use transition dynamics indicates that this system is globally integrated and that it has experienced functional shifts in the way it is organized, at this scale.

At the same time, the concepts of globalization, global economy, and globality bring to mind actions that impact the planet. They could be imagined as powerful forces which drop from the skies or are generated by processes that are ever present, all-encompassing, and inevitable. Globalization is often equated with products, capital, and ideas being everywhere all of the time. This conceptualization of globalization, however, is not accurate. No matter what the scale of the effect, all "global" processes originate in a specific place and are mediated by national, regional, and local conditions (Dicken 2003). The spaces of origin, even in regard to land use and land cover, are increasingly dense settlements. This should not be surprising as the human population is increasingly urban, and thus the sources of global flows are increasingly centered in dense settlements. The implications are significant. To understand the details of contemporary changes in land use and the role of various determinants, new models are needed. Those recently put forth as land teleconnections promise to provide a way forward to enhance our understanding of land-use dynamics in a global era. Equally important, "urban land teleconnections" (or how actors and processes originating or destined for dense settlements interact) must be part of these new models. As the world population concentrates in urban areas, activities associated with urban land use will be increasingly responsible for global environmental change, including land-use and land-cover dynamics.

7

Applications of the Telecoupling Framework to Land-Change Science

Jianguo Liu, Vanessa Hull, Emilio Moran,
Harini Nagendra, Simon R. Swaffield, and B. L. Turner II

Abstract

Over the past two decades, progress has been made in understanding and predicting land-use change in specific places, using frameworks such as coupled human-natural systems, coupled human-environmental systems, or coupled social-ecological systems. However, land-use change around the world is increasingly being driven by new agents and causes which emanate from distant locations, through forces such as trade, migration, transnational land deals, and species invasions. New conceptual frameworks are thus needed to account for such distant forces. This chapter applies a framework that explicitly takes distant forces into account in land-use change and builds on the concept of telecoupling (i.e., environmental and socioeconomic interactions among coupled systems over large distances). Telecoupling is a logical extension of coupled systems thinking; it draws insights from related concepts in different disciplines and serves as an umbrella concept to address and integrate various types of distant connections between coupled systems. The telecoupling framework includes five major and interrelated components: coupled human-natural systems, agents, flows, causes, and effects. An overview of the telecoupling framework is presented and two examples (transnational land deals and species invasions) demonstrate the application of the framework to global land use. Finally, challenges and opportunities in understanding telecouplings and their consequences are highlighted and calls made for new directions in land-change research.

Introduction

Over the past two decades, many advances have been made in understanding and predicting land-use dynamics at a global scale (Turner et al. 2007). In particular, land-use change has been extensively studied using systems frameworks, such as coupled human-natural systems (McConnell et al. 2011; Liu et al. 2007a), coupled social-ecological systems (Walker et al. 2004), or coupled human-environmental systems (Moran 2010; Turner et al. 2003). Coupled systems are integrated systems in which humans and natural components interact. These frameworks view land use as a function of interactions between socioeconomic and ecological factors within a coupled system (i.e., local or internal couplings). Although the frameworks for these systems are helpful in guiding the analysis of internal forces in driving land-use change, they fall short in their consideration of the increasing scale, extent, and speed of existing and emerging connections between coupled systems over large distances.

Distant connections between land-use systems are not new. They can be traced as far back as the third millennium BCE between areas now known as Iraq and India (Frank 1998). They were also present, for example, during the ancient Greek and Roman eras, along various trade routes in Asia, along the Silk Road between ancient China to Europe, and through the Columbian Exchange (following Christopher Columbus's voyage to the Americas in 1492, which led to widespread exchange of animals, plants, humans, food, culture, and ideas between the Western and Eastern Hemispheres) (Nunn and Qian 2010). Great Britain's rise to supremacy as an industrial power in the eighteenth century was also dependent on distant access to raw materials and markets. Modern connections progressed in the nineteenth century with advances in transportation, telecommunication, and economic industrialization (Headrick 1991). Today, even stronger connections have developed between coupled systems around the globe (Lambin and Meyfroidt 2011; DeFries et al. 2010; Seto et al. 2012b; Eakin et al. 2009; Haberl et al. 2009; Nepstad et al. 2006). These connections are related to many of the greatest and most complex challenges that face societies, such as food security, demands for energy, destruction of ecosystems, and biodiversity loss.

To address these unprecedented challenges, new conceptual frameworks are needed to guide analyses of these increasingly important distant interconnections, so that future land use can more accurately be projected and better governance and policies on land-use change can be developed. In this chapter, we present a multidisciplinary conceptual framework for such interconnections and present examples to illustrate key components of the telecoupling framework and their relations to global land use. We highlight challenges and opportunities in addressing telecoupling, and call for new directions in land-change research.

Overview of the Telecoupling Framework

Many disciplines consider interactions between distant systems. The idea that distant places and processes are connected is well established, as is the idea of

Applications of the Telecoupling Framework 121

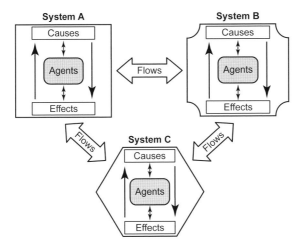

Figure 7.1 Model of the five main interrelated components of the telecoupling framework. *Causes* produce a telecoupling between at least two coupled human and natural *systems*, which generate *effects* that are manifested in one or more coupled human and natural systems. Telecoupling is conducted by *agents* that facilitate or hinder *flows* of material/energy/information between *systems*. Three systems are depicted, each of which can simultaneously serve as sending, receiving, or spillover systems, depending on the flow being analyzed. For further discussion, see Liu et al. (2013).

humans and the environment being connected (Linnaeus 1749/1964; Marsh 1864/1965). For more than a decade, the U.S. National Science Foundation has funded research to study coupled human-natural systems.[1] In atmospheric science, the concept of teleconnections relates to the environmental interactions between climatic systems over considerable distances (Glantz et al. 1991). This teleconnection framework has been applied to social science and land systems (Haberl et al. 2009; Seto et al. 2012b).

The concept of telecoupling used in this chapter builds on concepts such as land teleconnections (Haberl et al. 2009; Seto et al. 2012b), integrated human and environmental systems (Peterson 2000), and globalization (Levitt 1983). A significant feature of telecoupling between distant systems is the role of feedbacks; that is, reciprocal interactions among different coupled systems. The telecoupling framework presented here consists of five main interrelated components (Figure 7.1): systems, agents, flows, causes, and effects (see also Liu et al. 2013). *Causes* generate a telecoupling between a minimum of two coupled human and natural *systems*, which produce *effects* that are evident in one or more of the *systems*. A telecoupling is produced by *agents* that facilitate or hinder the *flows* of material/energy or information among the *systems*. Telecoupling studies differ from multisystem comparison studies in that they

[1] See http://www.nsf.gov/funding/pgm_summ.jsp?pims_id=13681

explicitly address socioeconomic and environmental interactions between the systems. Figure 7.1 illustrates the key components of telecoupled systems and the terminology used in this chapter; the process relationships between the systems have been further refined by Eakin et al. (this volume).

Systems refer to coupled human-natural systems and are defined as a set of human and natural components interacting to form a whole. Systems can be characterized as sending systems (origins), receiving systems (destinations), or spillover systems. Sending systems send *flows* out, while receiving systems obtain *flows* from the sending systems. *Flows* exchange material, energy, or information between the *systems*. Material or energy includes biophysical and socioeconomic entities (e.g., goods, food, natural resources, organisms, carbon) whereas information includes knowledge and agreements (e.g., trade agreements, land titles, agricultural techniques). Telecoupling *agents* include autonomous decision-making entities within sending, receiving, and/or spillover systems that are directly or indirectly involved in telecouplings, such as via the formation or dissolution of *flows*. They can be individuals or groups of humans or animals (e.g., herds of animals, households, government agencies). Spillover systems form as a byproduct of linkages between sending and receiving systems. Spillover systems may be linked to sending and receiving systems via three main mechanisms (Liu et al. 2013): (a) by serving as an intermediate stopover between two systems (e.g., a migratory bird stopover or airport layover), (b) by being located along the pathway between the sending and receiving systems (e.g., oil spilled during transport), or (c) by engaging with the sending and receiving systems (e.g., a third party in trade negotiations).

The *causes* of telecoupling are factors that generate dynamics (e.g., emergence, changes in strength) of a telecoupling. Most telecouplings have multiple causes which originate from a sending, receiving, and/or spillover system. Five broad categories of causes could include economic, political, technological, cultural, and ecological, although these categories interact with one another. *Effects* refer to environmental and socioeconomic consequences or impacts of a telecoupling. They can be manifested in sending, receiving, and/or spillover systems according to two main interrelated categories: socioeconomic and environmental. Types of effects observed in individual coupled systems (Liu et al. 2007b) may also be manifested in telecoupled systems, including indirect effects, feedbacks, cascading effects, and legacy effects. Effects are often nonlinear and may have time lags (i.e., they do not emerge for years or even decades after the telecoupling is initiated).

Telecoupled systems have a hierarchical structure. At the highest level, flows are transferred between multiple coupled human-natural systems. At the intermediate level, each coupled human-natural system contains agents that facilitate the telecoupling. Causes and effects are primarily manifested at this level. At the smallest subsystem level, each component has particular characteristics of interest to the telecoupling (e.g., individual agents operate within multiple, different types of institutions). In addition, cross-level interactions occur, such as when within-system agents facilitate cross-system flows.

Applications of the Telecoupling Framework

To illustrate the application of the telecoupling framework (systems, agents, flows, causes, and effects) to global land use, two very different examples will be used: transnational land deals and species invasions. We chose these two examples, and will provide a general description and specific case for each, because both are gaining importance as drivers for global land-use change through mediating interactions between multiple coupled systems. Additional telecouplings are listed in Table 7.1.

Transnational Land Deals

Throughout history, many communities and nations have exchanged land titles (i.e., rights to use and control land in different ways). Some notable historical exchanges include those between the classical Roman Empire and societies throughout Europe and Asia, the Louisiana Purchase by the U.S. Government from France, as well as numerous smaller land purchases by new immigrants and settlers from indigenous communities. Such "exchanges" of rights were frequently part of wider changes in relations between peoples, such as colonization, or postcolonial governance.

Over the past decade, there have been dramatic increases in different types of transnational large-scale land deals (exchanges of land titles, sometimes described as "land grabs" or a "global land rush") with an unprecedented number of countries involved (see also Hunsberger et al., Margulis, Gentry et al., all this volume). Between 2000 and 2010 a total of 2,042 deals with approximately 203 million hectares of land were reported to have been transferred (Anseeuw et al. 2012b). More than half of these deals (1,155), affecting approximately 71 million hectares, have been cross-referenced (i.e., confirmed via triangulation from different sources) (Anseeuw et al. 2012b). Most of these deals are long-term leases of government-owned land, although some are outright land purchases (Cotula 2012). These land deals increased abruptly in 2007/2008 and peaked in 2009. In 2010 (the most recent year of available data), the number dropped substantially; however, they were still well above the pre-2005 levels (Scherer 2012). The purposes of the recent land deals are more diverse than in the past, which focused mainly on agricultural food and fiber plantations, (Cotula 2012), with notable exceptions such as cacao, which served as currency in the time of the Aztecs in Guatemala. For example, biofuels are now a major factor in many deals. Of the land involved in cross-referenced deals for which the purposes are known, 58% is for biofuel, 17% for food crops, 13%

Table 7.1 Examples of telecouplings and hypothetical relationships to land use in sending, receiving, and spillover systems. Feedbacks may be common in telecouplings among different systems, although many of them may take a long time to emerge.

Telecouplings	Land Use in Sending, Receiving, and Spillover Systems
Trade of goods and products (e.g., food, energy, timber, medicine, and minerals)	Land is used for producing goods and products in the sending systems. In receiving systems, land may be used for other purposes (e.g., urban, residential) because of a reduced demand for goods and services fulfilled by the sending systems. In spillover systems, land use may be affected in various ways depending on the relationships with sending and spillover systems. In all systems, land may also be used for building facilities to store, transport and distribute goods and products.
Technology transfer (e.g. machinery, pesticides, tillage techniques)	Technology generation occupies land in sending systems (e.g., space for factories). Technology implementation may change land-use efficiency and intensity, and may promote land development (e.g., via powerful machinery) in receiving and spillover systems.
Tourism	Land may be devoted to tourism facilities (e.g., roads, scenic spots, restaurants, hotels) in receiving and spillover systems. Land use in the sending systems may also be changed (e.g., building something similar to what tourists have seen in other systems).
Development investment	Development investment (e.g., transnational land deals) may stimulate land use and land conversion (e.g., for agricultural production, manufacturing facilities) in the receiving systems, may or may not slow down or prevent land development and land conversion in the sending systems, and may influence land use in spillover systems in various ways.
Human migration	Human migrants may (or may not) abandon land in the sending systems and occupy land in the receiving systems (e.g., for housing and work/recreation space). Land use in the spillover systems may vary depending on their relationships with sending and receiving systems.
Knowledge transfer	Knowledge transfer (e.g., theories, techniques, innovations, governance and management approaches) may affect land-use efficiency and land expansion in the receiving and spillover systems. It is not clear whether land use in the sending systems is changed.

Table 7.1 *continued*

Telecouplings	Land Use in Sending, Receiving, and Spillover Systems
Water transfer	Facilities for water transfer (e.g., channels and reservoirs) may take up land in sending, receiving, and spillover systems, and may change land use in all systems.
Waste transfer	Transfer of waste (e.g., electronic waste) may conserve land in the sending systems, but occupy and contaminate land (e.g., landfills, pesticides in croplands) in receiving and spillover systems.
Conservation investment	Conservation investment (e.g., payments for ecosystem services) may conserve and restore land in the receiving systems, may or may not slow down or prevent land development and conversion in the sending systems, and may influence land use in spillover systems in various ways.
Seasonal animal migration	Seasonal animal migrants (e.g., migratory birds) may use land in sending, receiving and spillover systems (e.g., stopover systems) during specific times of the year.
Species invasions	Invasive species occupy and damage land (e.g., crops and native vegetation) in receiving and spillover systems. Land use in the sending systems can be changed through invasion control methods and policies.
Species dispersal	Species dispersal may or may not result in an animal, plant, or microbe occupying less land in the sending systems, but occupying more land in the receiving and spillover systems (e.g., dispersal corridors). Species may improve or harm land quality in each system depending on their specific relationships to land.
Air circulation	Circulation of air may change land composition and land cover (by transporting soil, organisms, and pollutants such as acid rain from one place to another) in the sending, receiving, and spillover systems.

for forestry, and the remaining 12% for other items (industry, livestock, mineral extraction, and tourism).

Telecoupling Components

Systems A total of 84 sending countries have provided land titles for land deals (Anseeuw et al. 2012a). Among the cross-referenced deals, Africa offered the largest share of land titles, with ca. 34 million hectares between 2000 and 2010. Asia supplied the second largest share (~29 million hectares), followed by Latin America (~6 million) (Anseeuw et al. 2012a). Other regions such as Eastern Europe and Oceania also offered land, but these offers were much smaller (1.6 million hectares).

There are a total of 76 land-title-receiving countries (or investor countries) (Anseeuw et al. 2012a). Of the cross-referenced deals, two-thirds (66.9%) of the land titles went to Asia. Europe, Africa, North America, and Latin America received 11.6%, 10.2%, 8.5%, and 2.7% respectively. The United States, Malaysia, United Kingdom, China, United Arab Emirates, South Korea, India, Australia, South Africa, and Canada are the top ten receiving countries.[2] A number of countries both offered and received land titles. These include Brazil, China, India, and South Africa. Although Brazil offered 27 times more than it received, the other three countries showed the opposite trend, that is, China, India, and South Africa received 9, 14, and 132 times more than they offered, respectively.

Some countries are important spillover systems, which facilitate investment from receiving countries to land-title-sending countries. For example, South Africa has played such a role due to its geographic proximity to land-title-receiving countries and expertise in agriculture throughout Africa (Cotula 2012). Mauritius is another spillover country because of its tax regime and its bilateral investment treaties with other countries in Africa, such as Mozambique (Cotula 2012).

Agents. Numerous agents have been engaged directly and indirectly in land deals. Private companies have been directly responsible for most of the land deals. They have also received diplomatic, financial, and policy support from their governments. Some of the agents are state-owned companies (e.g., China National Cereals, Oils and Foodstuffs Import and Export Company). In fact, for some land-receiving countries, land deals were initiated by their governments. The governments of land-sending countries are also deeply involved in land deals, from providing information to investors, to attracting investors through handsome incentives, to deal approval and implementation. National elites are also often key agents (Anseeuw et al. 2012a). Many international

---

[2] Data from http://landportal.info/landmatrix

organizations play facilitating roles. For example, the World Bank has promoted a good investment environment in some African countries to provide easier access to land by foreign investors (Daniel and Mittal 2010).

Flows. The primary flows involved in land deals include land titles (i.e., rights) transferred from the sending countries to receiving countries, and money to purchase the land titles transferred from the receiving countries to the sending countries. There are also consequential flows of people, machines, and techniques (e.g., crop varieties, pesticides, and fertilizers) transferred by the land title receiving countries to the sending countries to develop the land. Products produced on the land are then sent to receiving and spillover countries all over the world. These include timber, biofuel, food, nonfood crops such as rubber, and other raw materials such as minerals and gas (Anseeuw et al. 2012a). Hence the initial telecoupling (in this case purchase and transfer of land titles) acts as a catalyst for more complex flow patterns.

Causes. The key economic cause of transnational land deals is the interplay between global land supply and demand. The world is heterogeneous in terms of supply and demand. Large and fast-growing economies with relatively scarce land, such as China and India, create ever-increasing demands for land because of their economic growth and increased consumption of animal-based products and other land-based resources, most recently dominated by biofuels. When countries cannot meet these demands domestically, they turn to other countries for resources, especially those with abundant land area, relatively low population density, and weak land governance (Deininger et al. 2011). Rich but resource-poor countries, such as Saudi Arabia, also seek more land to feed and supply resources to their populations.

The international financial crisis in 2008 and the associated food crisis (e.g., hike in food prices) in 2007–2008 played a major role in the recent increase in land deals. These crises have made agricultural land a new strategic asset and an excellent opportunity for investors.[3] Global urban expansion has also played a role in encouraging land deals for two reasons. First, urbanization has significantly reduced the amount of agricultural land available. For example, one to two million hectares of cropland is being converted for housing, industry, infrastructure, and recreation in developing nations each year (Lambin et al. 2003), and this is typically better quality land close to the expanding urban centers. Second, the growing urban population has changing tastes and preferences (e.g., for animal products) and these trends create a belief that land will hold more value in the future (Smaller and Mann 2009), thus prompting investors to take advantage of cheap land in anticipation of higher prices for food and fuel (Anseeuw et al. 2012a).

---

[3] http://www.grain.org/article/entries/93-seized-the-2008-landgrab-for-food-and-financial-security

International conservation programs may also contribute to production land scarcity and subsequent land deals. For example, as indigenous forests gain new market values from programs such as REDD+ (Reducing Emissions from Deforestation and Forest Degradation in Developing Countries), a source of previously cheaper land for conversion to production is removed from the available supply. Ironically, it is thus also possible to displace indigenous people through land deals in the name of conservation (Lemaitre 2011). This illustrates the complexity of *effects* from telecouplings.

Effects. Land deals are a significant driver of land-use change and of consequential socioeconomic and environmental change. Some researchers argue that land deals provide opportunities to enable land-abundant but otherwise poorer countries to gain investment, employment, and technology (Deininger et al. 2011). Some of the land deals also set aside food produced for local communities or for the domestic market and receiving countries may also invest in the social infrastructure (e.g., hospitals and schools) of sending countries. Others argue that land deals threaten the livelihoods of the rural poor and promote social conflicts; they maintain that land deals could lead to the end of small-scale farming in sending countries, which could compromise local food production, and thus undermine food sovereignty. Land deals also promote an industrial model of agriculture that can lead to more rural poverty, since local job creation in most cases is very low (Deininger et al. 2011).

In terms of environmental effects, there is growing evidence that the land-use changes following land deals can generate large-scale and lasting environmental damage (e.g., loss of biodiversity, soil erosion, and pollution from pesticides and fertilizers) (Deininger et al. 2011). Furthermore, "land grabbing" is also a form of "water grabbing," because much water is embedded in land systems, and changing land use typically alters the use of water (e.g., in many regions, agricultural production accounts for most of the water consumption). The transport of food and fiber from countries that have sold their land titles also represents an export of nutrients. Socioeconomic and environmental consequences of land deals in the sending countries have therefore generated widespread attention in the international news media, which has led to political feedback. As a result, some receiving and sending countries have become more cautious in making new land deals (Agence France-Presse 2012; Perrine et al. 2011).

Insufficient attention has been paid to the net effects of land deals on countries that receive the titles and on spillover countries. For example, it is not clear what the opportunity costs and benefits of these land deals are, how many jobs will be transferred to the sending countries, how much water, fertilizer, and other types of agricultural input (e.g., pesticides) can be saved for the receiving countries, and how much loss of soil nutrients can be avoided. There are arguments that the supply of food and fiber from distant lands to cities in receiving countries in effect "exports" the environmental costs of urbanization.

The economic and environmental costs (e.g., CO_2 emissions) of transporting goods and people between sending and receiving countries are also poorly documented or understood. As the distances are large and amounts of goods and products are enormous, these costs could be high. Thus the issue of "food miles" and the carbon footprint of food supply has become a major consideration in debates over the sustainability of agriculture.

Laos As an Example

Worldwide, eleven countries, of which Laos is one, collectively account for 70% of the total land targeted in transnational land deals (Anseeuw et al. 2012a). In Laos alone, there have been an estimated 40 land deals involving 140,000 hectares (Anseeuw et al. 2012a). Transnational land deals are a relatively recent phenomenon in Laos, as they did not take place until the early 1990s, due in part to conflicts and political instability (Baird 2011). However, once they began, land deals increased dramatically. Currently 15% of the country's land system is under agri-business concessions (UNDP 2006).

The main *systems* involved are Laos (the sending country) and the investor (receiving) countries, primarily China, Vietnam, and Thailand. Within Laos, farmlands are converted to plantations (mainly for rubber) by investors. Deals were intended to be restricted to state-owned lands, but in practice, private lands have also been affected. Spillover systems include other countries that benefit from the sale of rubber, particularly those in Southeast Asia.

The primary *agents* include investment companies in other countries in Southeast Asia, for example the Asia Tech Company (a Thai firm) and the Dak Lak Rubber Company (from Vietnam). Other agents include the government of Laos, whose early policies opened the way for land to be acquired by foreign investors (Baird 2011), but whose recent crackdowns on overexploitation of the resource and threats to the livelihood of indigenous people have sought to curtail land deals (Agence France-Presse 2012). Local indigenous people in Laos also serve as agents, as they have shifted their livelihood strategies from farming to manual labor on plantations as a result of land deals (Baird 2011).

The main *flows* include money transferred from receiving countries to Laos to secure land, rubber production machines sent to Laos to develop plantations, and the rubber that is subsequently sent to countries all over the world for use in a variety of manufactured goods. Information flows include the land titles granted from Laos and the communications between the sending, receiving, and spillover countries that have interests in rubber development.

The primary *cause* of this telecoupling is economic. Investment companies seek developing countries where there is cheap land available to make profits (Baird 2010). The government of Laos, in turn, viewed the land deals as a means to increase state revenues and improve the country's export market (Baird 2010). Another cause is political: the transition of the political system in Laos from a socialist to a capitalist, market-driven system enabled the land deal era to emerge (Baird 2011). Successful implementation of land deals also required a politically stable environment, which was just emerging after decades of conflict at the time when the deals first became prominent.

There are a number of *effects* of this telecoupling. In terms of the natural system, effects range from degradation of ecosystems, loss of biodiversity, to loss of ecosystem services, all of which resulted when large-scale plantations were created on acquired land (Baird 2011). Effects on the human system include increased poverty among indigenous peoples, infringement on their access to natural resources (e.g., water, fodder, and fuel), and erosion of their social structures and family ties due to displacement for wage labor (Baird 2010, 2011). Feedbacks have also occurred between the sending and receiving countries, because the Laos government has instituted new limits and regulations on land deals made with other countries in response to emerging problems (Agence France-Presse 2012).

Implications

The implications of the recent surge in transnational land deals are potentially huge, but remain uncertain. These uncertainties exist because many deals have not been reported, almost half of the reports in the media have not been verified, many of the announced land deals have not been implemented, and there is relatively little empirical evidence of the environmental and socioeconomic effects. For example, a World Bank report found that farming had only begun on lands involved in 21% of the announced deals (Deininger et al. 2011). While the demand for land has been large, some researchers believe that the potential supply is also high. Some estimates suggest that the world still has over 445 million hectares of nonforested, uncultivated land suitable for rainfed cultivation of at least one of the five key crops (wheat, sugarcane, oil palm, maize, and soybean) (Deininger et al. 2011). Thus, the trend of large-scale land deals may continue in the future (Anseeuw et al. 2012a). More research is needed to confirm the actual amount of suitable uncultivated land.

Reducing uncertainties and generating reliable information is essential to understand the implications of land deals. Data on transnational land deals are challenging to obtain and verify (Friis and Reenberg 2010).[4] The Land Matrix Project, an international network of 45 organizations, has been able to establish the largest and most comprehensive database of land deals around the world. However, the data are still far from complete and must be considered conservative since, on one hand, they are mainly based on media reports and, on the other, many land deals have not even been reported in the media (Anseeuw et al. 2012a). Furthermore, even for deals recorded in the database, many of the details are unknown (e.g., size, date of contracts, boundaries), partly because much of the information is regarded as "confidential."

---

[4] See also the Land Matrix Project at http://landportal.info/landmatrix

Applications of the Telecoupling Framework 131

The telecoupling framework contributes a better understanding of this complex global issue to the land-change science community. It provides a means to answer key questions concerning the diverse causes of transnational land deals, the cascading and feedback effects of land deals on food security and land degradation across multiple coupled systems, and the diverse groups of agents involved in land deals, as well as their roles in facilitating flows among the systems. Such an approach is required if different institutions from multiple coupled systems are to work together to address the socioeconomic and environmental challenges that result from land deals.

Species Invasions

Worldwide, the number of invasive species has increased, as has the overall number of species introductions, which have greatly impacted global land use. Species introductions from one region or continent to another can be traced back to at least 1492 CE, corresponding to human exploration of the globe (Hulme 2009). Overall, however, species introductions were limited in scope and impact until around 1800, whereafter they began to increase rapidly due to the effects of the Industrial Revolution, which served as a catalyst for rapid expansion of the global economy (Hulme 2009). For instance, the annual number of introduced species in Europe increased by 300% for plant species and 600% for invertebrates and mammals between 1800–1850 and 1975–2000 (Hulme 2009) (Figure 7.2). Today, there are over 500,000 alien species worldwide—50,000 in the United States alone (Pimentel et al. 2007). A proportion of

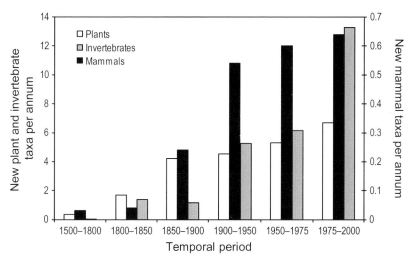

Figure 7.2 Dynamics of introduced plants, mammals, and invertebrates in Europe (after Hulme 2009).

these species have become invasive, with currently over 9,000 invasive plants and animals officially recorded (Pimentel et al. 2007). Even so, estimates of alien and invasive species are widely believed to be gross underestimates.

Telecoupling Components

Systems. Prior to the nineteenth century, invasive species spread from Europe (sending system) to other parts of the world (receiving systems) through human migration and trade. Over the past two centuries, globalization and associated advancements in trade, travel, and technology have brought invasive species to all corners of the globe. The receiving countries with the highest documented numbers of invasive species are the United States (3379), Australia (419), and New Zealand (410) (Sellers et al. 2010). Islands are at greater risk than large continents, and there is an increase in the number of invading species (per system) from North to South across the globe in the Northern Hemisphere, a sharp drop in the Tropics (where few invasions occur), with invasions then increasing again while moving southward into temperate lands (Vitousek et al. 1997).

Agents. Humans are agents of invasive species spread around the world, as they facilitate spread through travel, migration, and transport of goods and products. Human agency includes both production (e.g., the introduction of new species for agriculture or forestry) and consumption (e.g., the purchase of imported plants and animals for food and pleasure, such as gardens and pets). Invasive species themselves can also be considered agents because of their consequential actions. For example, many invasive species cause damage to indigenous ecosystems as a result of different competitive abilities over native species. The Invasive Species Specialist Group list of the 100 most destructive invasive species worldwide includes 36 plants, 26 invertebrates, 14 mammals, 8 fish, 5 fungi, 3 amphibians, 3 birds, 3 microorganisms, and 2 reptiles.[5]

Flows. The main flows involved in species invasions are the transfer of the invasive plants, animals, and microbes themselves. Other material flows include transport of control agents such as pesticides or natural enemies of the invaders. Information flows include the sharing of knowledge across systems on how to control the invaders. Knowledge is also transferred among groups of scientists and agencies when experts in receiving systems seek to understand the behavior of the invasive species in its native habitat (in the sending systems).

Causes. Invasive species may be transported intentionally by humans for the purpose of pet trade, horticulture, farming, or biological control of other

---

[5] http://www.issg.org/database/species/reference_files/100English.pdf

organisms. However, the majority of invasive species are spread accidentally by being present in vehicles, ships, and planes. Humans may also make purposeful land-use decisions that directly promote invasive species, even when they are mindful of conservation goals; for example, the recent increase in lands devoted to growing nonnative species of biofuels that display invasive properties in some systems (Danielsen et al. 2009). Land-use change may also create new niches or conditions upon which alien species can capitalize. For instance, land-use changes arising from the increased frequency of forest fires can subsequently promote the invasion of nonnative grass species that may be better adapted to higher frequencies of fire (Vitousek et al. 1997).

Effects. Some introduced species are beneficial for human welfare or ecosystem functioning, especially the crops and livestock that fuel agricultural economies worldwide, which are over 99% nonnative (Pimentel 2002). However, a large number of invasive species have had widespread and pernicious effects, such as the extinction of native species. As much as 80% of the world's endangered species have been threatened (Mooney and Cleland 2001; Wilcove et al. 1998; Armstrong 1995), through the collapse of whole animal and plant communities (Sanders et al. 2003) and through the disruption of large-scale ecosystem processes (Gordon 1998).

Species invasions are also a major cause of global land-use change and economic losses. Invasive species can directly bring about such changes by converting one land-cover type to another: invading trees convert grasslands to forests and invading grasses convert bare lands (that may have previously been forest) to grasslands. In addition, invasive species may indirectly cause land-use change. For instance, the Eurasian cheatgrass changed the fire frequency of shrub/steppe habitat in the Great Basin of North America from once every 60–110 years to once every 3–5 years, and then subsequently outcompeted local grasses not adapted to that fire cycle to take over five million hectares of land (Vitousek et al. 1997). The economic costs of invasive species are enormous (Figure 7.3). For example, the gypsy moth (a nonnative pest that originated from France) was transplanted to an urban forest in the eastern United States (Vitousek et al. 1997), where it has infected over 190 million acres of natural forest, requiring millions of dollars annually to manage (Mayo et al. 2003).

Besides receiving and spillover systems, invasive species may also affect the sending systems—feedbacks that are rarely discussed in the literature. Sending systems may gain benefits from selling invasive species or engaging in trade with other countries; this inadvertently results in invasive species spread. Sending systems may also incur costs due to invasive species prevention and control policies. Negative impacts on sending systems could occur if sending systems outlaw or issue fines within their own country or receiving systems implement international invasive species tariffs on sending countries to punish them (Touza et al. 2007).

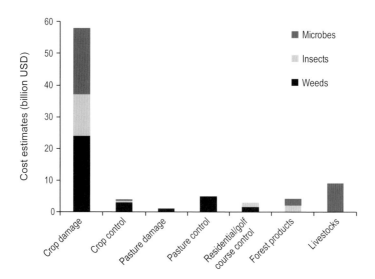

Figure 7.3 Conservative cost estimates of introduced weeds, insects, and microbes to select sectors of the U.S. economy (based on data from Pimentel et al. 2005).

Cheatgrass (Bromus tectorum) As an Example

Cheatgrass is an annual bunchgrass species native to Central Europe, southwestern Asia, and northern Africa. The species was accidentally introduced from Central Europe to numerous shrub-steppe ecosystems all over the globe via agricultural grain exports, contaminated packaging materials, and contaminated ship ballasts on ships in the 1800s. It was first discovered in the United States in the mid 1800s (Klemmedson and Smith 1964), and by the early 1990s it could be found throughout its current range. This species has the potential to alter entire ecosystems by outcompeting native species and has caused considerable damage to agricultural production systems (Knapp 1996).

The *systems* involved are countries in Central Europe (sending systems) and Russia, countries in Western and Central Asia, Japan, South Africa, Australia, New Zealand, Iceland, Greenland, Canada, and the United States (receiving systems).[6] Areas within the receiving countries that are affected include (a) sagebrush grasslands, particularly those that are already disturbed by human impacts which promote increased direct sunlight exposure (e.g., grazing, cultivation, frequent burning), (b) human-managed rangelands and pastures, and (c) wasteland and barren environments. Since cheatgrass invasions (and associated fire events and conversion of shrublands to grasslands) have had profound impacts on the global carbon balance, and have both contributed to and are affected by global warming, spillover systems include countries around the entire globe.

---

[6] Global Invasive Species Database (IUCN) at http://www.issg.org/database/welcome/

The main *agents* involved are the human traders, who inadvertently transported the grass seeds on ships, and farmers, who received the grass seeds embedded in agricultural imports. A number of agents have been involved in efforts to curb invasion, including livestock herders, farmers, fire control operators, and government agencies. Agents that have unknowingly contributed to invasive species spread are land users that degrade rangelands due to overuse (e.g., overgrazing), thus making them more vulnerable to invaders, as well as humans and animals that move long distances and can transport the seeds via their clothing and fur.

The main *flows* include the transport of the grass seeds themselves, in addition to information flow among different agents about techniques to control the spread of the grass (e.g., hand pulling, introducing grazing animals to suppress the grass, reseeding of native perennials, herbicide application, prescribed fire). Materials used for invasive species control (e.g., native seeds) are also transported among systems. Carbon may also flow into the atmosphere from receiving to spillover systems through cheatgrass conversion of shrubland to grassland and via fire events promoted by cheatgrass.

The initial *cause* of cheatgrass invasion was accidental—a result of global trade via ships in the 1800s. Subsequent spreading occurred when seeds were transferred on animal fur or human clothing, along roads and railways, or via wind. Successful establishment after spreading has been made possible by the competitive abilities of the species (i.e., their ability to monopolize available soil moisture at the expense of other plants through an extensive root system) and the species' resilience in disturbed environments (D'Antonio and Vitousek 1992). Humans have also promoted the spread of the invasive species through, for example, the practice of intensive grazing of livestock on rangelands in the Intermountain West of the United States, which depleted native grass species, thus paving the way for an explosion in the cheatgrass population (Mack 1981).

The *effects* of cheatgrass invasion are widespread and include reduction of biodiversity, destruction of habitat for wildlife, and destruction of agricultural crops (and resultant loss of income). For example, there is currently an estimated 31.5 million acres of cheatgrass in the Great Basin of the United States, where native species once dominated (Menakis et al. 2003). Cascading effects include the propensity of cheatgrass to increase the frequency of fires in habitats that it occupies, which can decrease soil nutrients. Managing for cheatgrass-caused fires in the Great Basin alone is estimated to cost nearly USD 10 million per year (Knapp 1996). Feedback effects occur between receiving, sending, and spillover systems as a result of global climate change due to CO_2 emissions from cheatgrass invasion. For instance, in the Great Basin of the United States alone, cheatgrass invasion has released 8 ± 3 Tg of carbon into the atmosphere (spilling over across the globe), and will likely release another 50 ± 20 Tg C over the next few decades (Bradley et al. 2006). Global climate change caused by cheatgrass invasion, in turn, affects further spread of the cheatgrass into new areas, because cheatgrass is sensitive to changing precipitation regimes and is predicted to shift its range dramatically to areas that receive lower precipitation under global climate-change scenarios.

Implications

The need for advances in predicting and managing invasive species is urgent, considering that one of the most critical aspects of invasive species biology is that there are significant time lags before the consequences of species introductions are detected on the landscape (Crooks and Soule 1999). In other words, the currently observed effects of invasive species may only make up a fraction of all long-term effects that may arise from telecouplings. For instance, the effects of the exponential increases in imports and exports from China on the country's invasive species load have not yet been fully realized and may not emerge for years (Jenkins and Mooney 2006).

Despite extensive research on invasive species, significant challenges remain in predicting the spread of invasive species and the management of invaded systems. Conceptualizing species invasion as a telecoupling can help propel this field forward, as it identifies how invasive species are linked in sending, receiving, and spillover systems. The telecoupling framework addresses these urgent needs in the context of land-change science by explicitly characterizing the relationship between the distant coupled systems involved. In addition, it addresses key questions:

- How do diverse agents change their behavior over time and space in response to an emerging invasion?
- How do flows of invasive species and control methods interact with one another across systems (including spillover systems)?
- How does the invasion create cascading effects that are not limited to a single system?
- How can the causes of invasion originate from multiple, different systems and at a global level?

The framework also provides a mechanism to address future policies on the control of invasive species spread, which requires collaboration between institutions from multiple coupled systems. Telecoupling allows for explicit characterization of how species invasions relate to other aspects of complex coupled systems, particularly those relating to dynamics in global human systems (e.g., socioeconomic and institutional drivers of trade and migration patterns). Species invasions are also closely related to other telecouplings, such as trade, since countries with higher rates of international trade experience higher rates of species invasions (Westphal et al. 2008) due to increased opportunities for species to be transported (either purposefully or accidentally) along with trade goods.

Challenges, Opportunities, and New Directions

Telecouplings offer new and unique challenges and opportunities for the land-change science community. As illustrated in the framework and the examples above, telecouplings are more complex than local couplings because the former involves multiple places, multiple flows, multiple agents, global causes, and global effects. Many crucial and complex questions can be addressed using the telecoupling framework. One question is how feedbacks between multiple coupled systems connect and propagate the effects of telecoupling widely across space. Another key question is what role spillover systems play in the telecoupled system. These spillover systems are rarely considered explicitly and have not even been recognized in previous research or management. The effects of telecoupling on spillover systems may sometimes exceed those on receiving and sending systems, as demonstrated particularly in the invasive species example.

Another major implication of understanding telecouplings is governance. Telecouplings create more challenges and complications for management and policy than local couplings. Because sending, receiving, and spillover systems are far removed from one another (and typically in different jurisdictions), the conditions within different parts of the telecoupled system are beyond the control of any single government or management agency. Managing one telecoupling is not easy, but managing multiple telecouplings simultaneously is a significant challenge for land-change governance (Table 7.1), as they may amplify or offset each other. Some international policies, such as the REDD program to combat deforestation, seek to manage the effects of telecouplings. However these endeavors have focused largely on individual telecouplings and have not fully considered the interactions of multiple telecouplings; they also have not considered the spillover effects in a systematic fashion or at all. Another challenge is how to manage the relationships between telecouplings and local couplings. For example, although there has been a push for consuming locally produced products (Desrochers and Shimizu 2012; MacMillan 2012), the increased trends brought about by telecouplings are difficult to reverse. Thus, current land-use policy and stewardship approaches which only consider local couplings need to adopt a new or revised structure that fully integrates telecouplings so that, in turn, positive effects can be enhanced and negative effects reduced with respect to sustainability.

A key issue to address in research and management is the data needed to characterize telecoupled systems. As shown in the examples above, huge data gaps exist in all telecoupling components. Thus far, work on telecouplings has focused on virtual resources used (e.g., virtual water, virtual land). For example, China is in the top ten countries engaged in the virtual trade of water (a global market involving a total of 625×10^9 m^3 of water exchanged around the globe per year) (Konar et al. 2011). Obtaining relevant data on telecouplings is more complicated, more time-consuming, and more costly than research on local couplings alone. This is because various systems (sending, receiving, and spillover) are in distant places and usually under different types of governance. As many driving forces originate from outside a coupled system and are difficult to predict, it is sometimes challenging to determine what kinds of data to collect and where to collect in advance.

It is encouraging, however, that new opportunities to address telecouplings are emerging. More researchers are realizing the need to address telecouplings in relation to global land use. More advanced tools for collecting, analyzing, and visualizing data are becoming available. For example, value-chain analysis, GPS collars, genetic markers, and barcodes are being used to track the long-distance movement of organisms, goods, and products. In the future, a relational database approach could be very useful for stimulating data organization and integrated analysis of telecouplings. For instance, information about transnational land deals could be entered into a relational database with a list of systems (sending, receiving, and spillover); each system would include a list of agents; each agent a list of other systems and deals in which the agent has also been active; each flow would include links to different systems and list each cause and effect that is related to each system. This could be linked with data about another telecoupling, say species invasion, thus providing insights into the relative complexity and interconnectedness of agents and systems involved in different types of telecouplings. Such an integrated analysis can help identify the types and interrelationships of telecouplings that have potential for the greatest socioeconomic and environmental impacts.

Rethinking and reexamining land-use dynamics in the context of telecouplings, therefore, requires new directions to be pursued in land-use change research. These may include (a) changes in the conceptual frameworks of land use, from a focus only on local couplings, to a combination of local couplings and telecouplings, (b) changes in research paradigms from site-specific and multisite comparisons to multisite linkages, and (c) changes from collaboration and dialog within the land-change science community to networking with experts in other disciplines and with various stakeholders around the world (e.g., the media and nongovernmental organizations, such as a Land Matrix for transnational land deals).

Conclusion

The telecoupling framework offers a useful analytical approach to integrate distant forces of land-use change across the globe. In the framework, agents, flows, causes, and effects as well as their relationships across multiple coupled systems are conceptualized as part of a broader telecoupled human-natural system. The examples provided in this chapter demonstrate the utility of the framework. Transnational land deals and species invasions vary considerably in terms of systems, agents, flows, causes, and effects, yet the framework consistently captured and connected all major relevant issues in both cases. Thus, the telecoupling framework can help researchers systematically analyze each of the system components and their relationships with one another.

Understanding telecouplings has enormous implications for managing and governing global land use at a time when the land-change science community faces unprecedented challenges and opportunities. New research directions are needed to meet the challenges of stronger and more widespread distant forces that drive land-use change around the world.

Acknowledgments

We acknowledge helpful comments by Ruth DeFries, Eric Lambin, Julia Lupp, Anette Reenberg, and Karen Seto. Funding for the preparation of this paper was provided by the National Science Foundation.

First column (top to bottom): Ruth DeFries, Karl Zimmerer, Eric Lambin, Karl Zimmerer, Birka Wicke, Eric Lambin, Ximena Rueda, and Jack Liu
Second column: Hallie Eakin, Suzi Kerr, Jack Liu, Group discussion, Simon Swaffield, Peter Marcotullio, Anette Reenberg, and Karl Zimmerer
Third column: Anette Reenberg, Simon Swaffield, Ximena Rueda, Group discussion, Ruth DeFries, Birka Wicke, Suzi Kerr, and Hallie Eakin

8

Significance of Telecoupling for Exploration of Land-Use Change

Hallie Eakin, Ruth DeFries, Suzi Kerr,
Eric F. Lambin, Jianguo Liu, Peter J. Marcotullio,
Peter Messerli, Anette Reenberg, Ximena Rueda,
Simon R. Swaffield, Birka Wicke, and Karl Zimmerer

Abstract

Land systems are increasingly influenced by distal connections: the externalities and unintended consequences of social and ecological processes which occur in distant locations, and the feedback mechanisms that lead to new institutional developments and governance arrangements. Economic globalization and urbanization accentuate these novel telecoupling relationships. The prevalence of telecoupling in land systems demands new approaches to research and analysis in land science. This chapter presents a working definition of a telecoupled system, emphasizing the role of governance and institutional change in telecoupled interactions. The social, institutional, and ecological processes and conditions through which telecoupling emerges are described. The analysis of these relationships in land science demands both integrative and diverse epistemological perspectives and methods. Such analyses require a focus on how the motivations and values of social actors relate to telecoupling processes, as well as on the mechanisms that produce unanticipated outcomes and feedback relationships among distal land systems.

Introduction

Over the last decade, connectivity between processes of land change and actors, decisions, and activities has accelerated across geographically distant places. The 2007–2008 global food crisis, the expansion of biofuel production, and the global emergence of niche and "green" markets have had widespread and often unexpected outcomes on land systems in disparate geographic locations. These connections are associated with accelerated urbanization as well as the development of new markets and are motivated by emergent demands of consumers with increased agency and an intensification of information and knowledge flows. On the basis of this observed "connectivity," we offer two propositions. First, nearly all land systems are now affected to some extent by these forms of connectivity, or *telecouplings*. Second, the increased significance of telecoupling for land change implies a need for integrating diverse epistemological perspectives, methodology, and analytical approaches that together complement the long-standing focus of land science on place-based research with a new focus on the networks and system interactions involved in land change. The telecoupling process links the diverse social, ecological, and economic outcomes of land change to specific, yet potentially diverse value systems held by different sets of actors—including scientists in disparate social networks. Research on telecoupling in land science is thus both embedded in the evolution of sustainability pathways for land systems as well as instrumental in the analysis of these pathways.

What Are Distal Land Connections and Telecoupling?

Teleconnections, as discussed in climate literature, have a specific reference to mesoscale atmospheric processes (e.g., ENSO) that have (concurrent) climatic consequences in geographically noncontiguous locations (e.g., Simmons et al. 1983; Trenbreth and Hurrell 1994). The idea of distal connections captures this essence of "acting at a distance": an action, phenomena, or process of change in one location has implications in a geographically distant location. In some senses this concept can be interpreted as unidirectional and linear, essentially reflecting the idea of an exogenous driver acting on a distant system. This concept has been applied to land systems (Haberl et al. 2009; Seto et al. 2012b).

The concept of telecoupling—preferred by this discussion group—captures the idea of two or more independently coupled, interacting social-ecological systems (Liu, this volume). In other words social-ecological interactions in one system generate mechanisms of influence over another. The process of telecoupling is different to the concept of coupling in that there is an element of social and spatial distance; that is, geographic separation between systems as well as a separation of social networks, institutions, and governance. The boundaries of the systems involved in the telcoupling are defined in terms of the placed-based social-ecological interactions as well as the potentially aspatial social networks, institutions, and governance structures that directly influence those interactions. There is no a priori assumption or understanding that these systems are integrally connected. They are assumed to be disconnected, and thus they are governed independently. Feedback processes, in some cases, may return the initial signal of change to the place of origin, provoking a change (in land use, policy, institutions, or behavior) in that place and causing a complete feedback loop. In other cases, differences in power and influence among the coupled systems may result in the implications to be essentially ignored, with potentially detrimental implications for ecological integrity and human welfare.

Because the governance of the linked systems is independent, the critical outcomes revealed in the telecoupling process tend to be indirect, emergent, and of a second or third order, such that they are more difficult to anticipate and to measure. Nevertheless, they may play a determinant role in land (social as well as ecological) outcomes in particular places. The interaction emerges essentially as an "ungoverned" process, such that the indirect outcomes of the interaction often appear unexpected or "surprising" because they lie outside the dominion of the existing governance arrangements. The disconnect between the problem origin and outcome challenge efforts at problem resolution. While existing governance and institutions may produce predictable supply and demand responses in a commodity market in two geographic locations, there may be "spillover" effects generating environmental change in a secondary resource in a third region as a result of demographic shifts provoked by the market. For example, an energy system is governed by specific suites of actors, energy policy, and regulations; while we know that many food production systems are highly energy intensive, such production systems are governed by separate policies, actors, and networks. The rapid interaction between oil prices, biofuel development, land use, and food security in urban areas that happened in 2007–2008 occurred in somewhat of a governance vacuum (Eakin et al. 2010). As these interactions and causal relations are made visible, they may be incorporated into governance, institutional design, and decision making if the volition and commitment exists, and if the problems generated through the interaction are tractable. What makes the concept of telecoupling so interesting for science is that it captures not only the "action at a distance" but also the feedback between social processes and land outcomes in multiple interacting systems. This creates both a need and an opportunity for a significant new research effort, focused on the question: How and where do telecoupled feedback processes influence global land-use change, and with what consequences?

Is the Concept of Telecoupling New?

The idea of connectivity between actions and actors in one specific geographic location and land outcomes in another is not new in the history of human environment interactions. Globalization—as understood as the increasing intensity and rate of capital flows and interdependencies across space (Held et al. 1999; Dicken 2003)—has several historical phases and has been implicated as the vehicle for "distal" linkages (primarily relations of production and consumption) for centuries. However, globalization as a diffuse, aggregated process of economic intensification and connectivity has not yet been specified in terms of particular causal social-environmental chains specific to a suite of actors, interacting noneconomic and economic flows and feedbacks, and place-based outcomes. Thus telecoupling as an analytical concept has the potential to transform land science and decision making at different scales.

The concept is novel in land science in several ways. First, there is an agreement that the spatial scope, intensity, and rate of connectivity is distinctly different now than in the past. The amount of land affected by the processes of interest, and the rate of change in land outcomes, is greater now than in recent history. Globalization, as a process, has served to accelerate the rate at which outcomes occur, as well as the scale and scope of outcomes.

Second, the context in which telecoupling occurs today is new and distinct. We now live in a time where many perceive that there are increasing claims to resources. In the near future, we anticipate a world with 10 billion people and are already experiencing significant constraints as we try to meet the land-based resource needs of our current population. Limitations on land availability and land-use options imply less flexibility, or fewer degrees of freedom in system response, such that the phenomena of telecoupling has potentially far more significant implications for system function than in previous points of history. The feedback linkages are "tighter," more rapid, and multiscalar; the potential for rapid acceleration to systemic transformation (surpassing thresholds) or crisis arising from multiple system interactions is potentially higher.

Third, the telecoupling of the current era is characterized by information-rich and information-intensive interactions, facilitated by the Internet, social media, and the capacities of communication that enable action (at a distance). While material flows are important in the process of telecoupling (flow of commodities, money, people), equally if not more important are nonmaterial flows. These flows are often in the form of information and knowledge and the social interpretation of that knowledge through the ideological lenses and values of specific social networks. The degree of information connectivity facilitated by globalization enables new forms of social contracts and empowerment, and constitutes an important feedback mechanism in telecoupled systems. Actors in one location can be informed of outcomes in a distant location, and their concern about the possible consequences of their actions can generate a response. There are imbalances in access to information and specific sources of knowledge, and this imbalance also translates into different degrees of agency (acknowledged or not) and positions in telecoupled systems.

Fourth, part of the ideological shifts that have occurred with the latest phase of globalization is an increased concern for sustainability and resource limitations, bringing awareness of telecoupling outcomes to new significance. These values, while not globally shared, now color more frequently how influential actors explain the ethical responsibilities for their actions and how they respond to new information (although often highly uncertain) about the (unintended) consequences of decisions in which they played a part. Globalization has enabled or facilitated "feeling (empathy) at a distance." In other words, today there is more concern about the consequences (good or bad) of actions that once would have been valued primarily in terms of a national context and national benefit. There are new moralities—new social contracts that are emerging as part of the globalization process—that imply new responsibilities for action. Thus there is a rising influence of the affective responses by specific social groups defined by specific values and preferences in relation to the nature of formal institutions and more informal social contracts that emerge in a globalized world (O'Brien et al. 2009). Actors who learn about outcomes (distantly) related to their actions may now be motivated to respond with behavioral or political change, creating an important feedback to the initial signals of change in the system.

Fifth, urbanization is playing a central role in creating the conditions of a telecoupled world. The process of urbanization, with the entailed rapid increase and shift in the nature of consumer demand, and an increased density of information, economic and political activity, social interactions, and knowledge creation (Seto et al. 2010, 2012b) have created a context in which telecoupling is more likely. Urban populations and places and the associated sets of values, activities, and consumption patterns have disproportionate agency globally, and thus are more likely be associated with the initial signal that produces the telecoupled effect. Urban centers, as a concentration of human activity and information, serve as nodes in telecoupled interactions and amplify signals to distant places. Urban processes also allow specific actors to obtain positions of greater relative legitimacy and facilitate their capacity to organize and acquire political influence. Urban areas thus have, potentially, an implicit if not explicit agency and responsibility in telecoupled processes.

An example of this is found in the indirect land-use changes caused by biofuel mandates. Several countries or regions have defined mandates for biofuels. The Renewable Energy Directive of the European Union, for example, specifies a 10% renewables content by 2020 across the entire membership. Other major blending mandates have been set in the United States, China, and Brazil. When environmental impacts of biofuels are evaluated, indirect land-use changes have become a central issue as they are caused by the competition for prime croplands, the international trade in agricultural commodities, and agronomic innovations facilitating crop substitutions under specific agroecological conditions (Lambin and Meyfroidt 2011). More specifically, when a bioenergy crop replaces a food crop in a field already under cultivation, or when crop production is diverted from the food market to the bioenergy market, the supply of the food crop decreases (e.g., for corn, sugarcane, potato, or wheat used for ethanol, or palm or rapeseed oil used for biodiesel). The market price for the replaced crop increases, thus causing more land to be allocated to that crop (Searchinger et al. 2008). This triggers a cascade of crop by crop substitutions, which eventually causes land conversion at the margins, a loss in ecosystem services, and could negate climate benefits from biofuels. The multiple crop substitutions and land conversion usually occur in places distant from the biofuel production site. As a result, there are additional environmental effects that are not immediately measurable and are difficult to attribute to the biofuel mandate. Estimating the magnitude of indirect land-use changes requires simulation experiments with global economic models, and results are sensitive to the modeling framework used and assumptions made. As quantification of their magnitude improved, indirect land-use change emissions were found to be lower than initially estimated by Searchinger et al. (2008), but still significant in the overall carbon budget of biofuels. As a result, the European Commission recently proposed limiting conventional biofuels with the risk of indirect land-use change emissions in contributing to the Renewable Energy Directive and instead encouraging advanced (low indirect land-use change) biofuels to contribute more to the targets, to decrease negative environmental impacts (EC 2012b).

Conceptual Framework

How Does Telecoupling Work?

For telecoupling to occur, two or more distinct social-ecological systems must exist, at some geographic distance from one another, such that the influence of one system over the other would not be expected or assumed. The systems may interact in a range of ways (e.g., through trade, through information exchange), but those interactions will be largely within the domain of existing governance and institutional arrangements.

Disturbance (e.g., a new technology, new information, a significant change in policy, or social mobilization in one context) to one of the systems (System A) causes the system to change rapidly, altering the type, number, or nature of linkages between that system and others (e.g., to System B, C…n) (Figure 8.1). The linkages or flows that connect the coupled systems may initially be economic, but the noneconomic and nonmaterial flows and linkages may be more instrumental in the telecoupled outcomes and feedbacks. These linkages may involve the movement of species (migration of people or species), environmental processes (dust movement, fire, carbon, nutrients), information, knowledge, ideas, and technology. These nonmarket interactions move through different media: through biophysical cycling and processes (the hydrological cycle, atmosphere, etc.), by people via social networks and migration, via the Internet or other communication media, and by banking networks and financial institutions. Each of these different forms of "flows" will have distinct modes of connectivity and nodes through which flows can be changed, amplified, or stopped. How different forms of flows are correlated in time and space is one way in which the potential for telecoupling is revealed. For example, in the EU biofuel policy example presented above, it is the process of crop substitution in places other than those producing biofuels, but which nevertheless are

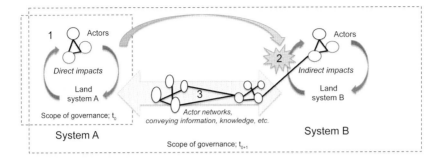

Figure 8.1 A schematic representation of telecoupling between two systems. (1) Coupled system A operates within a frame of governance; t_0 (e.g., via market transactions). (2) System A impacts on system B, producing a series of indirect outcomes and externalities, which may be unanticipated. (3) The externalities reach different thresholds of function, condition, attention, and mobilization, such that a feedback mechanism conveys pressure on the initial governance system to change its scope and address the externalities through new institutional arrangements (governance: t_{0+1}).

stimulated by changing relative prices (information) and the diversion of food stocks into fuel, that create the significant land-use change and eventually affect household food budgets in areas dependent on food imports.

Telecoupling as a land-change "problem" or concern frequently emerges when the institutions and mechanisms of governance in one system are unable to account for the consequences and interactions involving the more distant system, or they adapt too slowly to the new linkages that are formed following the disturbance. The systems' connectivities are thus "ungoverned": there are no higher-level governance structures to account for the new opportunities and risks that are entailed in the interactions. Typically, telecoupling also entails effects on livelihoods or land systems caused (indirectly) by spatially distant actors. These effects are essentially externalities (technological or pecuniary) that are unrecognized, inadequately compensated, or insufficiently mitigated. The temporal period that transpires until a telecoupled process is recognized and addressed formally in institutions and governance (something that is not inevitable) is particularly dynamic and critical. The way that ecological processes, economic systems, and social actors respond and value the resulting outcomes will potentially have a significant effect over the trajectory of development of the coupled systems and their future interrelationships.

The outcomes of telecoupling processes may be environmental or ecological (e.g., changing carbon emissions, reforestation, biodiversity loss) and/or social (e.g., an increase in well-being, loss of livelihood, population displacement). For our purposes, we are focusing on those processes that relate to land, in the form of outcomes for land-use managers and for land cover or land use, and are concerned specifically with outcomes that affect system resilience at different scales and sustainability of coupled systems and landscapes (Table 8.1).

Table 8.1 Features of telecoupling are illustrated for three examples: coffee production in Vietnam and Mexico (Eakin et al. 2009), British tourism in Portugal (Bell et al. 2010), and emigration in Bolivia (Zimmerer 2010). In each case, there is a trigger "event" or disturbance that alters the relationship between two distant systems. There are both direct and indirect impacts of this altered relationship, producing a feedback to the governance system. In some cases, institutional changes result.

	Coffee: Vietnam/Mexico	Tourism: U.K./Portugal	Emigration: Bolivia
Trigger	Development incentive for coffee production in Vietnam in context of deregulated market	Increasing disposable income in urban U.K. Portugal: emigration from coast	Changing opportunities in migration (including international migration and policies)
Direct impact	Increase in forest conversion to coffee Coffee enters global market	Change in residential ownership from Portuguese locals to U.K. citizens	Land-use intensification or disintensification depending on decisions (e.g., remittances)
Indirect impact	Migration, land degradation, and social conflict in Vietnam Depressed international prices	Increased interest in property amenity values but also resource degradation Urban expansion	Sustainability can increase (+) under viable land use, or decrease (−) where land use increases marginality
Feedback process	Coffee abandonment in Mexico International migration International coffee "crisis"	Temporarily vacant flats in U.K. foster in-migration in U.K. cities Mediterranean diets and new habits transfer to U.K.	Agrobiodiversity compatibility with intensification (+); soil degradation (−), e.g., terrace abandonment
Institutional change	Policy change in Mexico promotes economic diversification in coffee areas New attention in development agencies and coffee retailers to fair trade and alternative niche markets	None	Policy change and governance (*kawsay*) in Bolivia promotes investment for agrobiodiversity compatibility with intensification and food security improvement

The primary feedback from the outcomes to the international or "intersystem" context is often in the form of information, which will be mobilized through social networks into institutional change. For example, some group of actors (e.g., scientists, nongovernmental organizations, advocacy groups) observing the impact on System B will bring attention to undesirable or desirable outcomes. That information will, in some cases, be fed back to decision-making organizations who can then modify institutional arrangements to account for the unanticipated, secondary impacts of the system interactions. In other cases, the impacts may have such significant justice implications that affected populations are mobilized to protest, and this reaction can stimulate a policy response and the development of new governance arrangements to address the unexpected (indirect) outcomes. In yet other cases, teleconnections might remain ungoverned or uncompensated, which can be also a form of system response, exacerbating their intensity and potentially leading to thresholds of irreversible and potentially undesireable change.

A variety of new institutional arrangements might emerge from the recognition of a teleltoupled interaction: actors may adopt formal standards, laws about resource use or extensions to trade agreements, or compensatory mechanisms; they may motivate voluntary actions by private actors and NGOs in terms of certifications or "sustainability round tables"; consumer education campaigns may alter consumer behavior and preferences. The emergence of these new institutional arrangements can, in turn, have additional direct and indirect impacts. They may establish new examples of governance, which are then adopted by other actor groups who are anticipating potential analogous secondary impacts of their own activities. Alternatively, new institutional arrangements may produce new unanticipated secondary impacts and consequences, creating a new telecoupling process.

New Actors and Institutions

How the telecoupled system is defined, what outcomes are considered critical, and how the process evolves is in large part determined by the networks of actors, their activities, and their agency in the coupled systems. The processes of globalization and urbanization, and the inherent institutional changes, have accentuated the agency of some actors in the global system. In most cases the telecoupling process is characterized by asymmetrical relationships of material, capital, information, and ideology, such that the influence of one group of actors or one system over another can be instrumental. Similarly, the asymmetrical influence of different actors in the coupled systems also creates asymmetries in the responsibility for action and the nature of the response.

Global NGOs, for example, have assumed increased responsibilities and gained a new scope of influence in the wake of processes of state retrenchment and the declining influence of bilateral and multilateral investment and development agencies (Bebbington 2005). The World Wildlife Fund, for example, has been instrumental in mobilizing boycotts against the trade of bluefin tuna and pressuring trade organizations and companies to alter production practices which damage habitats and biodiversity. Global NGOs often have a physical presence in multiple countries and collaborate closely with local NGOs; in addition, they often have a strong political agenda and a geographically diffuse—but often highly urban and relatively wealthy—social constituency to which they are accountable.

Charitable foundations, typically defined by a specific set of issues and an agenda defined by the "high net worth" founding family or individual, are now exerting far more impact globally than they did even a decade ago. The Clinton Foundation as well as the Bill and Melinda Gates Foundation are just two organizations that have a new global presence and, by partnering with national government and bilateral and multilateral development agencies, are shaping the agenda of sustainable development globally.

Similarly, the consolidation of many commodity chains and commercial systems into transnational corporations has given a few specific corporate actors significant capacity to influence land outcomes globally. Cargill, ADM, Syngenta, Apple, international design firms, real estate agents, and retailers such as Tesco or Walmart represent just a few corporations with concentrated market presence. Their corporate policy decisions (and their shareholders) can have significant direct and indirect influences on resource allocation, land use, and even public policy in distant locations. With the intensified influence of corporate and commercial actors in resource decisions and management, voluntary round tables and consortiums of these actors (e.g., around sustainable foods or specialty coffee), in which key corporate policy decisions are made, assume greater influence over telecoupled processes. At a more local level, professional associations (e.g., producer organizations or commodity groups) can play instrumental roles in adopting new ideas and technology, responding to new opportunities, and in monitoring outcomes and creating knowledge about issues for which they are concerned.

In the public sector, institutional trends, such as deregulation and decentralization, have not only provided new spaces in which private and civil society groups can mobilize and act, they have also transformed the role of public sector organizations. Municipal governments often have new responsibilities and mandates, which may be poorly funded. In seeking ways to address these responsibilities, local governments have formed new networks and associations that allow them to exchange experience, lessons, and mobilize resources. By connecting directly to international donors, financial organizations or international NGOs, such local governments essentially "skip" traditional hierarchical relationships. They forge direct connections to other places and actors globally, accelerating the diffusion of ideas, technology, and institutional frameworks.

Improved social networking technologies and the Internet have also empowered social activists and associated social movements. The International Food Sovereignty movement, for example, and the World Social Forum represent an internationally networked coalition of disparate activist groups (often with different local agendas) which, among other activities, are monitoring the actions of international organizations and corporations in particular places, and advocating for change in international investment and international governance.

Bilateral development organizations and the private sector, together with global NGOs and global charitable organizations, are now also assuming new responsibility for reconstituting and enforcing failed international governance initiatives. The development of voluntary carbon markets, for example, reflects unenforced institutional designs developed in the 1992 UN Convention for Climate Change. The agency of these networks has significant system influence.

This brief overview of our conceptual framework suggests some key research questions concerning telecoupled systems in land science: How should a telecoupled land system be defined? Can we identify different types of telecoupled systems based on their functional characteristics and different pathways of emergence and development? Are there stages of emergence of telecoupling? What are the different processes through which telecoupling emerges? At what point can telecoupling be considered to be the primary force in the functional relationship between two bounded systems?

Conditions of the Telecoupled Systems

We still have much to learn about the nature of land-related telecoupled systems before a predictive model of system interactions and development can be developed. Nevertheless, it does appear that telecoupling emerges as a significant concern in land change after specific thresholds are crossed, all of which require further research. The sheer number and complexity of interactions in more complex telecoupled systems makes it particularly challenging to evaluate and anticipate when thresholds will be crossed. However, some indication may be provided by considering system functionality, condition, attention, and mobilization.

First, *functional* thresholds of influence and sensitivity are important. It is clear that one system must have characteristics that catalyze its potential for influence over others: it will be a price setter in international markets; it dominates the flow of information, capital, or technology; and perhaps it is a leader or example in the dissemination of ideas and knowledge. The other system, the "receiving" system, in contrast, will have characteristics that make it susceptible to the telecoupled signal: vulnerabilities exacerbated, for example, by the absence or reduction of domestic protections over land-use change or livelihood outcomes; the spatial extent or value of the resource affected; economic, infrastructural, or institutional conditions that trigger an elastic response of capital to new opportunities; or political conditions in which actors are well positioned to take advantage of the telecoupling and use it to their benefit.

Each of the interacting systems will be in some temporal process of change and dynamism involving social, institutional, and economic processes that are endogenous to the system as well as interactions with other systems. The state of the systems at the time of coupling will determine the implications of the coupling for the land outcomes of concern. Germany, for example, responded rapidly to the Fukushima nuclear disaster in Japan by changing its policy toward nuclear energy. This change in policy did not occur in other nations with similar dependence on nuclear energy. In Germany, however, the Green Party had been advocating for a change in energy policy for some time and used its recently acquired new political clout to catalyze the issue into national policy as a result of the ensuing environmental disaster in Japan. In other cases, industrial development motivated by wealthy immigrants with distinct cultural and aesthetic values may stimulate the growth of supporting real estate services and industry, altering the physical landscape and land use in a particular urban context. As such, urban design assumes symbolic status as it is transferred to other, unrelated contexts and becomes a model for urban development.

Second, *conditional* thresholds of impacts need to be crossed. In land science, these thresholds are reasonably well described and understood. Conversion of land from one use to another, or one land cover to another, typically depends on the combined influences of landscape factors (biophysical environment and land management attributes) and broader livelihood influences (market signals, migration remittances, demographic trends). Livelihood and human welfare outcomes are associated with land-use changes and, consequently, such outcomes establish thresholds in associated ecosystem processes. For example, migration may trigger environmental impacts in "sending" rural communities that are either negative or positive depending on conditions of information exchange, remittance-investment decisions, and other telecoupled factors. In the case of high migration communities of Bolivian smallholder farmers, these conditions have led in some cases to degradation (soil erosion) whereas in others it has led to conservation (agrobiodiversity) (Table 8.1). Thus while these conditions are critical in shaping telecoupling processes, more research is required to understand what degree of change is sufficient under which conditions to trigger the telecoupling relationship to become a prominent driver of land outcomes.

Third, for feedback to institutional design and governance to occur, similar to what occurs in the policy cycle, there needs to be an *attention* threshold of interest and concern. Institutional change is costly, particularly when it involves otherwise disconnected systems that are not operating within an existing governance system that can easily take on the emergent concerns of the telecoupling. For this reason, institutional change is most likely to occur when the implications of the telecoupling are of sufficient scale or consequence to trigger widespread concern and attention among a powerful community. Such consequences most likely occur when there is little possibility of substituting the affected resource, the resource is already contested or threatened, or when claims on the resource reflect multiple values and interests.

Finally, actors who have the agency and capacity to act must be concerned about the impacts and have influence over a system that is receptive to their actions: in short, there is a threshold for *mobilization* of resources to make a change in governance. In particular, those new actors, such as those described above, who have the capacity for cross-scalar action, play important roles. These thresholds raise a further set of research questions: What are the thresholds of functional change, condition, attention, and mobilization that are critical in different types of land-related telecoupled systems? Is it possible to identify characteristic features of such thresholds that typically lead to significant system change? Can they be predicted?

Analytical Approaches

The concept of telecoupling potentially offers a new heuristic from which to evaluate and think about land-use change. Analysis of telecoupling demands the integration of different epistemological perspectives on space and spatiality—one in which Cartesian space is the primary frame and point of departure, and one in which social space and its contingent aspects of agency and power are critical. In both science and policy, the tendency to view actions primarily within clear politically or ecologically bounded systems may need to change. As a heuristic, telecoupling shifts the focus from the processes and interactions occurring in one place or system to the processes and causal chain that links land parcels to land systems, to actors and actor networks, to institutions and governance, and ultimately to other land systems and places.

Telecoupling thus invites multiple points of entry for analysis and disparate lenses through which to understand its dynamics and implications. For example, telecoupling can be understood through the lens of environmental issues and sustainability concerns, as a way of bounding and framing the salience of the relationships. Research thus would tend to begin with a focus on a particular place-based problem and use that problem to define the system boundaries (e.g., deforestation in the Amazon or land degradation in the Sahel). The land parcel and its depiction in spatial analysis would be a critical starting point of analysis. Linkages are uncovered by working outward from the land-based focal problem of analysis.

An alternative approach might start with the telecoupled signal and the associated networks of actors and their activities, not the outcome on the landscape or parcel. For example, the EU Renewable Energy Directive, mandating a 10% renewable energy contribution to the EU energy portfolio, triggered a series of changes in cropping patterns and land use for the production of biofuels in the EU and in EU trading partners (Lambin and Meyfroidt 2011). In the analysis of telecoupling in this case, the signal might be a sudden rise in global oil prices coupled with a policy initiative in a particularly influential system to mitigate greenhouse gases. Here, the starting point of analysis might be the oil price signal and an identification of the social actors, and their networks, associated with that signal. By correlating the signal with other "flows" associated with land change (e.g., commodity prices, input prices, food prices, water, or labor demand), it may then be possible to anticipate where the telecoupling might emerge as a significant land issue. Alternatively, the entry point of analysis might be with the social networks involved: their agency, values, and capacity to instrument change.

The scope of observation and analysis consequently changes from what is often the case in land science. Multiple systems, separated in space, are typically involved in diverse possible causal and networked configurations. The emphasis is on the constitution of the linkages and nodes in the system and their implications rather than on hierarchy or scale of analysis. Networks of actors, of economic activity, and of environmental processes may occur at the same spatial scale or they may cross scales. For example, expansion of the imported quinoa market among high-end consumers in the U.S. market has led a variety of community organizations, indigenous rights groups, and food-security activists and public officials in Bolivia to use their connections through professional linkages, personal relationships, and funding sources to pressure international food, agriculture, and human rights organizations in Europe and the United States. Together, the actions of these different social networks have combined to make the role of quinoa in telecoupling processes an issue in an international FAO forum.[1]

Analytical approaches applied to telecoupled systems, therefore, need to be able to address several key questions:

- Who are the key actors? What are their activities and values? What is their agency and how is it instrumentalized (e.g., through which networks and linkages)?
- What are the institutions and governance arrangements in which the actors are embedded?
- What is the perturbation?
- What are the consequent direct and indirect flows, and how are they associated?
- What are the land system outcomes (social, environmental, economic)? What institutions are missing?
- What are the feedback mechanisms and associated consequences?

Many of the methods that could potentially be brought to bear on these questions are well developed in other disciplines in which geo-referenced space and place are not always prominent attributes of concern (see Table 8.2). Thus

---

[1] A proposal was made by the Government of Bolivia for an "International Year of Quinoa" at the 37th session of the FAO in Rome, June 25–July 2, 2011. C 2011/INF/18/ Rev. 1 (see FAO 2013).

Table 8.2 Analytical tools and lenses.

Analytical tool	Short description	Relevance for telecoupling
Actor theory-based approaches	Actor models comprise at least three nested components: • action as the dynamic interplay between activity and agency (means and meaning), • strategy of action as a combination of actions, and • institutions in which meanings of action are embedded.	An actor's capacity to influence land-use change can be discerned through his activity but also through the agency he can employ and how he is thereby influenced by external actors. Multiple interactions with other actors can be analyzed through direct influence on other actors (activity) or through shared institutions. This eventually allows indirect effects on land-use change to be identified that an actor exerts through actor networks.
Social network analysis	Allows the networked arrangements among many different actors to be analyzed in terms of various flows (e.g., information, money, material flows). The metrics permit network characteristics (e.g., density, centralization, bridging, and bonding ties) to be identified. Distinction of subgroups within networks is also an important measure.	Allows assemblages and hence system boundaries of telecoupled systems to be delineated, which then permit relevant socioecological systems to be identified and addressed. The identification of "brokers" (i.e., actors who ensure the connection between systems) highlights which actors and flows are decisive for enabling telecoupling, and hence represent leverage points for change and regime shifts
Process tracing	Identifies and decomposes all the detailed steps of the hypothesized causal chains that link some initial independent variables to the observed outcome of the dependent variable. When all steps of the causal chain, as well as their implications, are validated, and the counter-hypotheses are shown to be invalidated, then the causal link between the initial variable and the outcome can be established.	Well-suited to the longitudinal analysis of complex multidimensional cases, such as those which include nonmaterial flows and linkages that may be instrumental in land outcomes.
World city systems analysis	Scholars focus on two different aspects of global hierarchies: • differential attributes of global cities and the quantity, which involves ranking cities by the performance or level of specific traits, and • intensity of flows or linkages between them, which focuses on the interactions among cities within the hierarchy.	Explores partial "telecouplings" between cities in different nations. Theories, methods, and empirical results may be of use to those interested in land-use telecoupling.

Table 8.2 Analytical tools and lenses *(continued)*.

Analytical tool	Short description	Relevance for telecoupling
Assemblages	The concept of "assemblages" suggests the coming together and interaction of multiple things. This is used to theorize and study structural change in social relations and the global political economy. Assemblages do not privilege specific units of analysis or predetermined causal relationships.	May be a useful heuristic device to capture the complexity of analyzing the interaction among fragments of institutional forms, ideas, and actors across historical periods. Can be useful in examining the sets of actors and varieties of agencies across space that are then telecoupled to land uses.
Economic models	A wide variety of economic tools and modeling approaches are potentially applicable: • computable general equilibrium models capture linkages between input and output markets and can thus be used to study how changes in one market affect individuals in a distal market, • models of economic growth and migration capture flows of people, goods, capital, and firms across regions and feedback effects, and • models of social interactions capture the influence of groups and networks on individual choices.	Useful to capture the implications of interregional flows of people, goods, and services in specific places as well as the relative influence of particular flows on observed outcomes.
Diffusion theory	Addresses how innovations spread through space and time. Has been used in fields from communications to marketing to understand how information, knowledge, ideas, lifestyle, and consumer patterns "travel" through society and are adopted over time by an increasing number of people. The mechanisms by which this process of diffusion occurs relate to knowledge and interactions among people.	Can help identify the nature of the force (e.g., idea, policy), the communication channels, the time, and the systems (actors, linkages) that enable an innovation to flow through the system. Uses concepts such as critical mass and tipping points which should be considered when analyzing what makes two systems coupled.
Organization theory	Used to understand how organizations arrange actors and processes to obtain a goal.	Can help us identify the individuals, structures, processes, and motivations that guide the behavior of the key actors.

Table 8.2 continued

Analytical tool	Short description	Relevance for telecoupling
Trend analysis	An analytical approach in business (financial management, marketing) is employed to try and predict future behavior. Quantitative and qualitative methods are employed. To conduct marketing research, one tools consists of identifying "trendsetters" and following their actions, decisions, and interactions with others to detect the birth of a new trend.	As urban lifestyles become globally adopted, changes in patterns of consumption or interest of small but influential populations can be the origin of changes in behavior at a large scale and the telecoupling of distal systems.
Multisite studies	These are places of study that are connected by processes rather than comparable units of analysis or systems features. Such a study might involve all sites in which the activities along a commodity chain take place and use different methods for each site.	Is essential for analyzing telecoupling since, by definition, it occurs at multiple sites, with different actors, and activities of interest in each site.
Flow analysis	This general category of analytical tools incorporates commodity/value chain analysis, life cycle analysis, material flow analysis, and other methods that trace systematically the movement of material (energy, goods, capital) through a system.	Accounts for the movement of material and value between systems, and thus the feedbacks and linkages in telecoupled interactions.
Pathway analysis	Focuses on strategies that arise from decisions taken by individual actors, households, and groups of people. It stresses that opportunities and constraints on decision makers are imposed by other actors as well as higher-level institutions.	Is relevant for telecoupling analysis, which pays attention to the role of actors in dynamics and feedback.
Scenario and visioning	Different fields have developed strategies and procedures for scenario construction and analysis as well as participatory visioning activities. These approaches entail facilitating the construction of possible and plausible futures with key social actors, in which the indirect and direct consequences of actions and development trends can be explored.	May enable telecoupled land outcomes to be predicted and anticipated, and adverse outcomes avoided, through anticipatory actions and social learning.

integrating approaches requires innovation in analysis: How can social networks and their geographic influence be represented spatially? Are there ways in which tools common to land science can be employed to represent the material and nonmaterial flows critical in telecoupled systems? Can the different values that specific actors associate with land change be represented as attributes of parcels and places? Is it important to present evidence that specifically quantifies the degree of land system change associated with the initial telecoupled signal, and if so, what methods will enable this given the indirect and second- or third-order interactions observed in these systems?

Global Land-Change Implications: Vulnerabilities, Risks, Opportunity, and Adaptation

The undesirable impacts of telecoupling processes have drawn attention to these linkages. Nevertheless, it is important to recognize that the telecoupling process is driven by different—and possibly incongruous—values associated with diverse social actors, some of whom are in distal locations from where the impacts have materialized and may thus not have an accurate understanding of the causal relations or implications of the telecoupling outcomes they observe. For example, the concept of "food miles" gained attention in the popular press and media in the United Kingdom during 2005–2006, as a measure of the impact of food transport on the environment. The concept began to be used by activist environmental organizations to argue for the purchase of local food. This campaign threatened the viability of some types of food export from New Zealand to the United Kingdom, and hence potentially the livelihood of many New Zealand farmers. However, subsequent evaluation of the energy and emissions performance of a sample of food products using life cycle analyses (Saunders et al. 2006) revealed that locally supplied foods in the United Kingdom typically had significantly higher energy and emissions costs than the competing imports from New Zealand, due to the different production, storage, and transport systems and their cumulative performance. Here, the telecoupling which enabled U.K. NGOs to influence significantly the behavior of consumers unexpectedly prompted higher environmental impacts, due to misleading and incomplete information.

Thus, how telecoupling outcomes are valued will always be challenging and depend very much on assumptions and the framing of the process (e.g., "land grabbing" vs. "REDD+"), what attributes of a specific land system are affected, what social actors are present and mobilized to act, the spatial or temporal scale at which outcomes are evaluated, and the context in which the telecoupling occurs. One of the concerns with growing influence of telecoupling is that in a time of more acute resource scarcity, telecoupling may have redistributive implications, both enhancing and limiting the possibilities of action for different populations and actors in particular places. This redistribution may be positive (beneficial) or negative. Such outcomes are not only important to understand for development trends, but also have ethical implications for both science and governance.

For land science, the concern is particularly in relation to significant thresholds of change in land use and associated ecological and social processes. Telecoupling, for example, that results in sudden land-cover conversions, land abandonment, or shifts in land governance with consequent implications on resource management and use can potentially have significant global consequences for land-change trends. More broadly, regime shifts in land systems have implications for broader system resilience and capacity to manage shocks over the long term.

National governments may be more concerned with the potential threat to human and national security. Where telecoupling outcomes are undesirable in a particular place, social conflict may result. Land, as a scarce resource, is highly valued in many cultural and economic contexts. Telecoupling processes that result in marginalization of disenfranchisement of local populations to land and land activities has, in the past, provoked revolution and violence. For example, the French, Russian, and Mexican revolutions derived their strong impetus from exploited local land-working populations and resulted in transforming the social order. These experiences have been repeated in many colonial situations, where inequities in land-holding rights prompted civil unrest and eventual removal of colonial powers. Repeated land-reform movements, both local and national, have been prompted by inequalities provoked by changes in land use and market demand; they have resulted in less extreme transformations, but still transformatory realignment of rights to significant amounts of land. How such outcomes can be potentially anticipated is a potential subject for land science and governance in a telecoupled world.

Conversely, society in general may be better equipped today to manage these complex relationships through the same information and knowledge that has contributed to these accelerated linkages. People and organizations have the potential to anticipate consequences and mobilize collectively to act to avoid undesirable outcomes. The rising influence of social networks and information systems has given actors a capacity to "skip scale," redistributing agency to finer scales and enhancing the capacity of cross-space interactions. Through such connections actors can be both flexible and innovative in their responses. Telecoupling can also lead to other adaptive responses and opportunities via market specialization of actors newly linked to market signals, linkages between previously disconnected social networks operating at different scales, and the diffusion of institutional innovation and social mobilization to other systems that have yet to experience undesirable outcomes but wish to anticipate possible problems.

Outcomes of telecoupling processes that are valued as "positive" by some actors and communicated through knowledge networks can also be recognized and supported at higher levels of decision making. For example, various production standards associated with eco-certification have been adopted by other agents, even though they may not be seeking certification, given the benefits of the practices. This creates positive spillover effects to other producers, companies, investors, or governments. In some cases, sustainability standards become embedded in public policies or reinforce existing policies (Steering Committee of the State-of-Knowledge Assessment of Standards Certification 2012).

The new institutional changes that have been noted as an outcome in some telecoupled systems have in part been brought about by the recognition of responsibility by consumers, corporations, governments, or other actors in the systems that are the source of the telecoupled signal. Making telecoupling "visible" through information and knowledge networks can thus lead to enhanced agency and new forms of social contracts, mediated by sustainability certifications and standards or codes of corporate responsibility. These new institutions can "institutionalize" some of the externalities of the telecoupling process through the same linkages and flows that initially were the cause of the problem. This added value, then, can bring important benefits to telecoupled regions.

Research Needs, Opportunities, and Limitations

The development and application of the concept of telecoupled systems in land science opens up a stimulating range of challenging research questions: How can land science engage with other disciplines in a transdisciplinary project to analyze the nonmaterial dimensions of telecoupling, such as financial flows, social networks and values, and information and symbolic relationships? How can understanding of these procedures be most effectively integrated into established land science? Are there stages of emergence of telecoupling? Can we identify different types of telecoupled systems depending on their functional characteristics and different pathways of emergence and development? At what point does the functional relationship between two bounded coupled systems assume sufficient importance in the overall operation of the two systems for them to be conceived as primarily telecoupled systems? What tools and methods will enable tracing of indirect and second- or third-order interactions observed in these systems? What are the thresholds of functional change, condition, attention, and mobilization that are critical in prompting change in different types of telecoupled system? Most significantly, perhaps, where do telecoupled feedback processes critically influence global land-use change, with what consequences and potentials for improved governance?

It is important to recognize the limitations associated with this emerging conceptual approach. It could be argued that telecoupling is nothing new, and thus that it adds little novel insight. There have always been distal connections in human affairs, and so perhaps the current situation is different by degree, but not in essence. As is clear from the preceding account, we subscribe to the view that the social and nonmaterial feedback processes that have been enabled by modern technologies and new governance arrangements are substantively different in their nature and effect from anything that has gone before, and that they are potentially game changing in global land science. While it is by no means clear that the concept of telecoupling can be operationalized in an effective way *within* land science as it is currently constituted, the global significance of the phenomenon lays obligation upon the land science community to find ways to engage with the necessary concepts and analytical tools. This may require the development of a transdisciplinary land science, with profound implications for methodology and reporting.

Conclusion

The increased concern over telecoupling in land-change processes highlights the importance of system perspectives and integrated hybrid analyses in revealing the important drivers and feedbacks in land systems. The analysis of telecoupled systems also reveals the importance of social actors and their associated values, preferences, and networks on land change. Among those actor networks is the scientific community: science plays a significant role in making visible the relationships and consequences of value to the science community as well as to other actors. Telecoupled processes will be revealed because they affect attributes of value in a particular land system—attributes of the land itself, or social features and activities associated with land use. Understanding those configurations of values and the social systems in which they are embedded will define how the telecoupling process affects pathways of sustainability in land systems, and the relative influence of specific actor networks in defining the trajectories of those pathways.

9

Palm Oil as a Case Study of Distal Land Connections

Birka Wicke

Abstract

The production of palm oil is often associated with negative environmental and social impacts that are mainly related to land-use change (LUC), and more specifically to deforestation of tropical rainforest. However, most consumers of palm oil-based products are located far away from production, LUC, and its impacts so that they do not directly, nor immediately, feel these impacts. This chapter investigates the main trends and underlying factors that shape this connection and land use for the case of palm oil. It identifies possible entry points for minimizing undesired impacts of LUC related to palm oil production. Reducing the impacts of palm oil production generally focuses on the production areas and includes better land-use zoning as well as the use of degraded land for new plantations, increasing palm oil yields and applying production schemes that are more beneficial to local communities. However, to identify additional potential entry points for change, it is necessary to understand decisions that are not made at the local level of production; that is, those made increasingly by actors in distant places and through interactions between different markets for land, palm oil, and palm oil uses. Therefore, a global perspective on land use and land-use governance is needed. In addition to better global land-use governance, it is important to make producers, traders, processors, and consumers take more responsibility for the impacts. This could be made possible by requiring these actors to comply with international normative standards, by requiring multinational companies to take responsibility for adverse social and environmental impacts, and by creating consequences for not meeting standards. Consumers may influence decisions on where and how palm oil is produced by demanding certified sustainable palm oil or products produced according to normative standards for multinational companies. In addition, a global cap on LUC-related emissions for all countries, which should account for the emissions in all product chains, could provide better governance of land-use change and minimize the displacement of land use and associated emissions. Consumers and consuming countries could be made more responsible for LUC-related emissions by allocating these emissions to consuming countries or by placing a carbon tax on products with high LUC-induced greenhouse gas emissions.

Introduction

Immense increases in palm oil production in Southeast Asia, particularly in Indonesia and Malaysia, have come hand in hand with environmental, ecological, and social issues, mainly related to the land-use change (LUC) associated with the large areal expansion of palm oil production. Palm oil production has been linked to (a) forest and peatland fires which cause smoke pollution and related health hazards across Southeast Asia, (b) deforestation which causes losses in carbon stocks and biodiversity, and (c) peatland degradation. All of these lead to increased greenhouse gas emissions and thereby contribute to climate change (Gibbs et al. 2008; Koh et al. 2011; Koh and Wilcove 2008; Miettinen et al. 2012). Moreover, there are social problems associated with the rapid expansion of palm oil, including unevenly distributed economic benefits among stakeholders, land tenure conflicts, and labor rights issues (Colchester 2011; Obidzinski et al. 2012).

Most consumers of palm oil, however, do not directly nor immediately feel any impact from the LUC caused by palm oil production, as they are located all over the world, far away from where production is concentrated in Southeast Asia. Nonetheless, consumers and consumer countries are increasingly becoming aware and concerned about the impacts of their consumption. This concern has the potential to trigger a feedback mechanism that could cause, for example, less palm oil to be consumed or the demand for sustainably-sourced palm oil.

The relationship between the production of palm oil in concentrated areas and the widespread use of its products in distant places all over the world, particularly in urban centers, reflects the concept of distal land connections; that is "the linkages among land uses over large geographic distances" (Seto et al. 2010). This concept emphasizes that in order to examine the complexity and dynamics of LUC, its drivers, and its impacts, we need to account for the various links between distant places of production and consumption and possible feedback mechanisms. This chapter analyzes the main trends and underlying factors that shape these distal land connections related to palm oil, it examines linkages and feedback mechanisms between systems in distant places, and it explores the implications for local and global land use and society, including possible entry points for minimizing undesired impacts of LUC related to palm oil production. It begins with a brief description of the production and uses of palm oil. Thereafter, the palm oil system and its components are assessed through the lens of telecouplings. Based on an understanding of the system, entry points are proposed for reducing LUC associated with palm oil production and its impacts.

Palm Oil

The oil palm *Elaeis guineensis*, originally from Africa, grows in tropical regions within 20° of the equator and is now the dominant species for the production of palm oil. Currently, the largest share of global palm oil production (84% in 2010) takes place in Indonesia and Malaysia. However, other countries in Southeast Asia (Papua New Guinea and Thailand), Asia (China), Latin America (Colombia, Ecuador, Honduras and Brazil), and Africa (Nigeria and Ivory Coast) are increasingly producing palm oil (FAOSTAT 2013).

An oil palm starts producing the first palm fruit bunches after two to four years of planting. Although oil palms can live longer, the lifetime is normally limited to 20 to 25 years because, on one hand, the palms become too tall to be harvested easily and, on the other, yields decrease. Yields depend not only on the age of the oil palm but also on ecological conditions and management. In 2010, an average fresh fruit bunch yield of 17 ton per hectare was recorded for Indonesia and 18 ton per hectare for Malaysia (FAOSTAT 2013; MPOB 2013). Yields in other countries (with the exception of Colombia) are generally lower than those found in Malaysia and Indonesia (e.g., 14 ton per hectare in Thailand) (FAOSTAT 2013).

The palm fruits contain oil in the pulp (making palm oil) and in the kernel (making palm kernel oil). The oil extraction rate ranges from 17–27% for palm oil and 4–10% for palm kernel oil (FAOSTAT 2013). For Malaysia in 2010, the average palm oil extraction rate was 20%, with an average palm oil yield of 3.7 ton per hectare (MPOB 2013).

From 1990 to 2010, global palm oil production increased nearly threefold, increasing from 12 million tons in 1990 to 45 million tons in 2010 (FAOSTAT 2013). In 2010, the majority of production took place in Indonesia (which accounted for 45% of global palm oil production) and Malaysia (39% of global production) (FAOSTAT 2013). However, in recent years other countries have increased their palm oil production, most notably Nigeria, Thailand, Colombia, and Papua New Guinea (Figure 9.1). Large tracts of land are required to produce this amount of palm oil; in 2010, 15 million hectares of land were harvested for palm oil production globally (up from 6 Mha in 1990). Most of the increase in land use for palm oil over the last two decades took place in Indonesia (increasing from 0.7 to 5.0 Mha) and Malaysia (increasing from 1.8 to 4.0 Mha) (FAOSTAT 2013). Including the area allocated for immature oil palm—0.7 Mha in Malaysia in 2010 and a recently estimated 2.2 Mha in Indonesia (MPOB 2013; USDA 2009)—a total land expansion of approximately 9 Mha took place in Malaysia and Indonesia over the last two decades as a result of palm oil production.

Although palm oil is also consumed in countries that produce it, it is primarily a cash crop for export. For example, Indonesia and Malaysia both export over 80% of the produced palm oil. Palm oil is used for various purposes, most importantly in the food industry (e.g., margarine and vegetable fats) and

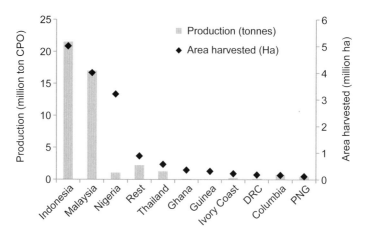

Figure 9.1 Production of crude palm oil (CPO) and area harvested for palm oil production in 2010, by country. DRC: Democratic Republic of the Congo; PNG: Papua New Guinea. Data from FAOSTAT (2012).

oleochemical industry (e.g., cosmetics, detergents, and plastics). The transportation fuel industry is increasingly interested in palm oil as a feedstock for biodiesel production (see section on Application of Palm Oil for Biofuels below).

The Palm Oil System through the Lens of Telecoupling

Let us explore the palm oil system in more detail using the lens of telecoupling and its key components (see also Liu et al., this volume): coupled human-natural systems, agents, flows, causes (including underlying and facilitating factors), and effects (including responses and feedbacks to these effects).

Coupled Human-Natural Systems

The following coupled human-natural systems are relevant for the telecoupled palm oil systems:

1. *Sending systems*: Malaysia and Indonesia currently produce 84% of all palm oil worldwide (Table 9.1), but it is increasingly being produced in other Southeast Asian countries (e.g., Thailand and Papua New Guinea), Latin America (e.g., Colombia), and Africa (e.g., Nigeria), with more plans for expansion in these areas in the future.
2. *Receiving systems* are located all around the world. While the number of large producing countries is limited (the top five countries account for 92% of global palm oil production), far more countries import and consume palm oil. Imports by the top five importers amount to 43% of all global palm oil imports (Table 9.1).

Table 9.1 Top five countries that produced, exported, and imported palm oil in 2010. Data from FAOSTAT (2013).

Production country	1000 ton	Import country	1000 ton	Export country	1000 ton
Indonesia	19,760	China	5833	Indonesia	16,292
Malaysia	16,993	India	3985	Malaysia	14,733
Nigeria	1350	Netherlands	1988	Netherlands	1168
Thailand	1288	Pakistan	1702	Papua New Guinea	521
Colombia	753	Germany	1434	Germany	232
Rest	3430	Rest	19,467	Rest	2374
Share of top 5 in total (%)	92		43		93

3. *Spillover systems* account for countries and regions that are heavily involved in palm oil trade or processing, such as Singapore, the Netherlands, and Germany (Table 9.1).

Agents

Palm oil production is strongly affected by new land agents and land-use practices. Many different agents are involved in the palm oil system including, for example, producers (ranging from smallholders to large international companies), the producer and palm oil industry associations, traders, retailers, investors, consumers, local and international NGOs concerned with the impacts of palm oil production, local people affected by production of palm oil, and certifiers. However, palm oil production is dominated by large multinational corporations. These corporations not only own oil palm plantations and mills, they also control the downstream activities (such as refining, trading, and processing of palm oil) that produce a wide range of products used throughout the world as well as activities in a wide range of other sectors (e.g., forestry, telecommunication, banking, and construction) (Wakker 2004). In addition to the multinational actors that are directly involved in palm oil-based supply chains, there are also domestic and multinational institutional investors that buy shares and bonds of palm oil companies, as well as foreign and domestic banks and other multilateral financial institutions that finance palm oil production, particularly the large initial capital that is needed to establish a plantation (van Gelder and German 2011). Another important actor in palm oil production is the government (both domestic and foreign); it provides subsidies, investment incentives, and loans and, in some cases, also owns palm oil companies (van Gelder and German 2011).

In addition to these agents, there are also actors from other industry sectors that must be accounted for as agents of land use and LUC associated with palm oil. In Indonesia, for example, it was reported that large tracts of land under palm oil concessions were clear-cut but never planted with oil palms (Casson 2000; Wakker 2004). This phenomenon is likely caused by getting a palm oil permit more easily than a logging permit so that companies from the timber industry apply for palm oil permits to get access to timber. This example illustrates the complexity of assigning shares to different drivers of LUC due to the interlinkages between drivers that are difficult to quantify with publically available data. Nevertheless, under the current system, forested land is financially more interesting for oil palm plantation companies than other types of land because the returns from the extracted timber help finance the establishment costs and maintenance costs during the immature phase of the plantation (i.e., the first two to three years).

Thus, the different types of actors influence decisions on land use for palm oil production in different ways and from different places, often far away from the production site. The influence of these various actors across country borders and jurisdictions points to the need for better governance of these actors. It also demands that different actors in different places take responsibility for minimizing, if not actually avoiding, negative local impacts (this will be discussed further below).

Flows

The distal land connections of palm oil are defined and further enhanced by globalization and the integration of economies. This is not only demonstrated by the multinational actors in different parts of palm oil-based supply chains (corporations owning plantations and downstream activities related to refining, trading, and processing palm oil; investors and financial institutions) as described above, but also by increased flows related to (a) palm oil products, (b) labor for palm oil production, and (c) technology and knowledge. Each of these aspects is explained in more detail below.

International Trade of Palm Oil Products

Increased international trade is a characteristic of globalization, but it also facilitates and enables globalization. International trade of palm oil has increased rapidly over the last decades (Figure 9.2). Global export of palm oil from producing countries increased from 8 million tons in 1990 to 35 million tons in 2010 (FAOSTAT 2013). More recent estimates by the USDA (2012) suggest a further increase of global palm oil export to 39 million tons for 2011/2012. In 2010, Indonesia was a net exporter of palm oil with 16 million tons (82% of its production) and Malaysia with 13 million tons (79% of its production) (FAOSTAT 2013). Leading importers are China and India (Table 9.1).

Import and export statistics also highlight the globalization and integration of the palm oil market. For example, the Netherlands and Germany are the third

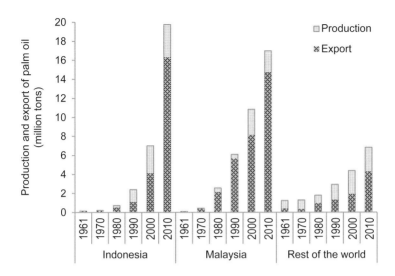

Figure 9.2 Production and export volumes of palm oil for Indonesia, Malaysia, and rest of the world, 1961 to 2010. Data from FAOSTAT (2013). Note: rest-of-world exports include exports from countries that do not produce palm oil but import palm oil. Therefore, the actual share of exports in total production for the rest-of-world countries is lower than shown in this figure. For example, in 2010, producing countries (excluding Indonesia and Malaysia) exported 2 million tons while the total global export volume amounted to 5 million tons.

and fifth largest exporters, respectively, worldwide although they do not produce any palm oil. Instead, they are also large importers of palm oil (Table 9.1).

Labor Migration

Another flow is related to labor requirements for palm oil production. Palm oil production is labor intensive, but in areas of recent expansion of palm oil production, population density is low and the amount of skilled labor is limited. In addition, plantation jobs, especially related to harvesting the fresh fruit bunches, are often not desired by the local population because of the hard physical labor that is involved. As a result, palm oil production is often reliant on migrant workers. Therefore, the governments of Malaysia and Thailand, for example, have migration policies that enable the immigration of foreign workers more easily (Kaur 2010). The case of Malaysia illustrates the importance of foreign workers in the plantation sector (most notably oil palm plantations). Currently it is estimated that nearly 70% of all plantation workers in Malaysia are immigrants, primarily from Indonesia (Abdullah 2011; Kaur 2010). But in Indonesia, plantations also rely on migrant workers, who originate mainly from more populous areas in Indonesia such as Java.

Diffusion of Technology and Knowledge

The exchange of technologies, knowledge, and information across and within regions and across plantations constitutes an important flow in the telecoupled palm oil system. Specifically, the diffusion of technology (e.g., high-yielding planting material) and knowledge of management techniques (e.g., precision agriculture, good harvesting standards, and early replanting practices) are important to optimize yields, which is a crucial component in reducing land requirements and new conversion of land to palm oil. Companies, which own large areas of plantations in different regions, can transfer this kind of information easily, as well as to emerging production areas outside of Indonesia and Malaysia. In addition, there is also a structural knowledge exchange through a collaboration of producer organizations such as the Malaysian Palm Oil Board and Colombian Oil Palm Research Center (CENIPALMA).

Increased and intensified information flows related to LUC and its impacts (such as information from satellite images on hotspots of deforestation and latest deforestation rates) also characterize the telecoupled palm oil system. This information flow is part of a feedback mechanism by which consumers in distant places become aware of the impacts of their consumption and potential to act upon them (for a discussion of the importance of information flows in telecoupled systems, see Eakin et al., this volume).

Causes

The main causes generating the telecoupled systems related to palm oil are ecologically favorable conditions in tropical regions near the equator and resulting high yields and low production costs compared to other vegetable oils. Thus, the ecological cause becomes an economic cause. In addition, various other factors drive the increased demand in palm oil and facilitate meeting this demand. These drivers include population growth, growing wealth and changing lifestyles (also related to urbanization), and new applications for palm oil, specifically for biofuels. Globalization and the integration of economies have facilitated the telecoupled palm oil system.

Population Growth, Growing Wealth, and Changing Lifestyles

Growing world population and wealth are attributed to be the two major underlying factors of increased palm oil demand. World population increased from 5.3 billion to nearly 7 billion people in 2010 and is projected to increase to 9 billion by 2050. In 2010, approximately half of the world's population lived in urban and peri-urban areas, and it is estimated that the share of urban population will increase to 60% by 2030 and to 70% by 2050 (FAOSTAT 2013, based on data from UN Population Division). Growing wealth is associated with changing diets, which is mainly characterized by a change toward more consumption of animal proteins, but also more vegetable oils, fruits, and vegetables (Kastner et al. 2012). Increasing demand for vegetable oils is to a large extent met by palm oil, which makes up the biggest share of vegetable oils (31% in 2010). Palm oil has seen the largest increase in production among all vegetable oils over the past two decades (average annual increase of 7%) and has a price generally lower than other vegetable oils (FAOSTAT 2013). Growing wealth is linked to, and also driven by, increasing energy consumption. In addition to growing wealth, increasing environmental awareness can become a factor that shapes land use. For example, concerns about deforestation may lead to a reduced consumption of food products that contain palm oil, while concerns about climate change may lead to the replacement of fossil fuels with biofuels (partially based on palm oil).

Application of Palm Oil for Biofuels

Increased demand for palm oil also comes from the biofuels sector, as palm oil can be transesterified to make biodiesel and used for blending with fossil diesel. Not only industrialized countries, such as those in the European Union, but also many developing countries, such as Indonesia, have become interested in biofuels as an alternative to fossil fuels, due to high oil prices, rising energy demand, and concerns about climate change (Caroko et al. 2011). Palm oil has the highest yield of all oil-bearing crops and therefore has low production costs and high land-use efficiencies. Both characteristics explain the interest of the biofuel industry in palm oil. Palm oil-based biodiesel is particularly interesting for the large palm oil producer countries such as Indonesia and Malaysia, because they can produce it domestically and thereby reduce their energy dependence on others. However, there is continued debate about the sustainability of palm oil-based bioenergy, especially related to deforestation of tropical forests in Southeast Asia and the associated greenhouse gas emissions, degradation of peatlands, and competition between food and fuel uses. This, together with high vegetable and palm oil prices in recent years, has slowed down investments in palm oil-based biodiesel plants or in low usage of existing capacity. However, when prices fall, future demand for palm oil from the biofuel industry is likely to increase.

Globalization and the Integration of Economies

Globalization and the integration of economies have been important facilitators of the palm oil telecoupled systems; both enable and simplify the global trade of palm oil, diffusion of technology and knowledge, and labor migration, as described above. In addition, free-trade areas and bilateral free-trade agreements have further facilitated the trade of palm oil. For example, both Indonesia and Malaysia are members of the Association of Southeast Asian

Nations' Free Trade Area, which has free-trade agreements with, for example, the largest importers of palm oil, China and India.

As Blanco (this volume) explains, advances in transportation infrastructures have shaped not only international trade but also facilitated increased trade at reduced costs. Thus, increased international trade is a characteristic of globalization, but it also facilitates and enables globalization.

A key factor and result of globalization is also the growth of multinational companies and conglomerates. The palm oil sector is largely defined by multinational companies and large agri-businesses. For example, one of the largest producers of palm oil worldwide is Sime Darby. This multinational company and conglomerate owns 880 thousand hectares of palm oil concessions (41% in Malaysia, 34% in Indonesia, and 25% in Liberia) and produces 2.4 million tons (equivalent to 6% of the global palm oil production) (Sime Darby 2013). Their downstream activities include manufacturing and distribution of oils and fats products, oleochemicals, and biodiesel. Their other core business activities relate to construction and property development; car dealerships; the purchasing, leasing, and selling of industrial equipment; energy and utilities (focusing on oil and gas exploration in Southeast Asia as a power provider), and health care (Sime Darby 2013). Multinational companies are able to finance the high upfront costs of the establishment and immature phase of palm oil plantations, exchange knowledge, information and human capital, and access resources more easily (and often at lower costs) than smallholders. Moreover, in Indonesia and Malaysia, large plantations have higher yields and thus lower production costs. However, the dominance of multinational companies exacts a major disadvantage: local communities receive hardly any benefits from palm oil production while they may even suffer from environmental degradation and forest resource losses, displacement from their homelands, or mistreatment as workers on plantations. These are just some of the effects of the telecoupled system that trigger responses and feedback mechanisms, as will be explained next.

Effects

In response to higher demand for palm oil, production has increased, as described above, primarily in Indonesia and Malaysia but also increasingly in other countries in Southeast Asia and elsewhere. Greater production is enabled mainly through land expansion but also partially by increased yields.

As already described, LUC in palm oil production has been linked to (a) forest and peatland fires which caused smoke pollution and related health hazards across Southeast Asia, (b) deforestation which causes losses in carbon stocks and biodiversity, and (c) peatland degradation (Gibbs et al. 2008; Koh et al. 2011; Koh and Wilcove 2008; Miettinen et al. 2012). Two recent studies have estimated the extent of palm oil production on peat soils. Koh et al. (2011) found that in 2000 an estimated 0.88 Mha of peatland was used for palm oil production in Peninsular Malaysia, Borneo, and Sumatra. For the same area, Miettinen et al. (2012) found that in 2010 2.1 Mha of peatland are used for oil palm plantations and that peatlands have been used at an accelerating pace over the last two decades for palm oil production. Moreover, social problems associated with the rapid expansion of palm oil include unevenly distributed economic benefits among stakeholders, land tenure conflicts, and labor rights issues (Colchester 2011; Obidzinski et al. 2012). Although most social problems and some of the environmental impacts (e.g., biodiversity losses) are local issues, other impacts (particularly forest fires and other losses in carbon stocks, which lead to increased greenhouse gas emissions and thereby contribute to climate change) have global influence.

Indirect Land-Use Change

Indirect LUC is a concept that has been applied mainly to biofuels, although it is also applicable to other land-based products. It is the market-mediated effect of biofuel production outside the feedstock production area and is caused by globalization and the integration of economies (Laborde 2011). Indirect LUC occurs when (a) bioenergy feedstock production displaces food production to another area in order to continue meeting the food demand and/or (b) the diversion of the crop to other uses triggers higher crop prices, which results in more land being taken into agricultural production elsewhere. In addition, infrastructure development for large-scale biofuel feedstock production and migration of people into and out of an area in response to bioenergy production can cause indirect LUC (Wicke et al. 2012).

Indirect LUC that is related to palm oil can occur as a result of increased demand for, e.g., the production of biofuels, but increased demand for palm oil can also be the indirect effect of the demand for, e.g., another biofuel feedstock. An example of the first mechanism is the increased use of palm oil for biodiesel production in Indonesia; this greater demand can be met by increasing production of either palm oil or other vegetable oils. The second mechanism is exemplified by increased rapeseed demand for biodiesel production in the European Union, which displaces rapeseed oil from other products or uses. To continue meeting vegetable oil demand, more vegetable oil from other feedstocks is needed. This is likely to be met by palm oil because it is lower in price than other vegetable oils.

Indirect LUC causes the same impacts as direct LUC, such as greenhouse gas emissions, biodiversity losses, and land-tenure conflicts. However, thus far, assessments have primarily investigated the greenhouse gas emissions of indirect LUC. Although all modeling efforts result in indirect LUC emissions being above zero, the results differ strongly and depend on many assumptions regarding, for example, the location and type of LUC, level of agricultural intensification, and agricultural crop yields (Wicke et al. 2012). More work is needed to improve the existing modeling frameworks so that they are able to analyze indirect LUC more appropriately and determine its extent and effects. In addition, there is a need to assess options for reducing the extent and impact of indirect LUC (see section on Reducing Impacts of Palm Oil Production).

Responses and Feedbacks

The first response to LUC related to palm oil production and its impacts has been increased information flows on the impacts. NGO campaigns have, in particular, directed attention in distant places to the impacts of palm oil production. Increasing awareness and concerns about impacts of palm oil production, as well as company responses to negative publicity, have resulted in various sustainability initiatives.

Sustainability initiatives include, for example, regulatory frameworks, voluntary standards, or certification schemes. Regulatory frameworks are being put in place in various countries and include, for example, the European Union Renewable Energy Directive, which affects land use in that it requires high biodiversity and high carbon land to be excluded from conversion to biofuel feedstock production. Voluntary standards or certification schemes are numerous, but the one most relevant to palm oil production is the Roundtable on Sustainable Palm Oil (RSPO). RSPO is an organization of various stakeholders, ranging from production companies and banks that finance palm oil companies to civil society groups, and was formed to promote the production and use of sustainable palm oil. This initiative can, in principle, shape land use for palm oil production as it promotes the use of degraded land and prohibits the use of highly valued conservation forest area (RSPO 2007). There are several important aspects, however, that remain to be solved before RSPO-certified palm oil can actually be considered sustainable. For example, while RSPO principles and criteria require that new plantings should avoid conversion of peatlands, there are no binding requirements, and thus peatland may still be used. Another example concerns the definition of high value conservation forest, which is defined by the individual member states and leaves room in some countries for forests to still be cut down for oil palm plantations and declared RSPO certified. Moreover, RSPO still does not have a greenhouse gas emission standard. An appropriate level would allow the exclusion of palm oil sourced from peatlands or previously forested land to be RSPO certified. An important additional concern is that, in 2011, only approximately 10% of all palm oil production was RSPO certified (WWF 2012). Thus, the still small market for sustainable palm oil can lead to leakage effects. That is, plantations that can easily meet the requirements are certified and provide the market with sustainable palm oil, while palm oil from plantations not able to meet the standards are shifted to markets that do not require sustainability standards.

In addition to international sustainability initiatives, there are also international incentive schemes and bilateral agreements that have the potential to affect land use and LUC positively. Important examples relevant to palm oil production are (a) the United Nations program on Reducing Emissions from Deforestation and Forest Degradation (REDD), which assists developing countries to implement REDD strategies, and (b) Indonesia's forest moratorium, which was initiated by a letter of intent signed by Indonesia together with Norway and covers a two-year ban on granting new concessions for primary natural forest and peatland areas. With any of these initiatives, but especially with Indonesia's forest moratorium, there are still many open issues related to, for example, the definition of forest land, potential loopholes, and government capacity to provide proper enforcement (Murdiyarso et al. 2011).

Reducing Impacts of Palm Oil Production

Various responses to the impacts of LUC for palm oil production have resulted in several approaches to reduce the impacts. Reducing the impacts of palm oil production focuses firstly on the production areas and includes better land-use zoning and the use of degraded land for new plantations, increasing palm oil yields, and applying production schemes that are more beneficial to local communities. However, the distal land connections described above highlight the global nature of land use for palm oil production, and thus local options can also have a global component. In addition, the concept of distal land connections, and particularly the telecoupling framework, provides entry points (originating outside the production areas) for reducing the impacts of palm oil production. These entry points focus on (a) better global land-use governance to account for the various interactions between different markets for land, palm oil, and palm oil uses, (b) producer responsibility, and (c) consumer responsibility. Options for reducing impacts, both from local and global perspectives, will be discussed in more detail next, including key questions that remain to be answered.

Better Land-Use Zoning and the Use of Degraded Land for Palm Oil Production

A key element for reducing the impacts of palm oil production is better planning and governance of land use, such as appropriate demarcation and protection of forest and peatland as well as improved monitoring and enforcement (Wicke et al. 2011). The use of degraded land is often proposed as an option for further land expansion for palm oil production. This is because it is generally assumed to be unused and has much lower effects, in terms of LUC-based greenhouse gas emissions and biodiversity losses, than other types of land. However, degraded land is not in all cases suitable and available for palm oil production. An example of degraded land for palm oil production is the *Imperata* grasslands, which are found on both degraded sites and soils with moderate to high fertility in Indonesia and Malaysia (Garrity et al. 1997).

Imperata grasslands are often considered degraded (even with moderate to high fertile soils) because *Imperata cylindrica* is an invasive species which inhibits the reestablishment of forests and is expensive to remove. On fertile land invaded by *I. cylindrica*, good yields could be achieved without additional agrochemicals if the grass is removed. However, on less fertile and degraded land, yields are likely to be lower, even with the addition of agrochemicals. This not only has implications for the profitability of and interest in palm oil production on degraded land, it also affects the total amount of land needed to meet certain production volumes. More research is needed to investigate what types and severity levels of degraded land are suitable for palm oil production, where this land is located, what yield levels could be obtained, and how to make it economically more attractive for palm oil production.

Besides the suitability, the availability of degraded land must also be investigated before degraded land is used. Degraded land is actually often an important resource for poor rural communities, particularly those with no formal land rights. Although the main competing uses and functions (i.e., food and feed production and high and unique biodiversity areas) can be assessed relatively easily, other land uses and functions are more difficult to capture accurately. Examples are the use of land for livestock grazing, hunting, and gathering as a result of seasonal and interannual variability, as well as the ecosystem and cultural services provided by degraded land. Not accounting for these uses and functions may result in palm oil production on degraded land displacing them and having similar impacts than using more productive land. Thus, when planning or investigating actual projects, an assessment of current land use, ownership, and functions needs to be conducted to minimize unsustainable land use, loss of ecosystem functions, and negative social and environmental consequences of palm oil production.

Increasing Palm Oil Yields

An important option for reducing the pressure of increased palm oil demand on tropical forests is increasing palm oil yields. This may be possible through, for example, earlier replanting with higher-yielding palms, applying fertilizer and other agrochemical inputs more precisely, practicing good harvesting standards, and quickly transporting the fruit to the mill. For both Indonesia and Malaysia, yields of smallholder production are generally much lower than for large plantations. Thus it is important to further develop production schemes and knowledge transfer to help increase smallholder yields.

Increased yields generally decrease the production costs of palm oil. This, in turn, can stimulate increased palm oil demand, because reduced production costs can trigger substitution of other vegetable oils by palm oil and/or make palm oil production more profitable and motivate an expansion of palm oil production area. This phenomenon is called the rebound effect (Lambin and Meyfroidt 2011) and is an indirect effect that calls into question whether increased yields would actually result in less overall land use for palm oil production. More research is needed to quantify the impact. The following questions need to be answered: To what degree can palm oil production be intensified, and what yield levels could be reached? What are the environmental and social impacts of this intensification? How large is the rebound effect of intensifying palm oil production? How does it change given different projections of demand for palm oil as food, fuel, or chemical? How can the rebound effect be controlled or minimized?

Production Schemes Beneficial to the Local Communities and Environment

The way palm oil is produced can greatly influence land use and the impacts of LUC. Several different types of production systems exist that can be characterized by scale (small scale vs. large scale or smallholder/outgrower schemes vs. large industrial plantations) and end users (local vs. national/international use) (Von Maltitz and Stafford 2011). The scale of production has a particularly large impact on the type and amount of land use, displacement of current users and uses, land productivity, and land-use efficiency (as well as impacts not directly related to land use, such as job generation, ownership, access to energy and many more). Large-scale production is generally favored by investors due to the lower investment risks, higher yields, and easier management. However, small-scale production may have more benefits for the local, rural population and the environment, but it will be more difficult to finance such projects (Von Maltitz and Stafford 2011). Options which help reap potential benefits for the local population while making financing possible are outgrower schemes related to plantation companies and small-scale producers becoming commercial farmers (Von Maltitz and Stafford 2011). Crucial for the success of either option are high yields, which reemphasizes the above-mentioned need for better knowledge transfer to smallholders and outgrowers and production scheme setups that facilitate high yields of these producers. Open questions are: How can local people benefit more from palm oil production? What is an optimal scale and type of production at which undesirable environmental and social impacts are minimized and desired impacts maximized?

Accounting for Market Interaction

The increased and intensified telecoupled palm oil system described above emphasizes the interactions between (a) different users of land, (b) different users of produced goods (fuel vs. food), and (c) different vegetable oils/substitutes of palm oil. These interactions make clear that interventions to reduce the impact of palm oil production must take a broader perspective beyond merely looking at palm oil to be able to account for potential indirect effects. For land use, it needs to account for developments in other drivers of LUC, their dynamics, and various interlinkages. For vegetable oils, it needs to account for developments in demand and production, including that of other vegetable oils, and possibilities for substitution. For applications of palm oil, developments in demand for different applications must be taken into account. Not accounting for these other uses of land or palm oil and substitution effects would risk indirect effects that may have detrimental effects in other places. Thus, governance of land use and LUC-related emissions is needed at a global level to account for market interactions. For example, a global cap on LUC-related emissions for all countries and accounting for these emissions in all product chains could help minimize displacement of land use and associated emissions. However, implementation of such a mechanism does not currently seem politically feasible.

While the interaction and integration of economies can cause rebound and other undesired indirect effects, it may also allow a faster diffusion of technology and knowledge for improving palm oil yields. Thus, Lambin and Meyfroidt (2011) suggest harnessing the benefits of globalization through better spatial management, increased land-use efficiency, and the use of degraded and low competition lands.

Producer Responsibility

Another option for reducing the impacts of palm oil production is to make producing, trading, and processing companies responsible for impacts of the production. Government-backed international normative standards exist (e.g., the OECD Guidelines for Multinational Enterprises and UN Guiding Principles for Business and Human Rights) which insist that companies are responsible for adverse social and environmental impacts (OECD 2011; UN Human Rights Council 2011a). These mechanisms are currently voluntary, but governments could make them stronger by creating consequences for not meeting the standards. For companies to take more responsibility, they could align company policies to these widely accepted standards, which implies accepting and acting on responsibilities. A key principle would then be due diligence, and companies should identify, prevent, and mitigate adverse impacts.

Consumer Responsibility

Just like producers, traders, or processors of palm oil, consumers of palm oil also share responsibility for LUC for palm oil production and its impacts. Consumers can influence decisions on where and how palm oil is produced by (a) demanding certified sustainable palm oil or products produced according to normative standards, such as the OECD Guidelines for Multinational Enterprises and UN Guiding Principles for Business and Human Rights, or (b) reducing their consumption of palm oil-based products. The latter option is, however, difficult due to labeling: often, only "vegetable oil" is marked as an ingredient and palm oil derivatives are not always clearly marked as having originated from palm oil. In the case of a global cap on LUC-related emissions suggested above, consumer countries should be given responsibility for the effects of their consumption by assigning the carbon emissions from LUC for palm oil production onto consuming countries. Consumer countries then would need to take responsibility for reducing their emissions by, for example, helping producing countries finance forest conservation and restoration.

Conclusions

Palm oil production in Southeast Asia is often associated with negative environmental and social impacts that are mainly related to LUC, and more specifically to deforestation of tropical rainforest. However, most consumers of palm oil-based products are located far away from the production areas and thus they do not directly, nor immediately, feel these impacts of LUC—although they are increasingly becoming aware and concerned about the impacts of their consumption patterns. This case study of palm oil illustrates the main contemporary trends and the many, palm oil-specific underlying factors that together (re)shape these distal land connections, and thereby land use locally and globally. Distal land connections are shaped by consumption in mainly urban and peri-urban areas because that is where diet changes and the majority of population growth take place. Distal land connections are enhanced by globalization and the integration of economies, as demonstrated by the large representation of international players throughout the palm oil supply chain as well as by increased flows related to (a) palm oil products, (b) labor migration, and (c) diffusion of technology and knowledge. An important aspect of globalization and the integration of economies is the displacement or indirect effects that can cause negative effects outside palm oil production areas. Important additional factors that shape distal land connections are related to new land agents and land-use practices. For palm oil, new land agents are not only the international corporations involved in production and trade of palm oil but also distant consumers who demand sustainable palm oil, international sustainability initiatives and incentive schemes, and new applications and users of palm oil (especially for biodiesel production).

Reducing the impacts of palm oil production focuses firstly on the production areas and includes better land-use zoning and the use of degraded land for new plantations, increasing palm oil yields, and applying production schemes that are more beneficial to local communities. However, distal land connections related to palm oil production highlight the global nature of land use for palm oil production, and thus local options also often have a global component. For example, increasing palm oil yields can stimulate increased palm oil demand because reduced production costs can trigger substitution of other vegetable oils by palm oil and/or make palm oil production more profitable, thus motivating an expansion of palm oil production areas. This rebound effect is an indirect effect that may result in lower than expected benefits from productivity increases.

In addition to local options for reducing LUC and its impacts having a global component, this chapter illustrates that decisions about land use are not made at the local level of production alone, but rather increasingly by transnational companies that invest in palm oil production, trade, and processing. Decisions by these international market players are also influenced by demands of consumers in distal places, by international sustainability initiatives or incentive schemes, and by the various interactions between different markets for land, palm oil, and palm oil uses. The influence of various actors across long distances, country borders, and jurisdictions as well as market-mediated, indirect effects point to the need for taking a global perspective on land use and governance of land use as well as for better governance of the various actors in international supply chains. It also demands that actors in different places—specifically producers, traders, and consumers of palm oil—take responsibility for minimizing negative local impacts. Consumers, in particular, need to be made more aware of and responsible for the impacts of their consumption patterns. This could be possible by allocating LUC emissions related to palm oil production (and others) to consuming countries or placing a carbon tax on products with high LUC-induced greenhouse gas emissions. In addition, producers, traders, and processors could be required to comply with international normative standards; this would encourage multinational companies to take responsibility for adverse social and environmental impacts while also creating consequences for not meeting standards. Consumers can influence company decisions on where and how palm oil is produced by demanding certified sustainable palm oil or products produced according to normative standards for multinational companies.

A global cap on LUC-related emissions for all countries, as well as an accounting for these emissions in all product chains, could help better govern LUC and minimize displacement of land use and its associated emissions. However, implementation does not currently seem politically feasible.

Acknowledgments

I would like to thank the reviewers and editors for their useful comments on an earlier version of the manuscript and the participants of the Ernst Strüngmann Forum, particularly the group on distal land connections, for the stimulating discussions.

Decision Making, Governance, and Institutions

10

Emergent Global Land Governance

Matias E. Margulis

Abstract

Land governance is currently the focus of many new global rule-making projects, marking a sharp break with past practices that sought to exclude land as an international governance issue. Wide-ranging concerns about land grabbing and its exclusionary and ecological consequences have driven this, prompting states and global civil society to devise new global land-governance instruments. This chapter offers a preliminary theoretical and empirical analysis of what is conceptualized as "emergent global land governance," focusing primarily on its international governance dimensions. A review of relevant land-governance policy instruments in the fields of investment, land tenure, and forestry suggests that emergent global land governance is likely to consist of multiple, overlapping instruments with diverging normative frameworks and objectives that are not closely coordinated instead of a singular, discrete international regime.

Introduction

Compared to past decades, land occupies a significantly higher profile in current governance deliberations at the global level. Driving this has been concern over large-scale acquisitions of agricultural land, which itself is part of the broader global phenomenon known as "land grabbing."

On the international level, land governance is currently at the forefront of new global rule-making projects at the Food and Agriculture Organization (FAO) of the United Nations (UN), the World Bank, and the Group of Eight (G-8)/Group of Twenty (G-20), the latter of which represent states with the largest economies. In terms of transnational advocacy, work is ongoing by prominent international nongovernmental organizations (INGOs), such as Oxfam, and global social movements, such as the food sovereignty movement, to establish global rules that will regulate foreign investment in agricultural land. In addition, private actors (e.g., transnational agrifood corporations and institutional investors) have expressed interest in global rules, as they perceive this will (a) ensure that their current and future investments in land are coherent with best practices on socially responsible and sustainable investment and thus (b) minimize risk to profits and reputation. Together, these developments attest to the rising interest in and demand for global forms of land governance.

The current flurry of activity surrounding global rule making for land makes a compelling case for scholarly analysis of new developments in land governance. In addition, because land has hitherto been treated as a local or national policy issue, rather than one of global concern, land constitutes an important case study for contemporary global governance practices. In this chapter, I discuss the theoretical basis and provide a preliminary empirical analysis of *emergent global land governance*; that is, the set of norms, rules, institutions, and practices that shape contemporary governance of land at the global level. Since much of the new global rule-making projects are primarily taking place within traditional international organizations (e.g., at the FAO), I focus on international modes of global governance. Still, the role of nonstate actors and private modes of global governance, and their overlap with international organizations, must be emphasized and is discussed by Gentry et al., Hunsberger et al., and Auld (all this volume). In addition, it is important to note that the distinction between international and transnational has become blurred; even though formal decision making remains the exclusive prerogative of nation-states,[1] global civil society and private actors are increasingly involved in policy making in so-called traditional intergovernmental bodies, such as the UN Committee on World Food Security.

I begin by framing the rising interest in land governance and demand for global-level forms of land regulation as a response to the contemporary phenomenon of "land grabbing." Next I present the concept of emergent global land governance in reference to contemporary debates. I then identify key policy instruments of emerging global land governance through a comparison across the fields of investment, food and agriculture, and forestry, and discuss major trends in this governance field. I conclude with a discussion of longer-term policy and political challenges relevant to emergent global land governance.

Rising Interest in Land and Demand for Global Governance

Land governance is by no means a new area of policy debate. The challenges posed by land governance factor into everyday politics in many jurisdictions at various levels (e.g., land redistribution to landless peasants in Brazil or farmland reclassification for conservation purposes in the United States). Since 2008, heightened interest in land and its governance has resulted from a

---

[1] This is evident, for example, in the formal participation of global civil society organizations and, increasingly, in the involvement of private sector actors in policy deliberations at the UN Committee for World Food Security.

massive wave of land grabbing worldwide. Contemporary land grabbing has significantly altered the political dynamics, discourse, and goals of land governance debates due to the challenges (both ethical and policy related) that competition for land has imposed on society in an increasingly teleconnected world.

Land grabbing is one of many terms that refers to large-scale acquisitions of land for the purpose of outsourcing agricultural production. There is no consensus definition and therefore a significant scholarly (and real world political) debate about what constitutes, or not, land grabbing (see Margulis et al. 2013). Rather than repeat that debate here, I take as a starting point the approach offered by Borras et al. (2012a:851), which establishes land grabbing as one manifestation of "control grabbing," defined as:

> [T]he capturing of control of relatively vast tracts of land and other natural resources through a variety of mechanisms and forms involving large-scale capital that often shifts resource use to that of extraction, whether for international or domestic purposes, as capital's response to the convergence of food, energy and financial crises, climate change mitigation imperatives and demands for resources from newer hubs of global capital.

This is an analytically useful definition of land grabbing because it is does not overemphasize the quantitative dimension of land grabbing, although scale and measurement remain important. This definition provides a framework to capture the potential diversity of localized forms of land grabbing, such as "green grabs" (Fairhead et al. 2012), "water grabs" (Mehta et al. 2012; Allan et al. 2012), and other forms of the "foreignization" of space (Zoomers 2010). In addition, control grabbing captures the idea that contemporary land grabbing is driven by changes in the global political economy, including the decision by distal agents to (re)valorize foreign land as an important economic commodity, which in turn, leads to what the globalization theorist Saskia Sassen (2013) argues is a nascent global land market.

Contemporary land grabbing echoes the great land rush during the era of imperialism and colonialism. In turn, this has prompted a scholarly debate on the similarities and differences between land grabbing today versus in the past (Alden Wily 2012; Ayers 2013). Depending on the theoretical framework and time scale used, different conclusions are drawn on what is novel about contemporary land grabbing and to what extent it is a continuation of historical patterns of capitalist world development. Postcolonial and world systems frameworks are useful because they shed light on how past (and contemporary) asymmetries in power among agents and local populations shape societal conditions, which in turn drive land-use decisions. Still, many characteristics do not fit well with neocolonial and world systems frameworks because land grabs, as currently practiced, differ in two key respects: First, contemporary land grabbing involves a more diverse set of actors, institutions, and governance practices that simply did not exist during the era of colonialism and imperialism (Margulis and Porter 2013). Second, many land grabs do not match earlier patterns of core periphery and North–South relations; polycentrism is a major feature of contemporary land grabbing as is the prominence of "land-grabbed land grabbers" such as Brazil (Borras et al. 2012b; Rulli et al. 2013). For a more extensive discussion on the process of land grabbing, see Hunsberger et al. (this volume).

Land Grabbing as Context

Recent acts of land grabbing have galvanized international attention and energy to the issue of land governance in an unprecedented manner. Global attention was first drawn to the phenomenon in the summer of 2008, following a report by the Spain-based NGO GRAIN, which documented a sharp increase in the sale or lease of very large units of agricultural land in developing countries by foreign investors and governments (GRAIN 2008). Subsequent research by academics and international organizations confirmed land grabbing as an extensive international phenomenon (Cotula et al. 2009; World Bank 2009; GTZ 2009; Friis and Reenberg 2010; Deininger et al. 2011; Anseeuw et al. 2012a, b; Oxfam 2012; Cotula 2011, 2012). To contextualize the unique features of contemporary land grabbing and understand why land grabbing has become such a prominent issue of global concern, let us review six key characteristics of the phenomenon.

First, global land grab is a recent phenomenon, having proliferated around the time of the "triple" food–fuel–financial crisis of 2007–2008. Although land grabbing took place prior to this crisis, its extent was relatively marginal. Table 10.1 illustrates this trend by drawing on recent data from the Land Matrix Project, which is considered to be the best available source at present.[2] Note the sharp increase in 2007–2009. This timing supports numerous theories which regard contemporary land grabbing as being driven by the attractiveness of rising returns in the agricultural sector and in farmland values, growing international demand for biofuels, and strategies for securing agricultural supplies by nations and corporations (Cotula et al. 2009; World Bank 2009; Borras et al. 2010). Since 2009 there has been a fall in land investments and it continues. The reasons for this are not clear and could be due to a lack of information about more recent deals and/or increasing wariness by investors about the risks of new deals given the heightened political attention and global resistance to land grabbing (Anseeuw et al. 2012b).

Second, contemporary land grabs have happened on a significant scale. Estimates of farmland that have been grabbed range from between 50 million hectares (World Bank 2009) to 227 million hectares (Oxfam 2012). It is

---

[2] Obtaining reliable data on land grabs remain a significant challenge because of the lack of transparency. To provide an empirical basis for this argument, Table 10.1 uses data obtained from the recent Land Matrix Project as an indicator of land grabbing activity. Table 10.1 displays only verified land deals in a given year; it is important to note, however, that at least 44 million hectares of land deals are in the process of being verified.

Table 10.1 Land grabs by hectares and number deals, 2006–2011, as reported by the Land Matrix Project (http://landportal.info/landmatrix/get-the-detail/by-year).

Year	Hectares	Number of Deals
2006	450,640	43
2007	1,123,275	29
2008	1,723,457	36
2009	1,879,108	39
2010	1,303,609	23
2011	422,200	3

generally acknowledged that data remains patchy due to the lack of transparency of most land deals, the lack of agreed upon methods (Cotula 2012), and the politics of numbers at play in global policy debates, given the centrality of numbers to informing actors' identities and preferences (Margulis et al. 2013). Despite this, all estimates clearly point to a substantial figure.

Third, the fact that land grabbing occurs in all regions of the world makes it a *global-scale phenomena*. Sub-Saharan Africa is certainly a hot spot for land grabs, but recent research has also shown intensifying activity in Asia (Hall 2011), Eastern Europe (Visser and Spoor 2011), in South and Central America (Borras et al. 2012b), as well as in Oceania (Filer 2012). There is growing evidence that land grabbing is not limited to developing countries but is expanding also in developed countries, as evidenced by the fact that Australia is currently a major site of foreign investment in farmland.

Fourth, the source of investors in farmland is highly diverse and extends beyond actors traditionally engaged in agriculture, including states (including sovereign wealth funds) and a range of private sector actors (e.g., agrifood corporations, commodities traders, and institutional investors) located in the North *as well as* in the South (Margulis and Porter 2013; Daniel 2012).

Fifth, available data suggest that a significant proportion of land grabbing has been used to produce "flex crops and commodities" (Borras et al. 2013; Anseeuw et al. 2012b:xi), such as palm, soy, and sugar, which are important to the expanding global biofuel complex. This suggests that demand for biofuels and the national policies which stimulate their greater use in transport fuels have turned out to be much more significant drivers of land grabbing than initially projected, when land grabbing was largely predicted to outsource food production (Robertson and Pinstrup-Andersen 2010).

Sixth, land grabbing is associated with the interaction between land and diverse sets of technologies. Examples include new seeds and genetic materials as well as advanced production technologies (i.e., automated mega-farms controlled by distal headquarters); new financial products, such as land investment funds, that are enabled through digitized trading and speculation in financial markets; and new ways of discovering, measuring, and mapping land with

Geographical Information Systems (GIS). GIS makes it possible, for instance, to identify "available" lands and facilitate land deals from a virtual distance; thus land investment transactions can take place without a buyer or seller ever having to meet or come together physically on site.

Demand for Global Land Governance

In response to the speed, scale, and extensiveness of land grabbing, demand for global land governance can be argued to be driven by two categories of concerns. For the purpose of the analysis here, these two categories are treated as ideal types. These ideal types are visible in the frameworks and ideas informing policy making and transnational advocacy on land governance, including the framing of collective demands for new and robust forms of global land governance.[3]

The first category falls under the label of *exclusion* and refers to situations where communities and individuals are displaced by land grabs. Exclusion can occur in legal, peaceful, and voluntary ways as well as under illegal, violent, and coercive circumstances. Exclusion is principally an issue of ethical or social justice framing. The principal problem here is that certain communities (e.g., small-scale farmers, agricultural workers, indigenous peoples, and other vulnerable or marginalized peoples) are more likely to experience a loss of livelihoods, violence, and be excluded from socially and economically valuable lands and resources. The actual forms of exclusion vary significantly, depending on the type of government in power and factors such as population density, the type of land-tenure arrangements in place, the type of agricultural practices, and the crops in question. The basic point is that most people who continue to live on the land are among the poorest in society who have few, if any, alternative economic opportunities. For many individuals, the loss of or diminished access to land can become a matter of personal survival.

The most visible expression of concerns over exclusion has come from global civil society. The 2011 campaign "Stop the Land Grab" led by the transnational peasant movement *La Via Campesina* brought attention to the importance of international regulation and proposed an international moratorium on land grabbing and to make states and investors accountable. In 2012, well-known INGOs, such as Oxfam and ActionAid, called for a moratorium on land grabs in Africa. Oxfam, for example, has made land grabs a key theme for their new high-profile international agriculture and food-focused campaign, GROW, calling for international good governance standards for land deals to prevent the displacement and further impoverishment of vulnerable people. Global campaigns organized around the idea of exclusion prioritize protecting the land tenure of vulnerable groups. This has resulted in the documentation

---

[3] These categories of concerns overlap with concerns about land grabs at the local and national level but play out differently in these contexts.

of cases of exclusion and has brought such cases to the attention of the global public and international bodies (Graham et al. 2010).

A related geopolitical and economic concern that echoes concerns about exclusion is the potential for escalating social conflict and economic risk arising from tensions between locals and land grabbers (e.g., access to land and resources, effects on land prices), citizens and their governments (e.g., political contestation over land deals, corruption and legitimacy of "complicit" governments, food insecurity), and among states contesting the legitimacy of land deals and their effects on commodity markets. Geopolitical and economic concerns have taken on great importance at interstate forums. At the G-8 summit in L'Aquila and at the UN, for example, debates about land grabs have been framed in terms of political instability in host states, especially after the fall of the Madagascar government in 2008 following a significant land deal there.

The second category of concerns can be termed *ecological risk*. This refers to concerns about the potential negative environmental consequences of land grabs. Such effects can occur in myriad ways, such as environmental damage caused by the intensification of extractive forms of production or the loss of biodiversity and traditional knowledge when industrial mono-cropping for biofuels supplants more traditional forms of agriculture. Only recently has agriculture been included more seriously in climate-change science, even though it has long been well known that agricultural production is the single largest use of land (Turner et al. 2007). A recent study by the Washington-based International Food Policy Research Institute (Nelson et al. 2010) calculates that agricultural emissions account for almost one-third of the global greenhouse gas emissions, a number significantly greater than estimated just a few years ago. Available data suggest that much land grabbing has been for the industrial production of agriculture commodities for global biofuel and meat production (Anseeuw et al. 2012a); both are carbon-intensive activities associated with net increases of greenhouse gas emissions. Therefore, land grabbing is a concern if it encourages the expansion of carbon-intensive modes of agricultural production.

Due to teleconnections, policies that are intended to promote positive environmental outcomes in one area can have negative environmental effects in another. Take, for example, fuel transport policies that promote the use of corn and palm-based biofuels, which are significant drivers of land grabbing. These national policies have spatially dispersed social and environmental effects when "green energy" that is produced in one country is consumed in another (Wicke, this volume; van der Horst and Vermeylen 2011). Land grabbing may also exacerbate the cycle of carbon-intensive agricultural practices and degradation of ecological resilience in light of the general variability of agriculture yields to climatic changes. From this standpoint, land grabbing of farmland is at odds with scientific consensus on the need to shift toward low-carbon forms of agriculture and sustainable intensification (see Fan and Ramirez 2012).

Ecological risk concerns are also evident in the work of major INGOs:

- ActionAid's campaign on land grabbing has called for reform of the European Union's biofuel policy.
- Friends of the Earth International (FOEI) has critiqued the World Bank's involvement in supporting land grabs to set up oil palm plantations in Uganda.
- Green Peace blockaded timber exports in 2011 related to land grabbing in Papua New Guinea.

These campaigns have generally portrayed land grabs as a threat to sustainable development. In addition, FOEI, for example, has highlighted the role of the World Bank in facilitating land grabs and has called for a reform of its lending policy. These concerns have made their mark, as agriculture has taken on a greater importance in the international climate-change talks since 2009.

Concerns about exclusion and ecological risk overlap considerably. However, in everyday real-world debates about global land governance, they remain disjointed. Madiado Niasse from the International Land Coalition argues that the debate about land governance "is typically ignorant of and disconnected from the discourse on the physical condition of the land, despite the fact that the latter affects the productive capacity of the land" (Niasse 2011:3). Similarly, in land-change science, land is disconnected from underlying socio-economic processes in distant places. Bridging this divide has been an overall objective of this volume. To this end, the concept of telecoupled systems (see Liu et al., as well as Eakin et al., both this volume) is particularly helpful as it provides both a conceptual framework and language with which to describe highly complex human-land interactions. The disconnect between exclusion and ecological risks in everyday politics results not only from a lack of knowledge or communication; many political actors are aware of these linkages, but must instead strategically focus on human indignities. Ecological risks and land-change science operate on much longer time horizons than can easily be framed into immediate political action. Nonetheless, the prospects for bridging the disconnect between exclusion and ecological risk in contemporary global land-governance debates are likely to improve once knowledge of telecoupled systems is advanced and translated to policy actors, and when interaction among actors engaged in articulating these concerns in global policy spaces becomes more frequent.

Emergent Global Land Governance

Rising concerns about land grabbing and resultant actions are producing emergent forms of global land governance. To date, issues of land tenure and investment have appeared to gain the most traction as evidenced by two recent global governance instruments: (a) the UN Voluntary Guidelines for Responsible Governance of Land, Fisheries and Forests in the Context of National Food

Security and (b) ongoing negotiations for principles on responsible agricultural investment. These instruments address concerns associated closely with exclusion. By comparison, the ecological risk appears to be less prominent in emergent global land governance.

Before plunging headlong into the analysis of the constituent parts of emergent global land governance, I now turn to a discussion of global governance. My purpose is twofold: to clarify the concept of global governance and situate land as a global governance issue.

Global Governance

Global governance is a term that is widely used to refer to the modern practice of governing transborder problems as well as to the institutions, rules, and actors that govern the global political economy. As an academic concept and field, global governance emerged in the 1990s in response to emerging global-level problems (e.g., HIV/AIDS, climate change, and migration) that were beyond the capacity of any single nation-state to manage on their own (Rosenau 1995). The field of global governance has also been deeply influenced by shifts in power at the global level, such as the fall of the Soviet Union and the resultant implications to U.S. unipolarity and multilateralism. More recent work has focused on emerging countries as new powers in multilateralism. In addition to this state-centric work, other research into power in global governance has highlighted the shift to nonstate forms of authority, with transnational business and global civil society increasingly taking on greater roles and influence in global governance institutions.

Several concepts from global governance research are relevant for the study of land as a global governance issue. The first is the concept of *authority*, which is closely related to the idea of governance: Which actors have authority to regulate a particular sphere of activity? Research into global governance has shown that authority has flowed in two principle ways over the past decades, and that these differ from the immediate postwar era:

1. Authority has shifted from state to international institutions and this has the capacity to constrain state sovereignty (e.g., EU, World Trade Organization (WTO), and International Criminal Court).
2. Authority has shifted to nonstate actors who have taken on governance functions in existing policy fields as well as in new areas of activity. The latter range from private actors (e.g., credit rating agencies) who have a significant influence on the state's financial affairs, to industry associations that create standards for self-regulation, to private international arbitration of financial and investment deals (Cutler et al. 1999b). Nonstate actors also include global civil society organizations (Smith 2007) that are involved, for example, in fair trade labeling and certification or in developing standards for humanitarian assistance.

Scholars have described the present state of global governance as thick and dense because of the proliferation of public and private global governance instruments (Raustalia and Victor 2004). The increasing density of global governance is important for the following reasons. It points to a significant change in the "architecture" of global governance. Much of the postwar system of international governance was founded on narrowly focused instruments to address what were perceived as discrete issue areas, such as security, finance, and trade. Global governance today, however, looks somewhat different: issue areas are no longer governed by singular instruments but often by multiple, overlapping instruments. Take, for example, intellectual property rights, which are governed by the 1994 Trade-Related Aspects of Intellectual Property Rights Agreement (TRIPS) at the WTO, and various standards negotiated by states at the World Intellectual Property Organization (WIPO). The trend toward greater plurality of instruments in global governance is partially explained by a greater awareness of, and acceptance by, policy makers and other policy actors that most governance challenges are multidimensional and complex, and that these require accordingly more extensive and flexible forms of governance. This is evident in various fields and is marked by greater attention to address the governance gaps that arise, for example, when it comes to trade and the environment, or security and migration. Multiple instruments that cut across issues also arise when states and nonstate actors engage in strategic and tactical approaches to global governance that assemble existing and new instruments into novel configurations (Young 2002; Raustalia and Victor 2004). Motives for this can vary highly, ranging from a desire to simplify and streamline governance, increase policy coherence, and reshape governance arrangements to make them more favorable to the interests of powerful or weaker actors.

Land As a Global Governance Issue

In contrast to other areas of global governance, such as climate change, HIV/AIDS, and terrorism, land has not traditionally been viewed or understood as a global-scale "problem." Several intellectual and political reasons contribute to this state of affairs. The foundations of the present international system are premised on the territorial sovereignty of the nation-state, of which land control is a defining feature. This is affirmed by practices such as international legal recognition of state borders and authority of nation-states to govern as they wish within their demarcated territory. In addition, politics and ideological struggles during the Cold War, in particular, kept land reform out of international deliberations and framed them instead as bilateral development issues (Margulis et al. 2013).

The current situation is, however, different: states and nonstate actors view land as a global issue that requires global-level forms of governance. This is primarily the case because today policy makers understand that land is affected by teleconnections and this is important given that land is the physical basis for telecoupled systems. Ever greater transborder flows of ideas, capital, and technology as well as transboundary challenges, such as climate change, are driving substantive shifts in land-use change as well as the distribution and access to resources (e.g., soil, water, genetic material, and subsoil minerals) which constitute "land" for those who live directly and indirectly off the land. However, two powerful paradigms of land are highly visible in these new global politics: land as a sovereign territory and land as a commodity. These are not, however, the only land-related paradigms of importance. The discourse on the use of land for environmental services under the climate-change regime has gained significant traction though remains problematic (Norgaard 2010). Also visible is an unfolding struggle to create new international human rights, such as the right of indigenous people and peasants to land (Jones 2011; Xanthaki 2010). The multidimensionality of land and the increasing plurality of relevant governance institutions and practices (mostly at the global level) suggest that land will less likely be perceived as a coherent, singular issue area. Thus, aspects of land governance will tend to cut across various levels and issue areas.

Policy Instruments of Emergent Global Land Governance

In this section, I provide an initial mapping of the policy instruments of global land governance at its current embryonic stage. Three policy fields—investment, land tenure, and forestry—are selected for discussion because they are specific and highly relevant to global land governance. Many other relevant instruments exist, of course; for example, those related to conservation (e.g., payment for ecosystem services, transfrontier conservation areas) or to the rights of indigenous and tribal peoples (Convention No. 169 of the International Labor Organization). These will not, however, be addressed here.

Investment

Investment in land by states, transnational corporations, institutional investors, and domestic elites is the principal mode of land acquisition. Investment in land raises several concerns, ranging from the lack of prior and informed consent of local communities, whose land is sold or leased, to the risks for investors active in states with weak regulatory regimes, where land deals may be revoked if the political climate changes.

The rapid rate of investment in land for agriculture elicited the creation of a new global governance framework for land. In 2009, the G-8 agreed to "work with partner countries and international organizations to develop a joint proposal on principles and best practices for international agricultural investment" (G-8 2009:113b). The first stage involved the consultations for the Principles on Responsible Agriculture Investment (PRAI), which was led by the World Bank in partnership with FAO, the UN Conference on Trade and Development (UNCTAD), and the Organization for Economic Cooperation and Development (OECD). In 2010, a first draft of PRAI was drawn up after initial expert consultation among international officials and experts.[4] The G-8 and G-20 have repeatedly affirmed their support for the PRAI, most recently at the 2012 Camp David and Los Cabos summits. However, many developed countries and global civil society refuse to endorse the PRAI because they feel that it was developed without sufficient broad-based consultation and participation. As a result, by the end of 2012, states agreed to reopen the negotiation on responsible agricultural investment at the UN Committee on World Food Security, a more inclusive and participatory forum. This process is to be completed by 2014.

From the onset, the idea of responsible agriculture investment has been to establish a set of standards for private and public investment so as to encourage investment in the agricultural sector and minimize negative social consequences. General objectives of the new regulation are to establish standards for fair consultations to ensure informed consent prior to the completion of land transactions, fair compensation for existing land users, and maximization of economic opportunities and protection of food security for local communities. The authors of the PRAI built upon various international standards and best practices relevant to investment in agriculture (and investment more generally); however, this process was tailored specifically to address the concerns associated with contemporary land grabs, with an emphasis on transparency and sustainability, but did not address many other important issues such as land tenure and the livelihoods of those living off the land without formal title. The push for an international code of conduct for investors is seen by policy makers as critical to maintaining consensus for a new international development model focused on increasing investment in agriculture of developing countries (Margulis and Porter 2013). For example, together the G-8 and G-20 have pledged 20 billion USD toward a multi-donor agriculture and food trust fund, the G-8 recently established the 3 billion USD "New Alliance for Food Security and Nutrition," and the Bill and Melinda Gates Foundation and other philanthropic actors have donated hundreds of millions to support this new development model for agriculture in the global South.

The effectiveness of responsible agricultural investment hangs on the assumption that increasing transparency and establishing criteria for best practices will lessen investment-related risks and increase the benefits of private investment in agriculture. This assumption draws on the success of other sectoral initiatives; in particular, the Extractive Industries Transparency Initiative (EITI) to which the World Bank is a major supporter. However, there are serious questions about the applicability of sectoral initiatives such as the EITI to

---

[4] The full text of the draft PRAI is available at: http://siteresources.worldbank.org/INTARD/214574-1111138388661/22453321/Principles_Extended.pdf.

agriculture, because the range of investors and producers is far more diverse than in extractive industries. In addition, land investment is highly variegated with respect to the range of the commodity being produced (i.e., food, feed, biofuels, and other industrial inputs), the methods of production, end use, and final market destinations. Another important consideration is that instruments such as the EITI are mainly concerned with corruption and monitoring payments by firms to governments. Corruption and "weak" governance in host states must be considered in land investments, but they have not been identified by actors as the most important policy issues. Thus, it is not clear whether the emphasis on corruption built into the EITI model can translate well, in terms of land investments, to address the problems of exclusion and ecological risk that frame the current debate and drive demands for global land governance.

The politics of responsible investment in agriculture are highly contentious. The very legitimacy of the PRAI is debatable, given the fact that it was developed by international organizations with the support of the G-8 instead of by a universal body, such as the UN; it was also created through a top-down, expert-led exercise that included little direct participation by the private sector, global civil society, and/or the populations most likely to be affected. Whereas much of global civil society was initially opposed to the PRAI, the decision to restart negotiation on principles at the UN Committee on World Food Security ensures that global civil society and the private sector will work together with states and international organizations to devise new standards for investment in agriculture.

Land Tenure

The most recent and politically significant policy instrument in global land governance is the Voluntary Guidelines for Responsible Governance of Land, Fisheries and Forests in the Context of National Food Security (henceforth referred to as "VGs"), which were adopted in May 2012 at the FAO. VGs constitute a nonbinding international agreement that provides a policy and normative framework to enhance the tenure and access to land and productive resources by poor rural households. Livelihoods as well as the prevention of political, social, and economic exclusion play a major role in this agreement.

Unlike PRAI, international cooperation for creating the VGs *preceded* contemporary concerns about land grabbing. The impetus for VGs can be traced back to the 2006 UN International Conference on Agrarian Reform and Rural Development (ICARRD), during which states and global civil society organizations called for a global framework to implement enhanced land tenure at the national level in order to address the perceived shortcomings of partial and unfinished land reform in developing countries. Land reform is an endemic, and often unresolved, political issue in many developing countries—one that raises the political stakes and salience of global land governance, for example, when actors use global-level instruments to influence national-level policies. Indeed, many of the actors (e.g., developing countries and transnational peasant movements) involved in the VG negotiations regard the guidelines as a continuation of the earlier ICARRD process, which was directed at strengthening national land tenure and land reform policies.

It is clear that contemporary concerns about land grabbing have significantly influenced the VGs' negotiations and subsequent outcomes. Some might claim that they put the negotiations on a new track and changed actors' perceptions of the importance of establishing the VGs. The official negotiation process for the VGs took place during 2010 and 2011. By this time, land grabs were a prominent political issue, and concerns about land grabbing were voiced at the VG negotiations. The zero draft (i.e., first draft of the negotiated text) of the VGs contained several proposals specific to addressing concerns about investment and large-scale land acquisitions. The states and global civil society actors that negotiated the VGs confirm that land grabbing was the most contentious issue in the negotiation (Seufert 2013). The final version of the VGs does not make direct reference to land grabbing; instead it refers to *large-scale transactions* (FAO 2010, 2012b), which is understood to refer to land grabs. The official text, however, used this term because many parties were not comfortable with the term land grabbing: some viewed it as being "too political" whereas others found it too vague for the specificity required in an international legal document.

The aim of the VGs, through its best practices-based approach, is to improve domestic governance of land tenure. Despite its emphasis on strengthening policies at the national level, there is much more to the VGs. First, the VGs include provisions specific to private actors and transnational corporations, including obligations for an investor's home state to prevent abuses of land tenure and human rights. In addition, they delineate the responsibilities of states involved directly (e.g., China, Gulf States) in large-scale land acquisitions to ensure that such investments are consistent with the protection of legitimate tenure rights and the promotion of food security in countries where investment takes place; in addition, they ensure that an investing state fully meets related obligations under national and international law (FAO 2012b:20). The VGs have also created a political space for transnational forms of implementation and monitoring of land-tenure practices. One option currently being discussed is the creation of a monitoring mechanism under the auspices of the FAO's Committee for World Food Security that would include the participation of state and non-state actors. Such a mechanism, if achieved, would potentially strengthen land tenure by providing a forum for affected parties, especially marginal groups, to bring their concerns directly to the international community (similar to the special procedures of the UN Human Rights Council, which allows parties who have had their human rights violated the ability to seek international justice when national institutions are unable or unwilling to address such concerns). By including provisions which focus on the obligation of investors and a transnational monitoring mechanism that goes beyond the national level, the VGs illustrate a form of incipient governance that seeks to address teleconnected processes and their effects on land use and control of land.

Forests

The global governance of forests is highly relevant for emergent global land governance because of the direct link between land grabbing for palm-based biofuels, land grabbing for pulp, timber and/or bioenergy production inputs, and other "green grabs" for conservation and carbon offset purposes—all of which occur on forest lands. Until now, international forest governance has not figured prominently in the global land grab debate. However, due to new environmental governance instruments such as REDD+,[5] which is purported to be a potential important driver for land grabbing in the future, future debate can be expected. In addition, forest governance has become very relevant to land-tenure issues now that the VGs explicitly apply to forests thereby linking land and forest governance policies in a new way.

In contrast to recent efforts to establish global governance instruments for investment in agriculture and land tenure, the global governance of forests is far more established. Although scholars continue to debate whether a singular international forest regime exists, all agree that it consists of multiple public and private institutions and rules, many of which loosely interact in a coordinated manner. An important feature of the global governance of forests is that it contains both international and transnational policy instruments; the latter is referred to under various labels, such as private authority (Cutler et al. 1999b), nonstate market-driven governance (Cashore 2002), regulatory capitalism (Levi-Faur 2005), and liberal environmentalism (Bernstein 2000). Many forest governance schemes are, in fact, hybrid forms of governance that involve both public and private actors in policy design and implementation. Private and hybrids forms of forest governance, such as the Forest Stewardship Council, have eclipsed international forest governance and are argued to rank higher in terms of quality and legitimacy than international forest governance (Cadman 2011).

Two international efforts particularly relevant to emergent global land governance are the UN Statement of Forest Principles (UNSFP) and UN Forum on Forests (UNFF). UNSFP is a nonbinding international policy framework (like the VGs) that was negotiated by states at the 1992 Rio Summit. It is important because its principles established norms of forest governance; namely, it framed the challenge as deforestation and nested forest governance under

---

[5] Reducing Emissions from Deforestation and Forest Degradation (REDD) is an effort to create a financial value for the carbon stored in forests. "REDD+" goes beyond deforestation and forest degradation to include the role of conservation, sustainable management of forests, and enhancement of forest carbon stocks; http://www.un-redd.org/aboutredd/tabid/582/default.aspx

the framework of sustainable development.[6] UNFF was established in 2000 and replaced several earlier interstate forestry governance forums. UNFF is an interstate body that serves to promote international cooperation on sustainable forest management. Nonstate stakeholders are consulted, but only states formally participate in decision making with respect to the negotiation of declarations and agreements. Although not a standard-setting body (Cadman 2011), the UN adopted the Non-Legally Binding Instrument on All Types of Forest in 2007. This is a set of voluntary guidelines intended to enhance international cooperation on sustainable forest management, monitoring of illegal logging, and promote political dialog and action to address issues affecting forest peoples. In addition, it established a framework for an international forest financing mechanism to support sustainable forest management and its implementation.

REDD+ represents a recent approach to forest conservation that has gained widespread support from states, international organizations, and the private sector. Currently under negotiation, REDD+ seeks to create an international market for carbon credits. Its design and governance modus has not been resolved, especially at the national level (Corbera and Schroeder 2011). Nonetheless, if, as expected, REDD+ emerges as a major pillar of the global environmental regime, it could become an important element of emergent global land governance. Because of its unique approach (i.e., economic incentives are used to control deforestation and sustainably manage forests), REDD+ could potentially link exclusion considerations (i.e., livelihoods) and ecological risks (i.e., reducing greenhouse gas emissions, preserving forest carbon stocks). However, given the low success rate of global environmental instruments reaching their goals (Jabbour et al. 2012), it remains to be seen whether REDD+ will be successful. Even if it is able to promote global sustainable forest management on the long term, there are significant academic and policy relevant questions related to how such an approach translates to land investments that produce significant carbon emissions and ecological degradation.

Conclusion

Currently, land governance is a significant issue in world politics and the subject of new global rule-making projects. Existing guidelines and ongoing negotiations on responsible agricultural investment represent efforts to establish global land governance in response to recent land grabbing, in addition to long-standing issues such as land tenure. Preexisting global governance instruments, such as those related to forestry, are highly relevant to this emergent global land governance but are not at the foreground of debate. A central feature of emergent global land governance is that its applicability is not limited

---

[6] This normative orientation diverges significantly from other international instruments in forestry, such as the International Tropical Timber Agreement, which initially sought to stabilize timber prices and supply through direct cooperation among major producers and consumers.

to any specific locality, region, or country; it is global in scope and its intent is to establish rights and obligations for actors inside and outside the state.

Global land governance is, however, a dynamic process. What are viewed now as the key policy instruments may change or expand over time. There is no reason to assume that global land governance will develop linearly or that path dependency is evident. Land grabbing may be the contentious political issue today that is driving the debate, but this can change rapidly, for example, if there are major systemic economic or climatic shocks.

Emergent global land governance is unlikely to exhibit the characteristics of a singular, self-contained international regime. Instead, it is likely to take place through several different types of global governance instruments that are loosely interlinked. Examples of linkages brought about by governance instruments (e.g., PRAI, VGs, REDD+) were presented in this chapter to show how they can span various issue areas (i.e., investment, land tenure, and forestry) and can diverge in their normative orientation and policy goals (i.e., risk management, land tenure and access to resources, and sustainable forest management) as well as the key policy actors involved (i.e., international organizations, agricultural ministries and transnational social movements, and forestry/natural resources ministries, respectively). The potential pool of global governance instruments (public, private, or hybrid) that may be relevant to global land governance is likely to be large since land issues are highly multidimensional, cutting across many different policy fields (e.g., gender, economic development, human rights). Thus, as it continues to develop, emergent global land governance will likely be a highly complex and congested domain. The fragmented nature is significant because it may exacerbate preexisting conflicts or resolve them; it could also create new policy challenges and related global political contests, including conflicts between and among states, international organizations, private actors, and global civil society.

Given the fragmented and conflicting characteristics of emergent global land governance, there is no agreement about which set(s) of norms to use to construct an overarching framework for global land governance. This situation can generate uncertainty about which course of action is most appropriate and may amplify latent political tensions among actors. Thus far, the most influential norms are those from the investment and land-tenure fields. These two sets of norms are complementary in some (but not all) cases. One can see that the goal of increasing investment in land may require the weakening of certain forms of land-tenure rights (e.g., collective, communal, or customary rights), which puts at odds the goals of protecting land tenure and reducing investment-related risk. Whereas such tensions are most likely to be solved at the national level on a case-by-case basis, global governance instruments are important because they can inform decisions on the ground, by providing actors reference to internationally accepted norms and standards. In addition, global rules alter perceptions about the legitimacy of certain land-grabbing and investment practices that can have reputational risks for states and investors alike. For example, if land is acquired through massive dislocation of local peoples, this could lead to institutionalized practices of naming and shaming as well as other forms of international condemnation that may change the behavior of investors and policy makers over the long term. As such, balancing competing sets of norms in global land governance may require further transnational policy-making efforts and consensus across a highly diverse set of actors with very unequal political and economic resources. This challenge becomes increasingly difficult when other normative frameworks are added to the mix, such as those found in global environmental governance. Sustainable development norms, such as those in global forest governance, and securing livelihoods do not always go neatly hand in hand. Conflicts between outsiders interested in resource conservation and destitute local populations dependent on the exploitation of resources for economic survival are inevitable and point to difficult ethical and social justice dilemmas which now must be increasingly negotiated on a transnational level, where the outcomes are highly uncertain for all parties involved.

These tensions, as well as the need to manage competing interests in land use and its governance, will shape the future course as global land governance continues to emerge. In preparation, in-depth research on land governance at the global level is required. So, too, is the need for greater dialog between land-use scientists and scholars of global governance; this interaction is requisite if we are to understand contemporary and future land-use challenges and provide broad-based solutions to them.

Acknowledgments

I thank two reviewers, Julia Lupp, Billy Turner, Sigrid Quack and members of Group 3 of the Ernst Strüngmann Forum, "Rethinking Global Land Use in an Urban Era" for helpful comments and suggestions on earlier drafts of this chapter. I acknowledge support from the Max Planck Institute for the Study of Societies for my writing time.

11

Large-Scale Land Transactions
Actors, Agency, Interactions

Carol A. Hunsberger, Saturnino M. Borras Jr.,
Jennifer C. Franco, and Wang Chunyu

Abstract

Large-scale land acquisitions, popularly known as "land deals" or "land grabs," bring together several important themes in global land-use change: competition for land, distal land connections, and governance across scales. Contemporary land grabbing occurs when actors with access to large-scale capital are able to capture control over vast tracts of land and other natural resources, often leading to changes in both land use and social relations. This chapter examines how the interaction of new actors and forms of political agency can help to explain the uneven character of contemporary land deals. By reflecting on cases where seemingly similar circumstances produced very different outcomes, it is proposed that the particular politics and histories of a given situation, the place-based characteristics of land and resources, and the interactions between mobilizations from below and the actions of state and capital from above all play a role in shaping the trajectories of contemporary land deals. Thus, greater integration between land-change science perspectives focused on the patterns of land-use change and social science perspectives focused on the dynamics of particular case studies could lead to an improved understanding of the variable processes and outcomes of land deals.

Land Grabbing, Global Context, and Actors

In recent years there has been an explosion of new land deals and land deal-making, raising questions about how these processes and outcomes can be understood and their implications for the governance of land resources. In this chapter we aim to contribute to the ongoing discussion by focusing attention on the political dynamics of large-scale land acquisitions, also known as "land deals" or "land grabs" (see also Borras et al. 2013). This phenomenon brings together several of the important themes in this volume: competition for limited arable land, long-distance connections between hubs of demand and sites of production, and complex governance arrangements that are based both on territory and on resource flows. To understand the links between land-use change and this particular form of changing control over land, it is important to complement a place-based perspective on where land grabs occur and the land-use changes they produce with a broader analysis of the political economic context of contemporary land grabs and the interactions between the actors involved.

Defining Land Grabbing

From the outset it is important to clarify what is meant by land grabbing. Defining land grabs too narrowly (i.e., large-scale land deals involving foreign governments which undermine food security in the host country and expel people from their land) is bound to miss significant aspects of contemporary land grabbing and possible trajectories of agrarian change. However, defining land grabs too broadly will miss what is distinct in the particular wave of contemporary global land grabbing. In this chapter, we exclude small- and medium-scale everyday forms of "dispossession by differentiation" (Araghi 2009). To avoid these problems, three interlinked features of contemporary land grabbing are emphasized.

First, as a fundamental starting point, land grabbing is essentially "*control grabbing*"; that is, grabbing the power to control land and other associated resources (e.g., water) to derive benefit from such control of resources. Control grabbing is inherently relational and political; it involves political power relations among state and nonstate actors. Control grabbing manifests itself in a number of ways, including land grabs, water grabs (i.e., capture of water resources, virtual as well as actual; see Woodhouse 2012; Mehta et al. 2012), and green grabs (i.e., resource grabs in the name of the environment; see Fairhead et al. 2012).

Second, the study of current land grabbing requires consideration of *scale*. This should not only be about the area of land involved, often defined using a particular benchmark such as 1000 ha. The *scale and character of the capital involved* are equally important. Bringing capital into the discussion helps to move the analysis beyond quantifying the amount of acquired land and describing the mechanisms of acquisition, so that information about motives and processes can be revealed. For example, a consideration of scale can facilitate greater understanding of why some investors prefer a specific type of renting land ("pools") in Argentina (Murmis and Murmis 2012), while other investors prefer contract-farming schemes in Indonesia (McCarthy 2010), and still others prefer land purchase where this is legally allowed. It can also help us see through layers of schemes, for example, in cases where foreign capital establishes national subsidiaries or forges alliances with domestic capital when land purchase by foreigners is not allowed (see, e.g., Franco et al. 2011; Murmis and Murmis 2011). This perspective can help us understand why some entrepreneurs grab control over land and implement land-use changes right away—as in the case of sugarcane plantations in Kampong Speu, Cambodia—while others do not, engaging in speculation or land banking instead—as demonstrated by various cases in Indonesia (McCarthy et al. 2012).

Third, current land grabs are occurring primarily *in response to* the convergence of multiple crises related to food, energy, climate change, and finance (McMichael 2012), as well as the emerging needs for resources by newer hubs of global capital, especially in BRICS countries (Brazil, Russia, India, China, South Africa) and in some powerful middle income countries (MICs). The key contexts for land (and water) grabbing, therefore, include food security concerns, energy or fuel security interests, climate-change mitigation strategies, and demands for natural resources by new centers of capital.

In short, *contemporary land grabbing is the capturing of control of vast tracts of land and other natural resources through a variety of mechanisms. It involves relatively large-scale capital and often shifts resource use orientation into a large-scale extractive character, whether for international or domestic purposes. Land grabbing reflects capital's response to the convergence of food, energy and financial crises, climate-change mitigation imperatives, and demands for resources from newer hubs of global capital.* This work-in-progress definition, anchored in the three defining features discussed above, sharpens the analytical lens, enabling us to navigate more realistically between a framework that is too narrow and a perspective that is too broad and includes everyday forms of dispossession by differentiation.

Land Grabbing As a Process

The working definition given above offers a useful counterpoint to a focus on "the grabbers"—their identities, declared intentions, assumed motivations, and apparent actions—as the main drivers of land deals. There are instances in the emerging literature on land deals where the grabbers have been presumed to be "successful" in acquiring lands unless proven otherwise; for example, when media reports about investor and government announcements of planned investments have been reported with confidence, even though many of these plans turned out to be abandoned later. In other cases the grabbers' "success" has been portrayed as inevitable, given their political power resources; in other words, their political power has been inferred rather than determined. As a corollary to this, and especially during the 2007–2010 period of initial reports on land grabs, host countries together with the people affected were often depicted as passive victims of a fait accompli rather than as active agents in an ongoing process marked as well by uncertainties and contingencies. However, this is now changing. Studies are coming out that emphasize the active role of national governments in land deals (e.g., Wolford et al. 2013). In addition, while the first major international academic conference on land deals at the Institute of Development Studies in Sussex in April 2011 yielded only a handful of papers on the political agency of the people affected on the ground, this topic received considerably more attention at a follow-up conference at Cornell University 18 months later. Our point is that while focusing on the grabbers has explanatory power with respect to many of the patterns associated with contemporary land grabbing, it also has some limitations that are already well recognized by the research community.

For one, history tends to be written by the winners; however, in land grabbing, history is still unfolding and thus subject to many kinds of potential hindrances, contestations, and setbacks which can, in turn, open up a range of trajectories. Land grabbing is thus better understood as an open-ended, indeed ongoing, political *process* (rather than outcome); consequently, explaining how the story ends—whether in specific cases or as a global trend—requires looking more closely at a range of process details. Our scan of emerging literature tells us that this is now being addressed, and it is on this emerging initiative that we want to build our discussion here. By adjusting the analytical lens to view land grabbing as an unfolding process—one that can result in different types and degrees of outcomes—we can begin to see how the political strategies and interactions of key actors in specific situations and contexts might make a difference to the process in terms of both its pace and direction.

To illustrate, a widely adopted explanation of the origins of contemporary land grabs (see, e.g., GRAIN 2008) traces them back to the food crisis in 2007–2008 and two related processes: (a) countries with limited arable land seeking to secure their future food supplies by directly sourcing food production in other countries and (b) high oil prices at the time which intensified investment in biofuel production, thereby increasing competition for arable land. The initial "food versus fuel" framing (see White and Dasgupta 2010) of large-scale land deals later expanded into other domains when land grabbing moved forward in the name of environmental protection (green grabbing) and financial security (linked to "agricultural" land investments for speculation and risk offsetting) (Fairhead et al. 2012). Resource-related grabs (for instance, water grabbing) also occur when land is acquired as a means rather than an end (Mehta et al. 2012). In this context, rising demands for food, energy, and other resources in newer economic hubs (especially Brazil, China, India, Korea, and the Gulf States) are also portrayed as driving the process of land grabbing (Margulis et al. 2013). These countries are seen as seeking to produce more food, fuel, and materials outside their borders to meet the demands of their growing populations, changing diets, and booming economies. They have been reported more frequently in research on global agricultural investments and labeled as much more active in this field than others.

This explanation of the processes driving land grabs places special emphasis on the grabbers, their reasonably assumed motives and declared intentions. It recognizes the convergence of multiple crises in the capitalist world and emphasizes the role played by the global movement of capital. Viewed from this perspective, doubts can be shed on some of the optimistic promises about large-scale land acquisitions; for example, that they will reduce hunger and poverty, revive agriculture and rural livelihoods, or produce a win-win situation when land deals have been properly controlled and managed (see Deininger et al. 2011). Understanding land grabbing as a further form of capital accumulation means that grabbers are seen to be seeking control of land (and other resources beneath, running through, or on top of it) due to the corporate hunger for profits, cheap labor, and rich resources, commensurate to the investor state's desire for food and energy security.

This argument explains quite well why Brazilian investors are actively buying land in Bolivia, Colombia, Paraguay, Uruguay, and Chile to produce soybeans, sugarcane, poultry, livestock, fruit, and forest products (Urioste 2012; Murmis and Murmis 2012), or why Chinese investors are seeking land in Africa, the Mekong River Basin, and in Southeast Asia (Hofman and Ho 2012). In other words, it explains why land-use change happens in certain ways when capital moves around the world in search of land and *succeeds*. In these cases, food production may be converted to cash crops and biofuels, or nonfood (forest lands, "idle" or "marginal" lands) to food and nonfood production (Borras and Franco 2010).

This narrative, however, cannot fully explain some of the other trajectories and outcomes that can be observed in global land grabbing. For example, a 1.24 million hectare investment by the Chinese government and private sector was reported in the Philippines, but was later cancelled after protests (GRAIN 2008). This was an outcome that would not have been predicted had the focus been on the intentions of land grabbers. Likewise, there are not many cases of transnational land grabbing reported in Vietnam, even though it is endowed with fertile land and a suitable climate for growing plantation crops (Sikor 2012b). In another unusual case, the Brazilian government has sought to control the "foreignization" of land within Brazil while encouraging outward investment into other countries; thus far, however, this control has not worked well (Borras et al. 2012b). Encounters with and reactions from actors at the local, national, and even international levels should be taken into account when the governance of land use is examined in the context of global land grabbing. For a broader discussion of the institutional context of land grabbing, see Margulis (this volume).

Land Grabbing and Land-Use Change

Land-use change is one of many different types of outcomes of contemporary land grabbing; others include changes in the particular crops grown, changes in the use of the crops grown, or changes in the social relations pertaining to a particular piece of land, to name a few. Indeed, land grabbing is occurring worldwide and producing many different types of land-cover and land-use change; for example, the conversion of forests, wetlands, and grasslands to agricultural production or forest plantations and the clearing of land for fuel or mineral extraction. However, land grabs also drive changes in the types of crops that are grown on land already used for agriculture, including food crops, biofuel crops and "flex crops" (discussed below). These crop changes can have important environmental and social implications even if they do not register in many discussions of land-use change because they involve shifts within, rather than between, land-use categories.

The concept of "flex crops" helps explain why particular crops are expanding so rapidly. Flex crops have multiple uses (e.g., food, feed, fuel, industrial material) that can be flexibly interchanged. They include soybeans (used for feed, food, and biodiesel), sugarcane (food and ethanol), oil palm (food, biodiesel, and commercial/industrial uses), and corn (food, feed, and ethanol). Production of these crops has increased substantially; the area cultivated with soybeans, for example, has nearly doubled—to approximately 100 Mha—over the past thirty years (FAO 2012a). The cattle sector, which involves flexible inputs rather than outputs, can also be considered to play a role in the flex crop complex. Flex trees and forests are also possible as demand for next-generation biofuels grows (Kröger 2012).

Flex crops resolve a difficult challenge in agriculture: diversifying a product portfolio to mitigate price shocks. With the emergence of relevant markets (or speculation of such) and the development and availability of new technology (e.g., flexible mills), diversification can, in theory, now be achieved—within a single crop sector. How much and how easily these crops actually do "flex" is a topic of ongoing investigation. While multiple and flexible uses of particular crops are not new, the contemporary character of flex crops is linked to the convergence of multiple crises already mentioned. Hence, the multiple and flexible character of these crops straddles the food, feed, fuel/energy complexes, and newer hubs of capital. Although changing the purpose for which a particular crop is grown does not constitute land-use change, understanding the nature of demand for flex crops can help explain why these crops are expanding in particular ways. From a governance perspective, it can also help explain the sometimes unexpected constellations of actors involved; for example, when auto manufacturers and oil companies take part in biofuels policy making, as some are doing by serving as members of the European Biofuels Technology Platform steering committee (European Biofuels Technology Platform 2012).

Much of the political economy literature on land grabbing is more concerned with social relations than with land-use change. The land "control grabbing" concept reflects this focus, as it considers a land grab to have taken place once control over land changes hands, whether or not land-use changes are actually implemented as a result. In the following discussion we emphasize the analytical distinction between process and outcome, examining some of the differences between cases where (attempted) land grabbing did and did not result in land-use change. Examples include cases where investors succeeded in implementing their plans as well as those where resistance or other interventions contributed to land deals being avoided, stalled, or cancelled.

Contested Processes, Unexpected and Uneven Outcomes

If we approach contemporary land grabbing analytically as an ongoing and open-ended political process—one that involves contingent interactions between new actors and new forms of political agency in specific contexts—this opens up the possibility of analyzing variable trajectories. This, in turn, can help to detect, and even explain, unexpected and unintended outcomes. Emerging patterns of the politics of land deals and deal-making demonstrate unevenness:

- Investors are targeting particular countries and regions in different ways and to varying extents.
- Government actors have cancelled or reversed land deals in some countries but not others.
- Investors have gone through periods of being more and less successful at implementing land deals within the same host country at different times.

The following examples illustrate this range of trajectories.

Similar Agroecological and Geographic Conditions, Different Trajectories

Vietnam and Cambodia are two countries with comparable biophysical potential for producing tropical plantation crops but they have very different patterns of transnational land deals. In Vietnam, private and foreign companies have not been able to gain a foothold easily in the plantation forestry sector. Instead, smallholders contribute to *two-thirds* of all productive tree plantations in the whole country (Sikor 2012b). The absence of transnational land grabs for fast-growing tree plantations in Vietnam seems to deviate from the general pattern in mainland Southeast Asia. Such an "anomaly" has been attributed by Sikor (2012b) to three aspects of the political economy of Vietnam:

1. Peasants gradually gained their access to land through occupation and decollectivization. Later their land ownership was formalized by local officials.
2. Fast-growing trees (eucalyptus and acacia) require relatively small monetary input and little extra labor, and loans with very low interest rates are available from the state-owned bank.
3. Local mills (wood markets) exist within a reachable distance for peasants. The Vietnamese state also runs state forest enterprises that are directly engaged in fast-growing tree plantations.

Adjacent Cambodia has broadly similar agroecological conditions for fast-growing tree plantations but has experienced land deals differently. In Cambodia, transnational land deals—including those by Vietnamese companies—have dominated the process. The state has encouraged land deals involving both domestic and transnational (mostly Chinese and Vietnamese) capital. Economic land concessions reportedly cover just under 20% of Cambodia's total protected areas (UN Human Rights Council 2012). Recently, Cambodia has witnessed widespread conflict around these land deals, as local communities resist expulsion from their lands (UN Human Rights Council 2011b).

Same Host Country, Different Investors and Outcomes

Land deals can play out very differently even within the same country, where one would expect the institutional context and geographical endowments to be relatively similar. Two cases from Kenya's Tana Delta illustrate how different investors experienced contrasting outcomes even within the same district. In one case, domestic state and private sector actors—Tana and Athi Rivers Development Authority (TARDA) and Mumias Sugar Company—together pursued a 33,000 ha sugar project, which met with considerable opposition and was eventually stalled. In the other case, a foreign company—Bedford Biofuels, based in Canada—initiated a 64,000 ha jatropha plantation for biodiesel, which received a more favorable (though still mixed) response from local residents, and it has begun to be implemented in stages (Smalley and Corbera 2012).

Smalley and Corbera (2012) identify four crucial factors that differed between these cases: land tenure, the reputations of the investors, perceptions about the value of the land in question, and the political prominence of local responses. In terms of land tenure, in the case of the proposed sugar plantation, the government gave the investors access to the targeted land. In the jatropha case, the investor had to negotiate with ranch owners to obtain leases, creating a situation where the investor may have felt compelled to adapt their plans to reflect some of the ranch owners' views (Smalley and Corbera 2012). Regarding the investors' reputations, local residents had had prior negative experiences with TARDA related to previous development schemes, thus biasing them against TARDA's latest proposal. By contrast, Bedford Biofuels had not previously been active in Tana Delta and thus was not hampered by negative preconceptions on the part of residents (Smalley and Corbera 2012). In terms of the value of the land, the area covered by the sugar proposal included lands of high seasonal importance to pastoralists, which opponents of the project claimed the government had undervalued, whereas the drier ranches covered by the proposed jatropha project area were not perceived to be as important to many land users. Finally, opposition to the sugar project was amplified through local, national, and international awareness campaigns and this crystallized in a lawsuit that temporarily stopped the project in 2006. In the jatropha case, there was "no unified opposition to Bedford Biofuels from squatters and other threatened resource users across the six ranches" involved (Smalley and Corbera 2012:1063).

Although these two land deals are in close proximity to each other geographically, context-specific factors and interactions between investors, residents, government officials, and campaigners produced remarkably different processes and outcomes.

Same Investor, Different Countries and Outcomes

China has long been identified as a main actor in the contemporary land grab picture. Hofman and Ho (2012) argue that there was a rapid rise in the incidence and scale of such land investments after China's "going global" policy was initiated in 2000; however, China has been more successful in some countries than others when it has outsourced its development.

In Cambodia, for instance, around 25 land deals (over 42% of the total) were owned by foreign investors according to a ministerial investigation in 2007, of which 13 entities belonged to Chinese companies (UNOHCHR 2007). With a total of six confirmed Chinese investments over 1000 ha, China acquired over 105,000 ha of land, producing agricultural products such as palm oil, fruits, livestock, cassava, corn, vegetables, castor oil, and trees (UNOHCHR 2007). Among a total of 18 reported land deals between China and Cambodia (including those not confirmed), 13 have reportedly aroused protests from civil society (Hofman and Ho 2012). Many protests of local villagers have been brought to the district, provincial, and national levels, but UNOHCHR (2004, 2007) reports that these protests were ignored or suppressed by the state, or met with empty promises and bribery.

Land deals involving Chinese investment, however, have undergone a slowing down process in the Philippines, interwoven with resistance from local (and national) social movements, as well as shifts in related national policies. China's recent land investments in the Philippines have been impeded by various protests. The 2007 Memorandum of Understanding signed between several Philippine government actors and various Chinese investors, which covered about 1.4 Mha of lands for planned agricultural production, has been suspended. Another package of land deals granted to Chinese companies such as telecom giant ZTE, involving 1.24 Mha of land, was also reportedly suspended after social protest (GRAIN 2008).

These examples concern the same investor (China) and similar social protests in Cambodia and the Philippines. However, the trajectories were quite different: Chinese investors gained a foothold in Cambodia but experienced more difficulty in the Philippines, at least for the time being.

The Importance of Interactions

Contemporary global land grabbing has ushered in new patterns of alliance and contestation between actors from local to global levels. We argue that the interactions between actors and (new) forms of political agency (especially at the ground level) in their specific political and institutional contexts are potentially the most important factor in explaining the differences between cases where land deals proceed and where they do not, as well as how they proceed when they do. We contend that this factor deserves more focused and systematic attention in the current literature on the politics of global land grabs.

In the next section, we introduce three analytical tools to highlight more clearly the political and institutional factors that shape land-grabbing processes and account for the widely variable and uneven outcomes, in terms of pace and direction, that is currently being observed globally. The first draws on the work of Fox (1993) to highlight the dual, and contradictory, role of the state—particularly the central state elites, who must balance efforts to facilitate private capital accumulation with efforts to maintain political legitimacy. The second emphasizes the specific national historical-institutional setting in which land grabbing occurs, and which structures the choice of political strategies of key actors as well as their interactions. In many countries this includes a potentially plural-legal landscape (e.g., national land, forest, and water laws, in addition to customary laws and arrangements that may or may not have formal legal standing). The third draws insight from Tsing (2005) to stress the "frictional" encounters between global and national actors and local actors and forces, especially as they unfold in perhaps unexpected and unintended ways on the ground. Together, these perspectives underline the role of contending political choices and strategies around land grabbing and, ultimately, the crucial importance of tensions, frictions, and synergies among and between actors.

Contradictory Role of the State

Fox (1993) identified two foundations of state power: facilitating private capital accumulation and maintaining a minimum level of political legitimacy. These are contradictory tasks, and an understanding of them can help to interpret the central state's role in contemporary land grabbing. The state is ever-present in large-scale land acquisitions; its intervention partly shapes the character and trajectory of land deals. The various case studies in the collection by Wolford et al. (2013) demonstrate this. In different circumstances, the actions undertaken by central states can gravitate toward one or the other of these two poles.

Actions to facilitate land deals on the part of investor states as well as host country governments can be seen as favoring the capital accumulation/economic growth objective. Sometimes "open for business" rhetoric is part of the announcement of such deals. For instance, the Cambodian state opened its land market in 1991 to foreign and private investors in pursuit of economic growth soon after the end of two decades of civil wars and conflicts (UNOHCHR 2004). The state leased out approximately one-third of its most fertile territory to corporate investors within seven years, referred to as "land concessions for economic purposes" under its 2001 Land Law (UNOHCHR 2004). Similar cases can be observed in China and India, where special economic zones provide almost free land, favorable taxation, and cheap water and electricity to investors. Fox (1993) also argued that over time the state develops its own capital accumulation interests that are distinct from promoting private capital accumulation. This may fit with what is happening in Vietnam: although transnational actors are not engaging in land grabs, the state is apparently doing so by acquiring land for its own forestry schemes.

In cases where host country governments consented to particular land deals but later stalled or canceled them, one can interpret these actions as efforts to restore damaged political legitimacy. Kenya's refusal to issue the required permits for the planned Tana Delta sugar plantation following local, national, and international opposition provides one example. Proactive measures to discourage land grabbing in general also fit this pattern, as in the case of Brazil and Argentina's anti-foreignization measures, or Tanzania and Mozambique's moratoria on large-scale biofuel production. In Madagascar, opponents of the Ravalomanana government heavily criticized its reported land deal with the Korean company Daewoo in 2008, using the issue to attract support for Andry Rajoelina in the lead-up to the military coup of 2009 (Ploch and Cook 2012). Discourses of lost legitimacy and the prominence of public opposition to proposed land deals played an important role in these cases (discussed further below).

National Historical-Institutional Setting

No actor, however "new" or powerful, is able to shape the political field and determine outcomes by sheer will alone. As Marx (1852/2008) famously observed in the *18th Brumaire*:

> Men make their own history, but they do not make it just as they please; they do not make it under circumstances chosen by themselves but under circumstances directly encountered, given and transmitted from the past.

This observation highlights how historical and institutional factors structure encounters between and among different actors, including those who may be involved in and affected by land grabbing. Within broadly similar national historical-institutional contexts, there are still differences, however slight, that can contribute to producing variable and uneven outcomes.

Here "institutions" refers to formal-legal, official institutions as well as those that may be considered informal and customary (e.g., not necessarily codified by national legal frameworks). This distinction is particularly relevant with regard to the rules and laws in a given country that play a part in determining who has what rights to which land, for how long, and for what purposes.

Specific historical-institutional settings, stretching between the national and local levels, have a key role to play in shaping the choices different actors make with regard to land and water resources, including their choices of political strategy either as would-be land grabbers or in response to attempts at land grabbing by others. They also play a role in structuring the particular interactions of different and contending actors around land grabbing processes.

To illustrate, investors seeking to capture control of land for large-scale industrial agricultural ventures are very likely to target prime agricultural land. Whether they aim to do so via purchase (as in Polochic Valley, Guatemala, where one family purchased 5400 ha for sugarcane; see Alonso-Fradejas et al. 2011) or lease (as in the Northern Philippines, where Green Futures/Ecofuel leased agrarian reform land where its sale is restricted by law; see De la Cruz 2012) is determined by the set of laws governing land tenure that prevail in a particular country. People on the ground in areas targeted by these land deals, who are resistant to losing control of the land and water resources they occupy and use, may indeed try to resist such deals from taking place. However, what form their resistance will take depends, at least in part, on whether or to what extent their preexisting relationship to these resources is recognized by national law as "legal." For instance, we can expect that resistance against grabbing in China, where collective land ownership is widely recognized and protected, would be very different from that in Brazil, where private ownership dominates. In an exceptional case, an indigenous group (B'laan) in the southern Philippines, which had decades earlier been forcibly expelled from their ancestral land by a local landlord, formed what could be considered a tactical alliance with South Korean investors, who wanted to gain control of the land for a jatropha project, in order to regain access to the land (Borras et al. 2011). Here, there is an opportunity to draw further insights from the literature on legal pluralism and customary resource rights regarding access and use of land and water.

Political Reactions from Below As a Source of "Friction"

As suggested above, land grabbing is best understood as an ongoing and open-ended process—one that involves attempts by would-be land grabbers to capture control of land and water resources in order to control the benefits that can be derived from them. Land grabbing is a dynamic process—one that transpires through real people and across a variety of boundaries and differences in terms of aspirations, cultures, social histories, practices, and knowledge. Land grabbing involves "encounters across difference," which can generate tensions and conflicts, and bring about unexpected twists, turns, and outcomes. These can be conceptualized as "frictional encounters"; that is, "the awkward, unequal, unstable, and creative qualities of interconnection across difference," according to Tsing (2005).

This is a potentially very fruitful way to think about why and how land deals generate variable and uneven impacts within and between communities and countries, showing differences across social class, gender, ethnicity, and nationality. It is these existing social fault-lines that determine the uneven benefits of land deals among affected communities. In turn, these differentiated outcomes influence the shape of political responses to land deals from one case to another. Research into land grabs that refers to "the local community" without acknowledging the differentiation and power hierarchies that are involved can overlook this important aspect.

McCarthy (2010) analyzed peasants' adverse incorporation into oil palm plantation in four villages in Sumatra, Indonesia and the divergent outcomes of these processes. Even within one village, he found that some peasants may be enrolled in the smallholder scheme; some may be forced to sell their scheme entitlements, while still others may become a provider of cheap labor. Those adversely incorporated into the scheme may become poorer than they were before. Palm oil requires expensive inputs and creates a threshold of inclusion and exclusion for smallholders (Fortin 2005). It can be concluded from the McCarthy study that the variability of outcomes of land deals was determined by localized livelihood strategies, which in turn are shaped by economic, social, and political relations.

For instance, McCarthy (2010) reports that some Melayu farmers found that they could not cope with the oil palm schemes because the land allocated to them was located in remote communities, or because they had to sell them for loans or food. Under these circumstances, their land entitlements were bought up at a relatively low price by village elites, officials, in-migrant businessmen, and teachers. The KKPA scheme ("Primary Cooperative Credit for Members") focused on integrating local populations, although the three villages applying this scheme that McCarthy investigated exhibited very different results: the first village was the most prosperous; the second experienced both positive and negative impacts; the third remained poor. Those who benefited from the palm oil boom were primarily Melayu landowners. Meanwhile, below-poverty-line smallholders were excluded from livelihood gains and faced a rapid mechanization of village lands. Wage laborers were generally better off than these smallholders, and some even managed to save enough income to buy a piece of land (McCarthy 2010).

These accounts from Sumatra (and many other similar ones) show that differentiated outcomes and land-use changes may happen *within and between* communities when they are confronted with large-scale land investments. As the affected rural community is not a monolithic entity, political responses from below can take different forms and involve different alliances, including land sovereignty and land peasant movements as well as openness to such land investments (McCarthy 2010), as well as mixed attitudes and groups in between.

Concluding Remarks

When political reactions and mobilizations from below interact with state and capital actions from above, the trajectory of a particular land deal is shaped. Here we have proposed that the particular political configuration and interactions—not only the original intention of land grabbers—strongly influence the character and trajectory of *land deals as a process*, from one case to another. While this insight can add value to investigations that focus on the place-based land-use outcomes of land grabbing, the inverse is also true: research on political interactions can similarly be enhanced by taking into account the role that place-based attributes, as elaborated by land-change science, can play in influencing actor interactions and decisions.

Have land grabs revealed new actors in global land-use change? Indeed some have emerged, including investors from newer economic hubs, financial actors drawing on new forms of capital including pension funds and private equity, and civil society groups—some facilitating and others opposing land deals. However this wider range of actors cannot alone answer the two questions posed in this chapter: Why are some actors more successful at implementing land deals than others? Why do land deals in some situations proceed through to the implementation stage but not in others? To understand these patterns better, research must shift from considering the *range* of actors involved to examining more closely their *interactions*. Scrutinizing these interactions could provide a missing link between understanding the drivers and motives involved, and interpreting the actual outcomes of land deals for land-use change and social relations.

This chapter therefore proposes adopting interaction as a unit of analysis to help understand land deals and their relationship to global land governance. This presents some analytical challenges related to complexity and scale. First, the convergence of actors from multiple sectors produces a high level of complexity: land deals can involve unlikely alliances and tensions between actors from the fields of energy and agriculture, conservation and finance, resource extraction, and the military. Compounding this diversity of actors, land grabs play out within a complex and shifting array of governance arrangements, some of which focus on land as territory while others concern the flows of land-based products and resources (see Gentry et al., this volume). To adopt an analytical approach that is broad and yet flexible enough to track the interactions between this wide variety of actors, and sufficiently nuanced in its treatment of governance instruments and institutions, remains a key challenge.

Second, the cross-scale nature of these interactions complicates the analysis. New analytical perspectives need to be able to address complex power relations involving local and national networks and hierarchies, state politics, international relations, transnational economic (and civil society) alliances, and international governance efforts. Finally, greater integration between land-change science perspectives focused on patterns of land-use change and social science perspectives focused on the dynamics of particular case studies could lead to a stronger understanding of the linkages between the variable processes and outcomes of land deals.

12

Private Market-Based Regulations

What They Are, and What They Mean for Land-Use Governance

Graeme Auld

Abstract

Private market-based regulations are becoming prevalent features of governance across economic sectors. From programs, such as the Forest Stewardship Council, that set rules for management practices in forests to the e-Stewards programs that addresses the recycling of electronic waste, a diverse range of private regulatory initiatives have potential significance for land use around the world. This chapter examines how these private regulators contribute to problem-oriented attention in land-use governance. To make sense of the complex array of private regulators, the chapter reviews the variation in their characteristics. From this descriptive foundation, it assesses what we know about the construction, evolution, and consequences of these initiatives on micro, meso, and macro levels, and assesses, in a preliminary way, how various private regulators may affect pressures on land use. The chapter closes by discussing three key problems that are presented by an incomplete field of private regulators. One is the mismatch between the scale and rate of the processes of land use and the scale and rate at which private regulations are being adopted and exerting influence over land-use practices. A second challenge is the spatial mismatch between where private regulations have gained the greatest inroads and where the greatest land-use practice concerns reside. A third challenge is the problem and importance of institutional fit. Land-use governance is an encompassing field that extends well beyond the policy focus of any given private regulator. Hence, for private regulation to attend to this challenge, coordination and co-operation will be necessary to address influences on the land that are at the intersections of economic sectors and stages of global supply chains.

Introduction

Private market-based regulations (hereafter private regulations) are becoming prevalent features of governance across economic sectors. Their significance and potential import as governance mechanisms which operate alongside or in complement to governmental processes and rules has spurred a great deal of discussion and academic analysis.

This chapter scrutinizes how private regulators contribute to problem-oriented attention in land-use governance. Considerable complexity exists below the surface of this issue. One primary challenge is posed by the differences among the many forms of private regulators. Different initiatives address different problems, from narrow attention to a single environmental concern (e.g., shade production on coffee farms) to multidimensional assessments of the ecological footprint of a product's production process. Some are consumer-oriented, providing compliant operators the opportunity to sell products with an on-product label that communicates the value of their program participation. Others are business-to-business oriented. Some may require external verification, but not all do. Finally, initiatives have varying decision-making procedures that have implications for the engagement of stakeholders and their perceived legitimacy as global rule setters. This chapter reports on current knowledge about the array of private regulatory initiatives that now exist and how they operate individually and in concert.

To do this, I employ a threefold analytic structure and begin by assessing what we know about the construction, evolution, and consequences of these initiatives on three levels. First, it examines the micro level, connoting the level of individuals. Individuals play several roles in the operation of private regulation, most obviously as consumers, but also as key entrepreneurs within organizations, companies, government agencies, foundations, and other actors that promote, steer, or resist the development of these initiatives. Second, it examines the meso level—the organizational level—where companies, non-governmental organizations (NGOs), governments, and other actors interact. Third, it examines the macro level, or the background conditions against which private regulation has formed, evolved, and ultimately may or may not have any significant effects on policy problems.

From these individual strands, a larger picture of how, and the extent to which, a field of private regulation is being constructed comes into focus and can be assessed. The patterns and themes that emerge reflect the contemporary trends of urbanization, economic integration, and new land agents and practices that are reviewed by Seto and Reenberg (this volume). Applying their framework, private regulation exemplifies the nexus of "distal connections" and new "decision making and institutions," since many of the programs reviewed below are, at least in part, the product of "transnational coalitions and advocacy agendas" and since these programs are also creating new venues where decisions about appropriate land-use practices are being made. Early examinations of these initiatives noted the potential benefits that come from a reorientation away from the territorial units of states as hierarchical nodes in a global system of governance to an issue or supply-chain centered governance system (Cutler et al. 1999a). As Cashore (2002:512) explains, in the context of nonstate market-drive governance, "the location of authority is grounded in market transactions occurring through the production, processing, and consumption of economic goods and services" (see also Bernstein and Cashore 2007). One potential benefit of this reorientation is a lessening of the need for cooperation among states to secure action on global public problems, such as land-use change (Webb 1999; Cutler et al. 1999a).

This optimistic view has been fraught, however, with challenges surrounding the scale, rate, and place of changes associated with the recent surge in private regulations. The scale and rate of private regulation has not alone been sufficient to manage the scale and rate of pressures associated with land-use change. Similarly, there has been a mismatch between the place where the activities of private regulation have flourished and the places where land-use pressures are most severe and demand the greatest policy attention. Finally, and beyond the focus of the Seto and Reenberg framework, the construction of these new regulatory institutions face the standard problem of institutional fit (Young 2002). That is, the policy focus that private regulators choose creates issue-area boundaries which programs must coordinate across to address cross-issue problems (Auld 2013). As a consequence, land use, as a problem affected by the operations of many economic sectors, is a particularly challenging issue space for current private regulation to address.

Private Market-Based Regulations

To lay the foundation for the analysis which follows, I begin by reviewing the various ways that private regulation has been defined and what characteristics exemplify the programs that fall within its categorical boundaries. I then introduce the three-level analytic framework to review the way these initiatives are currently understood.

Characteristics

The most common theme in existing work on private regulation has been to delineate *which actors set the rules* (Abbott and Snidal 2009) (Table 12.1). Broadly defined, the private category includes those initiatives where governments are not part of the processes established to set standards or verify compliance. How the state is involved matters as it affects why operators choose to comply with a program's rules. At one extreme, private governance operates in opposition to the state, such as mafia or criminal networks (Williams 2002); adherence to rules, in these cases, is not directly the consequence of the threat

Table 12.1 Descriptive characteristics of private market-based regulations.

Features	Key Concerns
Decision making: Who decides the rules?	Role of the state Implications for compliance incentive Degree of multi-stakeholderism—has implications for democratic legitimacy
Monitoring and enforcement: How are the rules monitored and enforced?	Resulting implications for performance when monitoring and enforcement are conducted by: • a first party (the company checks itself), • a second party (an industry association checks compliance), or • a third party (an external organization or body)
Policy focus: What is the character of the problem the standards seek to address, and what form do the standards take?	Types of behavior that standards seek to induce: social and environmental performance, or technical quality Actors that are targeted to regulate: all actors in a sector or just some, such as small producers Types of standards used: procedural vs. performance, discretionary vs. nondiscretionary
Market features: What are the program's market features?	Voluntariness and market-based compliance incentives

of state sanction. Cashore's nonstate market-driven governance is a middle category where the "state does not use its sovereign authority to directly require adherence to rules" (Cashore 2002:509). Nevertheless, this form of private governance is embedded in a system of state rules, including contract and property law, which are not directly challenged by the programs (Meidinger 2006; Gulbrandsen 2006; Bernstein and Cashore 2007). Then there are instances where states "delegate" authority or partner with the private sector and thus act, either implicitly or explicitly, to cede authority to private actors (Cashore 2002; Clapp 1998; Börzel and Risse 2005).

Private regulators are also differentiated based on which nonstate actors hold decision-making power (Fransen and Kolk 2007). At one extreme, a broad range of stakeholders is given direct influence over the policy decisions of an initiative. At the other extreme, a single private actor—be it a company, a research organization, or an NGO—establishes the rules of an initiative. Whereas the role of the state garnered attention for those interested in compliance incentives, the degree of "multi-stakeholderism" has been of interest more for scholars concerned with the democratic legitimacy of private regulators (Tollefson et al. 2008; Dingwerth 2007; Raynolds et al. 2007). It is important to recognize that "private" and "public" are broad terms with many different meanings, and that often the distinction between the two in these governance processes is blurry. Moreover, many of these "private" regulators deal with concerns of public and collective good and could thus, in this respect, be termed public regulators (Sikor 2012a; Cutler 1997). Not surprisingly, this ambiguity has led scholars to use many different terms to connote these emerging forms of governance, including "private authority" (Cutler et al. 1999b; Hall and Biersteker 2002), "civil regulation" (Bendell 1999; Zadek 2001; Vogel 2008a), and "regulatory standard-setting schemes" (Abbott and Snidal 2009), among others.

A second characteristic builds on the interest in compliance. It focuses on differences in programs' *approaches to monitoring and inspection*. As private regulations are voluntary, an interest in why operators choose to comply is evidently important. Monitoring can occur through first-, second-, or third-party verification (Garcia-Johnson 2001). Existing research identifies the perils of weak verification procedures. Without some form of external, or third-party, verification, private regulators are likely to induce processes of adverse selection whereby poorer performing operators will join a program to reap any benefits participation provides, but will do little to improve their performance (Prakash and Potoski 2006). Work on voluntary programs and industry codes of conduct in the United States and other countries have shed important light on this matter by identifying the relationship between the strength of a program's compliance and enforcement provisions and the performance of participants (King and Lenox 2000; Rivera et al. 2006; Rivera and de Leon 2004; Rivera 2002).

A third characteristic is the *policy focus of the program*. This has three main characteristics. First there is the focus or scope of the rules, defined as the range of activities the program seeks to regulate (Cutler et al. 1999a). Some programs focus on reducing market transaction costs or increasing efficiencies by generating technical standards for products or production processes (see, e.g., Salter 1999; Botzem and Quack 2006; Mattli and Büthe 2003; Büthe and Mattli 2011). Other private regulators focus on social and environmental externalities (Lipschutz and Fogel 2002; Cashore 2002; Cashore et al. 2004; Knill and Lehmkuhl 2002; Bernstein and Cashore 2007). The policy focus of private rule makers is important as it comes with different processes by which actors assess whether or not to adopt the rules. A focus on externalities means that private rule-makers face a public goods problem. Targets of the rules will need some reason to participate or they will face strong incentives to shirk their compliance responsibilities. Technical standards, by contrast, can be self-enforcing once in place, because they reduce transaction costs and increase market efficiency (Young 1999b). There may be distributional conflicts over what the technical standard ought to be, since different producers are likely to gain relative to their competitors with the adoption of one quality standard over another (Knight 1995, 1992; Salter 1999; Krasner 1991; Mattli and Büthe 2003). However, there will be strong reasons to adopt the standard as it gains market dominance.

Although beyond the scope of this chapter, it is important to note that the distinction between these two problems is more complicated in a dynamic and multi-actor world. For instance, when actors know they will be interacting over a long period of time, the willingness to cooperate to address externality problems can grow. By contrast, a long time period can mean technical standards become more contested, the logic being that when actors face relative losses from a coordinated agreement, the longer the agreement is to last, the higher the incentives are to find ways, at some point, to argue for a new arrangement or to construct one (Snidal 1985, 1996).

A second criterion is the domain of actors regulated. Programs vary in how open they are to the participation of different operators within an economic sector. A program may, for instance, restrict access to only small operators or operators that are organized as cooperatives. Others do not impose any such eligibility restrictions. Cutler, Haufler, and Porter (1999a) also discuss this domain criterion in relation to who sets the program's rules. With some private regulators, those that must adhere to the rules are involved in developing the rules. This category captures many corporate codes of conduct, where companies set rules to govern their own industrial practice (Cashore 2002; Andrews 1998). With other private regulators, rules can extend well beyond the domain of those who developed them. For instance, a large retailer that develops a set of social, environmental, and/or quality requirements for its suppliers has established rules that may affect many interconnected global supply chains.

The final policy-focus characteristic concerns the nature of standards. Procedural versus substantive and discretionary versus nondiscretionary are two dimensions of variation used to describe the types of standards which programs develop to govern and delineate appropriate practices in a given issue area. Although these distinctions have been applied to classify entire programs (e.g., ISO 14001 environmental management standards are considered procedural because operations are only required to have plans in place to achieve their own goals), they are also valuable in characterizing individual rules (Auld and Bull 2003; McDermott et al. 2008).

The final characteristic concerns the *market-based nature of private regulations*. This has been variously construed depending on what scholars view as a "true" market-based instrument (Elgert 2011). A key distinction to draw is that private regulations are not equivalent to market-based instruments, such as pollution pricing via taxes or charges and pollution permit and trading schemes, which are considered beneficial because they are more cost-effective means to achieve a given environmental target as compared to performance standards or technology standards (Keohane and Olmstead 2007; Goulder and Parry 2008). These instruments give operators flexibility to choose how best to achieve a given target, which can drive down costs of compliance. Though there are exceptions (e.g., the now defunct Chicago Climate Exchange), private regulations are neither pricing nor permit and trading schemes. Rather, they are performance standards and sometimes technology standards that are set by private actors (Auld 2010b). They are considered market based for at least two other reasons: they are voluntary and they rely, to varying degrees, on market incentives to drive compliance; that is, the incentive for participation is generated within the market place rather than via government sanctions (Cashore 2002).

This distinction between voluntary as the basis for a market-based instrument as opposed to flexibility is important. Flexibility is useful for finding the least-cost means to achieve a given environmental target; voluntarism is plagued by the problem of adverse selection, as noted above. At worst this means that operators that are already performing at a high standard will select into a program that has high standards and strict enforcement. Hence, voluntary private regulation may only serve as a sorting tool, allowing those with the desire to buy high-performing products the opportunity to do so, but not broadening the effects of private rules to a sector as a whole. This has been one of the dominant concerns for scholars considering the effects of private rules, and one that the discussion returns to below. Technical standards, as noted above, are slightly different given that positive network externalities can create self-enforcing mechanisms, making it in the interest of companies to adopt and adhere to the standard in the short term.

Building on existing work, Table 12.2 offers additional characteristics of these regulators that deserve future consideration. First, finances are an area that requires more careful analysis. Unlike states that have taxation powers,

Table 12.2 Characteristics of private market-based regulation that have received limited research attention. Examples were drawn from www.ecolabelindex.com

Features	Potential Variation	Examples
Finances: How are the activities of the private regulator financed?	Public funding License fees from use of label Foundations Membership fees Other	Forest Stewardship Council (FSC), membership fees, and foundations Marine Stewardship Council (MSC), more funds from label licensing
Focus of label: What products can carry the label?	Single product Multiple products	Chlorine-free paper Organic product
Focus of standard: What characteristics are assessed in determining whether an operator can use the label?	Single criteria Multiple criteria Life-cycle, single criteria Life-cycle, multiple criteria	Green-e LEED Green Building Carbon neutrality Ecological footprint
Geographical scope: In which geographic units can the label be used?	Subnational National Regional Global	Organic labels for individual Canadian provinces National organic labels EU organic label MSC

private regulators must generate revenue to cover operating costs via grants, fees for service, membership fees, donations, and other revenue sources (Cutler et al. 1999a). The implications of different funding models have received little examination. Second, the character of a program's label, if it has one, has received some attention, but not in a way that is well linked to the overall problem-oriented effects of private regulations (Teisl 2003; Teisl et al. 1999, 2002). Third, discussion of standards have not carefully considered the trade-offs and implications of programs that focus on single or multiple social and environmental criteria and/or whether they do so at a single point in a supply chain or seek to assess a product's life cycle. Finally, the geographic scope of private rules has received insufficient attention. There are two extremes of the literature: one side examines programs that operate within a single country's domestic public policy context (Lyon and Maxwell 2007); others focus only on the transnational manifestation of private rules (Gulbrandsen 2010). There are, however, reasonable arguments to be made that the geography and place of these standards will matter in ways that may be path dependent. These possibilities have been given only initial consideration (Auld 2009).

Analytic Perspectives

What is the significance of these new governance and rule-setting institutions? Building from the previous section, the rise and potential of private regulations will be explored, drawing on existing work.

Micro Processes

Individuals are implicitly or explicitly a critical part of how academics and policy analysts explain the rise and potential of private regulations. Individuals are conceptualized most readily as citizens or leaders/entrepreneurs. These are addressed in turn.

There are at least three ways citizens feature: as a consumer, as a civil-society supporter, and as a rights holder. First, as consumers, citizens are seen as an instrument; they make purchasing decisions. If these decisions can be shaped such that the consumer citizen acts on their political values—say concern for the environment or social equity—then private regulations can be tools to encourage these practices among economic actors. Great interest in this possibility emerged in the late 1980s in the United States and Europe. Consumer boycotts begot tools designed to inform consumers of good and bad purchasing decisions; these, in turn, begot more and more formalized mechanisms establishing standards and verifying claims that one company or product was preferable over another (Bartley 2007b). The high-water mark of this early interest included the publication and wide distribution of the Council on Economic Priorities' book, *Shopping for a Better World*. Within five years after its release in 1988, it sold over four million copies (Dickenson 1993).

Behind the enthusiasm, however, a great deal of academic work has examined and raised questions about the extent to which consumer values can lead to significant market demand, given the understood tendency of stated preferences (i.e., willingness to pay) to exceed those preferences revealed by purchasing behavior (Loureiro et al. 2001; Fuerst and McAllister 2011) and the different norms and institutional context in which citizen voters versus citizen consumers operate (Green 1992; Sunstein 1993). Moreover, there is an understandable tension between using consumer power as a vehicle for addressing environmental problems, when levels of consumption are partly a cause of the problems private regulations seek to address. Making products appear more ethical can have the unintended effect of justifying more consumption rather than less (Horne 2009).

Second, citizens are a focus for the support they provide to NGOs via donations or through public opinion. Here, instead of directly supporting private regulation through consumption, citizens are an indirect source of legitimacy and, related, they provide funds that are important for the operation of some groups that promote private regulation. Put another way, they provide legitimacy to these organizations since citizen backing helps build the perception of popular support for a set of values and activities (Cashore 2002). Businesses may also be concerned about the general perception of citizens for reputational reasons, which can include a fear that they will lose their social license to operate (Gunningham et al. 2003).

Third, citizen as a rights holder draws from the idea that the information that private regulators produce should be rightfully disclosed to citizens. Here the act of disclosure serves a normative end. It is not merely an instrumental disclosure of information with the aim of changing behavior. Rather, private regulators generate information that should be available to citizens. What citizens do with this information is less material. The "right to know" has been most prominent as a focus for food-related labels particularly in debates about labeling products that contain genetically modified organisms (Degnan 1997).[1]

Aside from citizens, the role of individuals within private regulatory initiatives has received less explicit attention. An exception has been the role they may play as entrepreneurs in the development of programs and the diffusion of private regulations across different sectors. These roles, however, have typically been analyzed through an organizational lens (see Bartley and Smith 2010), not based on networks among individuals divorced from their organizational affiliations, even though anecdotal evidence indicates that some individuals have played pivotal roles as they have moved from one organizational affiliation to another.

---

[1] With each of these citizen roles, a major concern has been the limited capacity of individuals to sort through a confusing and potentially conflicting array of product labels. For discussion of this concern with U.S. public labels, see Noah (1994).

Meso Processes

Groups, networks, coalitions, and movements constitute a second key analytic focal point for the study of private regulations. At the most general level, this analytic focus is captured by those who seek to understand how the interactions of actors shape the rules of the game that facilitate and affect market activities (Overdevest 2010; Pattberg 2007).

Two general perspectives dominate the understanding of these interactions. One places firms and the market at the center of the analysis. Bartley (2007b), examining the rise of labor and forest certification programs, termed this the "market-based approach." It, according to Bartley (2007b), centers on three proposals:

1. Private regulations are designed to protect the reputation of companies.
2. Companies are willing to create private regulations that include external verification in order to provide credible information to consumers, something they struggle to do alone.
3. Private regulation can be a tool for companies to limit competition (Bartley 2007b).

Others add the shadow of hierarchy: companies self-regulate to preempt or fend off the possibility of government intervention. For all this work, companies are the central unit of analysis. It is their calculus—reacting to government, competitive pressures from other companies, and pressure from NGOs and broader public interests and consumer demand—which shapes the character of private regulations. These labels are one more instrument available to individual companies and collectives of companies to facilitate the maintenance of their autonomy and ability to seek profit (Bartley 2007b).

Another perspective broadens the group of relevant actors at the meso level. For Bartley (2007b), the institutionally embedded political struggles among states, companies, NGOs, and social movements are what matter to a "political-institutional" perspective.[2] Many authors have examined private regulations in ways that align with this perspective. Dingwerth and Pattberg (2009) argue that an organizational field of transnational rule making formed from 1990 onward as a product of the social interactions of leading private regulator initiatives, such as the FSC and the International Federation of Organic Agriculture Movements (IFOAM). From here, isomorphic pressures worked to drive convergence in the general organizational form of private regulators that seek to govern different policy problems. Providing much greater detail, Bartley (2007a) demonstrates the role of philanthropic foundations as institutional entrepreneurs: they did more than channel the activities of social

---

[2] He goes on to explain: "The driving forces in this account are conflicts among states, NGOs, social movements, and [companies] about the legitimacy of various ways of regulating global capitalism" (Bartley 2007b:309–310).

movements (or NGOs) toward more professionalism or less radicalism, but rather used the latter as leverage to help create clearer incentives for compliance with the emerging private rule-making bodies. Funds were, on one hand, provided to support the development of the FSC, MSC, and other programs, and, on the other, to campaigning organizations, such as ForestEthics, which were targeting large, brand-sensitive companies near the end consumer to develop purchasing policies which demanded certification from their suppliers of wood and paper products, among others (Conroy 2006; Sasser 2003). It has been these corporate campaigns that have supplanted much attention to consumer demand for the labels of private regulators, as, the logic proposes, if large corporate buyers make company-wide commitments to certified products, the decision of the end consumer will be moot (Gulbrandsen 2006). In this way, economic integration, as discussed by Seto and Reenberg (this volume), has been a necessary condition for a key process generating compliance incentives for private regulation. It is this integration that has made the campaigns of transnational coalitions much more potent.

Within this vein of work, others have examined these meso-level interactions to assess how industry and NGOs interact to affect whether private regulators can achieve political legitimacy (Bernstein and Cashore 2007), why support for private regulations vary across companies (Sasser et al. 2006), across countries (Gulbrandsen 2005; Espach 2006; Cashore et al. 2004) as well as across industrial sectors (Espach 2005), and the implications of support for the balance of power among domestic interests (Ponte 2008) or existing domestic regulations (Besky 2011). To date, however, fewer studies have examined the role of other actors, such as consultants, certifiers, academic institutions, or the role of epistemic communities. In addition, few works have conceptualized how the boundaries of groups operating at the meso level have effects on how the political-institutional approach understands the rise and evolution of private regulators, in general, and for land-use governance specifically. For instance, professional groupings, such as planners, agronomists, architects, engineers, and foresters, are agents within the development of private regulatory processes as much as they are constraints defining the boundaries of a sector and the appropriate policy focus of any given private regulator.

Macro Processes

Both meso- and micro-level interactions are generally seen to operate within broad institutionalized rules, norms, cultural scripts, and ideas (Bartley 2007b; Dingwerth and Pattberg 2009; Bernstein and Cashore 2007). Virtually all of this work notes the importance of neoliberal institutions and ideas as being causally related to the rise of private regulation (Bartley 2003). A key interest of this perspective is then to understand whether private regulation, once in place, can have transformative effects on the very same macro processes that led it into being. A number of scholars, for instance, have approached private regulation with an interest in whether this form of governance can serve as part of a Polyanian countermovement to re-embed markets, offsetting the negative effects of neoliberal globalization (Mutersbaugh 2005; Taylor 2005; Guthman 2007; Bernstein and Cashore 2007; Fridell 2007; Klooster 2010). Perspectives on this question are mixed, and most scholars conclude that it is too early to know what the ultimate potential of private regulation is or can be.

Uptake and Potential of Private Market-Based Regulations

By many measures, it is clear that private regulations have been on the rise over the past several decades, particularly those programs that seek to address social and environmental concerns with production processes (Bartley 2007b; Dingwerth and Pattberg 2009; Abbott and Snidal 2009). One measure is offered by the recently launched eco-labeling index project, which collects information on eco-labels from around the world.[3] By its count, there are over 430 eco-labels operating worldwide. This is likely to be an underestimate, due to the difficulty in surveying labels confined to single countries. However, it could also overestimate the number of private regulators by some degree, as many of the eco-labels are government led. In additional, many of the labels focus on common concerns or standards. For instance, over fifty of the eco-labels reported using some form of private or government organic standard. Further programs also build from organics, using it as a component in a multi-dimensional standard that specifies organics as one requirement for a certified process or product.

Only some of the private regulators behind these labels are relevant to land-use governance. Figure 12.1 provides a simple schematic of pressures on land use mapped onto a stylized product life cycle. The narrow arrows represent the stages of the stylized life cycle. The four thick arrows represent separate sources of pressure on land use exerted by a product's life cycle. Arrow "A" represents the effects on land use of extraction and production practices as well as the provision of services. Arrow "B" captures the direct effects of consumption via the purchase of products or services that have implications for land use. For instance, someone traveling to a tourism destination will exert impacts through the use of fuel and due to the direct effects they have on the geographic place they are visiting (e.g., displacing turtles from nesting beaches). Arrow "C" represents the implications of recycling practices on land use. This may be associated with the location of recycling facilities, the inputs they use, and the outputs they create. Finally, arrow "D" captures disposal and waste management activities.

In practice, these are not independent sources of pressure. Still, they are useful to separate for the purpose of this chapter, since they highlight how

---

[3] See www.ecolabelindex.com.

Private Market-Based Regulations 229

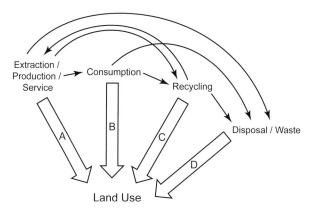

Figure 12.1 Simple conceptual diagram of a product life cycle that moves through the stages of extraction, production, and service provision to consumption before going to recycling and/or disposal and waste. The thick arrows labeled A, B, C, and D capture the sources of pressure which these product life-cycle stages exert on land use.

private regulations must find ways to govern this system as a whole as well as the individual sources of pressure.

Table 12.3 reviews the uptake of private regulation and provides a look at how different private regulators affect the four sources of pressure captured by arrows A, B, C, and D. An important caveat is that the table is neither a tally of all private regulatory programs nor is it a representative sample. Hence, it is difficult to draw concrete conclusions about how the field of private regulators relates to pressures on land use. Still, it does point to some important research questions about the additive effects of private regulators, which I will return to below.

The uptake of private regulations has been varied across programs, sectors, and countries. Some programs have gained sizable toeholds in individual sectors. In the forest sector, the FSC and Program for the Endorsement of Forest Certification (PEFC) together have certified 10% of the world's forest area, which is estimated to produce 26.5% of the global industrial roundwood supply, or 469 million cubic meters (UNECE and FAO 2012). The Better Sugarcane Initiative (BSI) reports that certified sugarcane now covers 1.67% of the world's surface. Based on FAO data, sugarcane annual harvests comprise around 24 million hectare (Mha), so if the BSI data is to be believed, they have certified more sugarcane farmland than is in annual production, which means that we can assume that the crops operate on some rotational basis. In coffee, Utz Certified has more than doubled the quantity of coffee certified by the program between 2008 and 2011, but it still represents around 1.7% of world coffee production (Auld 2010a).[4] What Table 12.3 does not capture, and

---

[4] World coffee production totaled 7,875,180 metric tons in 2011/2012, http://www.ico.org/historical/2010-19/PDF/TOTPRODUCTION.pdf

Table 12.3 Sample programs (and the date they were established) that identify how private regulation may affect the pressures on land use coming from the extraction and production of goods and services (Arrow A in Figure 12.1), the consumption of goods and services (Arrow B), recycling processes (Arrow C), and disposal and waste processes (Arrows D). Data from Bartley (2007b); Riisgaard (2009); Auld et al. (2010); Auld (2010a, 2011); the eco-label index (www.ecolabelindex.com); and program websites.

Sample Programs (date)	Uptake	Potential Regulation of Pressure on Land Use
FSC (1993)	163 Mha forestland certified	Arrows A and C; encourages improvements in the extraction or production of forest products, with some attention to encouraging recycling
PEFC (1999)	240 Mha forestland certified	Arrows A and C; encourages improvements in the extraction or production of forest products, with some attention to encouraging recycling
BSI (2008)	30 Mha of sugarcane production certified	Arrows A and D; encourages improved production of sugarcane, including attention to waste effluents from production facilities
Utz Certified (1997)	Cocoa: 42,704 metric tons sold Tea: 48,142 metric tons sold Coffee: 136,752 metric tons sold*	Arrows A and D; encourages improved production of coffee, including attention to waste effluent from processing facilities
BCI (2008)	n/a	Arrows A and D; encourages improved production of cotton, including attention to waste runoff from farms
GlobalG.A.P. (1997)	Area of top 10 cover covered crops (e.g., tomatoes): 123,000 ha Area of open field crops (e.g., bananas): 2.7 Mha*	Arrows A, C, and D; encourages "good" agricultural production for various crops, including a focus on recycling, waste, and disposal practices
RJC (2005)	Member sales of USD 45 billion, with 39% of members certified*	Arrows A, C, and D; encourages improved mining extraction/production, with attention to recycling and waste disposal in processing facilities
BPI (1999)	132 certified compostable products	Arrows A and D; encourages the production of compostable plastic and fiber-based products

Table 12.3 *continued*

Sample Programs (date)	Uptake	Potential Regulation of Pressure on Land Use
USGBC (1998)	159,000 registered and certified projects; certified buildings cover 2 billion square feet	Arrows A, B, and C; encourages building design and construction that considers environmental effects of materials, location, and usage of building
BREEAM (1990)	245,000 buildings certified, mostly in the U.K.	Arrows A, B, and C; encourages building design and construction that considers environmental effects of materials and usage of building
Carbon Trust (2011)	Certified 600 organizations with combined emission reduction of 5.5 metric tons	Arrows A, C, and D; encourages GHG emissions accounting and emission reductions from extraction, production, services, recycling, and disposal
VCS (2005)	873 registered projects; 101 million metric tons GHG emissions	Arrows A, C, and D; encourages GHG emission reductions from extraction, production, services, recycling, and disposal
Earth Check (1997)	1,300 members involved	Arrow A; encourages improved practices of tourism services and operators with a focus on GHG emissions and use of natural resources
e-Stewards (2003)	3 recycling companies certified, 12 in the process**	Arrows C and D; encourages improved disposal and recycling of electronic products
EPEAT (2006)	47 participating manufacturers	Arrows B, C, and D; encourages electronic product designs that are more energy efficient in use, easier to recycle, and less harmful when disposed
BEST (2006)	n/a	Arrows A and C; encourages improved production practices for batteries and appropriate recycling to reduce emissions.

* Data for 2011; ** Data for 2010

Abbreviations: BCI: Better Cotton Initiative; BEST: Best Environmental Sustainability Targets; BPI: Biodegradable Products Institute; BSI: Better Sugarcane Initiative; FSC: Forest Stewardship Council; GHG: greenhouse gas; PEFC: Program for the Endorsement of Forest Certification; RJC: Responsible Jewellery Council; USGBC: U.S. Green Building Council; VCS: Verified Carbon Standard what is discussed further below, is the spatial location and scale of uptake relative to the pressures on land use.

Table 12.3 also illustrates the ways different private regulators seek to affect the different pressures on land use. Though it is difficult to generalize, for the reasons noted above, two patterns deserve further research attention. First, returning to the point about the rebound effect identified by those interested in citizens as consumers, consumption as a driver of land use is something that private regulations only deal with indirectly. They seek to promote consumption that internalizes environmental and social considerations. They do not deal with levels of consumption as a problem. The U.S. Green Building Council (USGBC) and BREEAM (an environmental assessment method and rating system for buildings) do seek to promote reduced impacts from consumption (e.g., by making buildings more efficient). However, these are exceptions rather than the rule, and they may be susceptible to the rebound effect, whereby knowledge that consumption has reduced effects leads to higher levels of consumption. Second, the coverage of the pathways of influence appears incomplete and unevenly distributed. Moreover, there are apparently many ways in which the problems addressed by one private regulator may interact with problems addressed by other private regulators; the actions of one private regulator may have positive or negative spillover effects for the actions of other private regulators (Auld 2013). Neither of these problems has been adequately addressed, given the patchwork of programs that have so far emerged to contribute to land-use governance. This second issue will be discussed in greater detail below.

Problem of an Incomplete Private Regulatory Field

Linking the previous discussion back to Seto and Reenberg (this volume), two issues connect directly to their framework, and one extends beyond it.

Problems of Scale and Rate of Uptake

The data on uptake in Table 12.3 provides some indication of the scale to which private regulation has begun to make inroads in various economic sectors. The growth of some programs has exceeded expectations. However, in the forest sector, the current rate of certification means that it will take 80 years for 50% of the world's forests to be certified (UNECE and FAO 2012). For other sectors, private regulation has oversight over an even smaller fraction of the total production and will likely grow in prominence at an equally slow rate.

These sectors, however, do not stand still. For instance, in coffee, Vietnam rose from an insignificant producer in the early 1980s to overtake Colombia as the second largest producer by 1999. In the United States, major forest product companies, such as International Paper, sold off sizable areas of industrial forestlands in the last decade, shifting their role in the forest product supply chain in a way that has potential implications for the operation of private regulations. Rising oil prices, likewise, had implications for agricultural production practices, the forest industry, and many other sectors. The challenge, therefore, is for private regulation not only to grow to a scale commensurate with the global challenges of land-use governance, but also to keep pace with the rate of change in various economic sectors. The limited uptake also points to the need to consider carefully how private regulation interacts with the rules set by governments to govern land use within their territory.

Problems of Spatial Location of Uptake

In addition to the scale and rate of uptake, the spatial location of where private regulation gets adopted has received considerable attention and has raised notable concerns for those interested in the problem-oriented effects of private regulations (Ebeling and Yasué 2009; Ponte 2008; Vandergeest 2007; Thornber et al. 1999, unpublished). Though the founders of private regulatory programs may not have realized the degree to which this would be a challenge, there have been strong self-selection effects at play due to differences across countries in the levels and enforcement of laws (Cashore et al. 2006), technical know-how (Pattberg 2006), and access to international networks and markets (Nigh 1997; Mendez 2008). In addition, larger, vertically integrated companies also benefit from lower per unit costs of being independently audited to prove adherence to private regulatory rules (Cashore et al. 2004). Put another way, those operators that face the least cost to achieve the requirements of a voluntary private regulatory program are the most likely to participate, all else being equal.

As a consequence, the adoption of certification has been skewed toward those countries and operators which do not necessarily pose the most concern for the problems of a given sector. In forestry, for instance, only 2% of the forestlands in tropical regions have been certified, whereas in Western Europe 56.7% of the forestland has been certified and in North America 32.2% has been certified (UNECE and FAO 2012). Previous analyses have shown that those countries reported by FAO to have had annual losses of forests have participated very little in certification, even though the FSC had made slightly further inroads in these countries than the PEFC when these analyses were conducted (Auld et al. 2008; Gullison 2003).

These spatial patterns pose two main concerns. First, and again, they indicate that private regulation alone may struggle to address the core problems that face individual sources of pressure on land use, as identified in Figure 12.1. Second, they may create unintended negative consequences for equity by erecting barriers to entry for producers in developing countries (Delzeit and Holm-Müller 2009; Pattberg 2006; Fuchs et al. 2009).

Problems of Institutional Fit of Programs with Land-Use Issues

Aside from the questions of uptake, little attention has been placed on how private regulators fit with broad governance problems, such as land use. Private regulatory programs focus on many different problems, and do so by targeting different actors within global supply chains, from the primary production stage through to recycling and disposal. Figure 12.1 and Table 12.3 illustrate that the overlay of these programs onto stylized sources of pressure on land use is patchy, and that additive affects—those coming from the interactions of different problems or policy initiatives of different governance initiatives—are not necessarily tackled directly. That is, there are no private regulatory programs that address all of the factors affecting land use, which means that to address these problems systematically, coordination among different private regulators will be necessary.

One way to think about this challenge is to consider a trade-off between specialization (i.e., focusing on a narrow policy problem) and comprehensiveness (i.e., trying to address all the causes of a problem and their interactions) (Auld 2013). Specialization may be useful where additive effects are small and where the relative costs of coordinating across programs is lower than the administrative costs involved in establishing and running a comprehensive program. Hence, the policy focus of different private regulators will affect how they act individually and in concert to address land-use governance challenges.

Programs appear to be aware of the challenge this presents and have been developing individual strategies and coordination strategies to help address interactive effects (Auld 2013). First, private regulators can change the scope of their rules to encompass new problems or include new eligibility requirements so more operators can participate. The Fairtrade Labeling Organization and IFOAM each followed this strategy by adding environmental and social criteria to their respective standards. Second, private regulators can modularize their requirements so as to apply them to new products or sectors. By doing so, these regulators may be better able to regulate farm systems or land-use practices that involve more than one crop or activity. A number of initiatives have taken this approach. GlobalG.A.P., for instance, has various standards that cover different scopes, some general and some very specific. Within crop-based farming, it has thus far developed standards for fruits and vegetables, combinable crops, coffee, tea, and flowers and ornamentals; within livestock, it has pigs, poultry, turkey, and ruminants, which are even further broken down into cattle and sheep, diary, and calf/young beef. GlobalG.A.P. also has general standards for transportation and feed manufacturing.[5] The USGBC has taken a similar modular approach by expanding to different forms of development and building use, including new construction of commercial and institutional

---

[5] http://www.globalgap.org/cms/front_content.php?idcat=176

buildings, commercial interiors, schools, retail, health care, residential homes, and neighborhood development.[6]

Three additional strategies require private regulators to work together. The first involves one-on-one partnerships to address some common challenge or problem. Starting in 2009, for instance, the FSC and Fairtrade International began working on a joint initiative to apply Fairtrade standards to FSC-certified forestry operators; in doing so, the programs addressed the perceived problem that small forest owners faced in fairly competing in global timber markets. Second, programs can work together through coordination mechanisms to address mutual areas of concern. The International Social and Environmental Accreditation and Labeling (ISEAL) Alliance, formed in 2002, brings together a group of certification initiatives to advance their common interests. Some of its codes of good practice provide guidance on how programs manage overlap or interactions among their areas of policy focus. The third strategy involves the creation of a supra-sectoral mechanism, which develops new criteria and draws on existing programs to address certain environmental or social performance criteria (Auld 2013). The USGBC LEED standard has done this to an extent by using the FSC as its benchmark for appropriately managed forest products; however, more efforts in this direction appear to be a clear challenge at the global level, for private regulation to have a greater ability to address the various individual and interactive pressures on land use.

This last issue raises a key point: Governance approaches for addressing a broad challenge such as land use will need to develop ways to balance the benefits of specialized attention to particular pressures on land use while also attending to the interactions among all the different sources of pressure.

Conclusions

Private regulation comes in many forms with the intention of addressing different policy problems. In this chapter I have provided an overview of the characteristics of private regulators as well as the analytic perspectives, from micro to macro, that seek to explain the rise and dynamics of these institutions. In addition, I have sketched, albeit in a preliminary fashion, how various private regulators map on to four different pressures of land use and discussed three key problems that result from an incomplete field of private regulators. In their multidimensional framework, Seto and Reenberg (this volume) focused on the importance of scale, rate, and place. This lens sheds light on the first two challenges; that is, the mismatch between the scale and rate of the processes of land use and the scale and rate at which private regulation is being adopted and exerting influence over land-use practices. Added to this is the challenge that land-use practices are not stable. Hence, private regulators need to both

---

[6] http://www.usgbc.org/DisplayPage.aspx?CMSPageID=222

increase the scale of their effects as well as remain adaptable to address the changing geography and agents of land use.

The second problem involves the spatial mismatch between where private regulations have gained the greatest inroads and where the greatest land-use practice concerns reside. Those operators that face the least cost to achieve the requirements of a voluntary private regulatory program are the most likely to participate, all else being equal. This means that private regulation alone may struggle to address the core problems facing individual pressures on land use. In addition, private regulations may create unintended negative consequences for equity by erecting barriers to entry for producers in developing countries.

Finally, and beyond Seto and Reenberg (this volume), the analysis in this chapter emphasized the importance of institutional fit. Land-use governance is a broad challenge that extends well beyond the policy focus of any given private regulator. Hence, for private regulation to attend to this challenge, coordination and cooperation are necessary to address influences on the land at the intersections of economic sectors and the stages of global supply chains.

Acknowledgments

This chapter benefited from many useful comments from the participants at the Ernst Strüngmann Forum in Frankfurt, Germany. The author also thanks two referees for their useful and constructive comments, as well as Constance McDermott and Karen Hebert for their early advice and input. Thanks are also due to Benjamin Cashore, Jessica F. Green, Stefan Renckens, Luc Fransen, Lars Gulbrandsen, and Jennifer McKee for their ongoing research collaborations on private regulations, all of which have informed this chapter.

First column (top to bottom): Brad Gentry, Heike Schroeder, Group discussions, Heike Schroeder, Graeme Auld, Anthony Bebbington, and Caroline Upton
Second column: Thomas Sikor, Tor Benjaminsen, Brad Gentry, Caroline Upton, Anne-Marie Izac, and Matias Margulis
Third column: Carol Hunsberger, Matias Margulis, Graeme Auld, Tobias Plieninger, Thomas Sikor, Tobias Plieninger, and Anne-Marie Izac

13

Changes in Land-Use Governance in an Urban Era

Bradford S. Gentry, Thomas Sikor, Graeme Auld,
Anthony J. Bebbington, Tor A. Benjaminsen,
Carol A. Hunsberger, Anne-Marie Izac, Matias E. Margulis,
Tobias Plieninger, Heike Schroeder, and Caroline Upton

Abstract

Land use is being fundamentally transformed worldwide. Governance mechanisms that manage land use are changing from territorial organizations to global institutions anchored to specific resource flows between urban and rural areas. This shift reflects an underlying change of values attached to land, from the creation of new monetary values to the assertion of social values. Such a revalorization has, in turn, fueled global competition and led to governance arrangements that may appear fragmented from the vantage point of any particular land plot. In addition, rising urbanization impacts and reflects governance arrangements for land use. This chapter addresses the governance of land use in an urban era, with a focus on the emergence of global arrangements to address land competition and the telecoupling effects that arise between coupled multiscalar systems.

Introduction

Land use is undergoing fundamental transformations worldwide (see Seto and Reenberg, this volume). It is becoming a global issue in that land-use changes in particular places tend to originate from distant locations or are driven by actors operating around the world. As a result, new global-level governance initiatives seek to address land-use changes worldwide. Urbanization is increasingly influencing these transformations through the demands for resources and environmental amenities articulated by urban consumers, the involvement of urban-based actors in rural production operations, and the changing lifestyles of urban residents. New actors have emerged around the world, extending from private financial companies to civil society organizations.

In this chapter, discussion focuses on the governance of land use, with a particular interest in the emergence of global arrangements. Governance is about collective decision making; that is, the rules that guide collective decisions on land use, as well as the actual decisions. Governance plays an important role in the transformations of land use, because collective decision making influences these transformations and because changes in land use affect governance arrangements. Decision-making procedures mediate the globalization of land use, the influence of urbanization, and the emergence of new actors.

The key proposition advanced in this chapter is that land governance has expanded from its traditional focus on territorial governance, such as through nation-states and local communities, to include more global institutions anchored to specific flows of resources between urban and more rural areas, as illustrated by timber exports from forests certified under the Forest Stewardship Council's forest management standards. This shift in global land governance has come about as a consequence of underlying changes in the values attached to land, including the creation of new monetary values (e.g., through forest carbon markets) as well as the assertion of social values (e.g., in the form of biodiversity habitats, aesthetic beauty, recreational potential, territorial sovereignty, and so on).

These global revalorizations of land have fueled global competition over land and led to governance arrangements that look fragmented from the vantage point of any particular plot of land. Land use is no longer under one single territorial institution—if it ever was—but rather the subject of multiple, flow-anchored governance arrangements. This suggests that instead of pinning its hope on the establishment of all-encompassing land institutions at the global level, future land governance needs to accomodate a central role for flow-related, functionally differentiated governance arrangements, while still working in combination with territorially based institutions.

The rising intensity of land competition and its increasing recognition as a global issue puts the spotlight on the global-level governance arrangements available to respond to this competition, as well as on ways to improve such arrangements in the future. Governance plays an important role in "telecoupling" because it helps to transmit—or prevent transmission of—signals between coupled multiscalar systems. Finally, the ascendance of urbanity affects and simultaneously reflects governance arrangements, as the latter helps connect urban centers to other areas and is linked to urban lifestyles and forms of decision making.

This chapter builds on the discussions among the working group at this Ernst Strüngmann Forum. We began by focusing on five questions, covering a range of current issues and perspectives around the governance of land use:

1. How do we define "land grabbing"?
2. Given the multiplicity of actors that affect land use at different scales, what governance arrangements are emerging?

3. How is the production and use of knowledge affected by and how does it affect different land-use governance structures?
4. How are the relationships among individual and collective interests in land across scales changing?
5. How will moves toward a "green economy" affect land use and its governance?

To analyze changes in land-use governance that are currently underway, we developed a conceptual framework (described in detail below). We used the framework to analyze several illustrative types of land use in an effort to identify areas that appear ripe for future research.

Our discussion on governance connects to three core premises of this Forum:

1. Competition for access to land has increased over recent decades and will continue to intensify for the foreseeable future.
2. Much of this competition is driven by nonlocal sources of demand, involves nonlocal actors, and transforms land use in ways that can have collateral and often nonlocal effects.
3. These shifts in demand for land and its potential products are largely driven by changes in lifestyle and resource use that derive from changing patterns of urbanization.

Taken together, these premises suggest that substantial changes are underway in the processes and flows that link territories across space, as well as in the actual dynamics of land use within given territories. This, in turn, demands that the research community first seeks to understand the factors that have made these new governance processes possible. Other chapters in this book point to some of these factors: new levels and sources of demand, new trade relationships, new flows of investment capital, etc. However, each of these factors only becomes effective through particular governance relationships: relationships that transmit demand to certain lands and not others; relationships that make new trade and investment relationships possible and that define the rules of these trade relationships; tenure and policy arrangements that make land available for new forms of use and investment.

A second set of governance questions relates to how current competition over land is being governed. Once again, there are many possible avenues of enquiry. Land-use competition might be governed through changes in tenure arrangements, the imposition of policy edicts, or through the establishment of new settlements that are arrived at via processes that can range from negotiation and dialog to the use of force. Indeed, competition and conflict are frequently different sides of the same coin, and many of the processes outlined by Wicke, Margulis, Hunsberger et al., and Auld (all this volume) involve conflict that arises as different stakeholders seek to secure their access to the same land, water, and forest, or seek redress and compensation if they lose such access.

A third area of inquiry concerns how the collateral impacts of these changes in land uses are also being governed. When prior land uses and livelihoods are displaced, or when land-use change produces new environmental impacts *in situ* or downstream, what rules, regulations, authorities, relationships of power, and decision-making processes determine and contain the final effects of these displacements and impacts? If these collateral impacts are not being governed in any formal sense, the question remains: What broader governance arrangements may lead to a situation in which such impacts are not regulated?

The challenges laid down by the premises that underlie this book are, at their core, ones of governance. These challenges must be addressed by research (i.e., to understand what has happened and what is happening) as well as policy (i.e., to determine the proper response).

Conceptual Framework

Figure 13.1 illustrates the conceptual framework that emerged from our discussions of changes in global land governance. We address these changes in relation to revalorizations of land, differences in knowledge about land, and changes in politics, as well as local and national dynamics. These, in turn, are negotiated by differentially positioned social actors, as will be discussed below.

Land use involves a series of social actors in different positions. The positions of these actors differ in the levels of assets they have, including natural, financial, social, human, and physical assets. Natural assets of critical importance concern not only the size and quality of land in the actors' possession, but also access to complementary natural resources, such as water. Important financial assets include personal wealth and access to credit and other sources of finance. Social assets with particular relevance to land are the networks

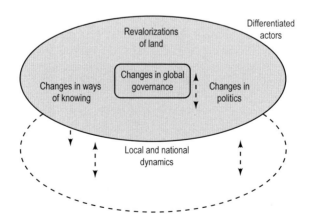

Figure 13.1 Conceptual framework of factors that effect changes in global governance of land.

available to actors as well as the social identities attributed to or claimed by them, including their power to influence the decisions of other social actors. Human assets refer to educational attainment, technical skills, and more broadly the significance of actors' knowledge. Physical assets include machinery and other similar capital stocks.

Governance emerges from the interactions among these differentiated actors. It includes not only formal organizations, such as elected governments or international bodies, but a variety of collective decision-making procedures. For example, private market-based agreements are part of global land governance (see Auld, this volume). Voluntary guidelines for private investments are equally part of global land governance, as are the social movements organizing against "the global land grab" (see chapters by Margulis and Hunsberger et al., this volume). The rules governing collective decision making may not even be formalized in written agreements or institutionalized in specific organizations, as illustrated by the routinized rituals (e.g., demonstrations, blockages) enacted by social activists and host governments on the occasion of World Trade Organization (WTO) negotiations or G8 summits.

A key driver of changes in global land governance are the global revalorizations of land currently underway. The term "revalorization" refers to changes in the values that social actors attribute to land, including the emergence of new values and dimensions of value, loss of previous values, as well as increases and decreases in existing values. Values may be expressed monetarily, allowing a quantitative ranking. However, it also refers to nonmonetary aspects of value, in the sense of what forms of land use are considered desirable and good. For example, a particular revalorization may take the form of emphasizing the contributions of land use for the conservation of biodiversity over its productive potential for agriculture. Revalorizations thus highlight some of the values attributed to land (e.g., biodiversity, productive potential), making some more visible than others.

Revalorizations of land work in close association with "ways of knowing" about land; that is, how actors understand the relationships between nature and society, as well as the types of nature-society relations that they consider desirable. Ways of knowing frame the generation of particular knowledge about land and, more broadly, the relations between nature and society that are pursued in practice. For example, the ecosystem services framework is informed by particular understandings of how nature operates—a predominantly anthropocentric, instrumental perspective which focuses on nature as a provider of services for human well-being. This view of nature-society relations is radically different from one that conceives of nature as a complex, constantly evolving system, of which humans are but one part.

Revalorizations of land also operate in tandem with politics. Politics concerns the material and discursive practices employed by socially differentiated actors to achieve their own objectives by influencing other actors. It takes place in formal venues (e.g., global negotiations between nation-states) and informal spaces (e.g., dinner meetings between government officials and lobbyists). A suitable example comes from the UNFCCC Conferences of Parties, where political negotiations are not limited to the negotiation room and party delegations, but rather extend into the corridors and squares populated by activists, scientists, and others. Recent work on the Convention on Biological Diversity offers similar insights (e.g., see Corson and MacDonald 2012).

The last, but equally important, category in our conceptual framework is "local and national dynamics." By including this category in our framework, we want to highlight the close interactions between the dynamics of land use, governance, ways of knowing, and politics across scales. For example, global governance arrangements reflect both the influence of collective decisions, land-use practices, particular knowledges, and political negotiations at the local level, and they affect them in return. Revalorizations do not operate at the global scale in the abstract, but are based on changing practices at local and national scales. An illustration of this comes from the "Principles on Responsible Investment in Agriculture" promoted by the World Bank and G8 (World Bank et al. 2010). These principles do not originate merely from global-level decisions within the World Bank or in relation with other global-level actors; they reflect, among others, the influence of local-level land-use decisions and dispossessions and local- to national-level social mobilizations and regulations worldwide.

Next we turn to a deeper, but by no means complete, examination of the different components of this conceptual framework, drawing extensively on our discussion of the five questions listed in the Introduction.

Components of the Conceptual Framework

Differentiated Actors

Many actors—on local, national and global scales, from the public, not-for-profit, and for-profit sectors—have shown increased interest in land-use decisions. While there is some debate about how "new" this phenomenon is, awareness of and engagement by groups beyond the local level (e.g., owners, mayors, managers, workers, residents) is certainly heightened in the following groups:

- National governments, as owners or developers of land within their own boundaries, acquirers of control over productive land in other countries, and bilateral funders of capacity-building projects in other countries.
- Global private sector actors: from private equity investors seeking opportunities in "real assets" (land) around the world to multinational

companies that manage their global supply chains, as well as real estate developers and designers who erect new buildings in urban areas.
- Global social movements and nongovernmental organizations (NGOs), as they work to protect underrepresented interests around the world, including both the local human communities affected or threatened by the actions of the national and private interests described above, as well as the natural, nonhuman communities that are also affected or threatened by human activities.
- International governmental bodies, which offer forums for resolving conflicts or for developing standards to be followed to reduce conflicts about appropriate land use across the actors described above.
- A wide range of other potential actors: from those pursuing the regulation of the global financial sector, to those promoting the Olympic Games or the World Cup.

The difficulties facing efforts to track and understand these multiple actors are compounded by the fact that actors' interests in particular pieces of land or even land with particular attributes vary over time—with changes in technologies, markets, prices, and other relevant factors—and across scales (see Table 13.1). These changes in interest are reflected in changes in their relationships/configurations on particular governance issues.

Multilevel networks are used to connect the work of national and global level actors to on-the-ground decisions affecting land use (e.g., from a bilateral donor to a global NGO operating in a particular country to local governments or landowners/users). Shared outcomes or goals are negotiated and pursued along and between such networks. The enhanced "telecouplings" made possible by the vast improvements in digital communication have facilitated the rapid expansion of these networks.

A central goal in the mapping of such social networks is to identify the key "brokers" of conflicts in or between the different interests in a piece of land. So too is finding the creative agents who are asking the "what if" questions about how land use might best be enhanced or conflicts over land use might best be addressed. For a detailed discussion of the interaction between new actors and forms of political agency, see Hunsberger et al. (this volume).

Changes in Global Governance

Global governance has emerged as an important research focus in recent decades. It is a broadly encompassing term that is not confined to the global scale. Rather, it refers to "systems of rule at all levels of human activity—from the family to the international organization—in which the pursuit of goals through the exercise of control has transnational repercussions" (Rosenau 1995:13).

Governance does not equal government, and global does not simply mean supranational institutions. Governance can involve multiple

Table 13.1 Examples of actor involvement in agriculture and their spatial scales of influence.

Actors	Spatial scale of influence
Individual farmers	Farm and village lands
Village heads/chiefs and communities	Land under village "control"
Provincial/local state governments	Land throughout the province/state

NGOs from various sectors:
  Agriculture
  Environment (water, biodiversity)
  Forestry
  Community organizations
  Indigenous rights
  Fair trade, eco-certification, etc.
	Local/farm to province/national/international

Scientists:
  National research institutes
  International research institutes
  Research institutes in the North involved in projects in developing countries
  South–South cooperative research
	Local to global

Farmers' organizations	Local to national
National governments (i.e., ministries of agriculture, forestry, fisheries, the environment, and finance)	Local (?) to national

Transnational production companies:
  Voluntary agreements
  Land grabs
  Investments in large-scale plantations
	Local to global

International public and private financial sector:
  Futures markets for agricultural commodities
  Carbon markets
  Debt for nature swaps
  International systems of agricultural subsidies
  Other related activities
	Local to global

International public organizations:
  FAO
  IMF
  World Bank
  WTO
  G20
  Conventions and treaties (e.g., Convention on Biological Diversity, International Treaty on Plant Genetic Resources)
	National to global

actors—governments, businesses, nonprofit organizations, social movements, and others—operating across scales that set rules to control and steer their own behavior and/or the behavior of others (Rhodes 1996).

Increasingly, global governance involves more than formal rules negotiated by government. Private actors alone (such as when firms dominate a particular sector and enjoy market power) and through multistakeholder processes (e.g., private standards, industry codes of conduct) are becoming rule-setters in many economic sectors. They are operating alongside, and interacting with, the intergovernmental processes of the United Nations.

Following these definitions and applying them to the field of land-change science, we suggest that there is something that can be identified as global land governance. Global land governance can be said to consist of all actors across scales that contribute to or influence the rules that are set, including both public and private rules, formal rules (such as binding intentional law), and soft, voluntary arrangements (such as certification).

One implication of this expansive definition is that scholars seeking to produce an empirical map of the actors and rules involved in the global governance of land will have to look beyond what is self-evident and recognized as global land governance. For example, they will have to look across the "Voluntary Guidelines of Responsible Land Tenure, Fisheries and Forests in the Context of National Food Security" put forth by FAO, to the less obvious governance arrangements around REDD+, which is self-identified in terms of forestry and carbon storage, but also provides a vital framework for land-use decision making.

As part of global land governance (and similar to any global governance issue), it is well recognized that actors have different capacities to set rules and that they experience the effects of rules in asymmetric ways. In other words, actors are endowed with differentiated capacities to create governance for land; we refer to this as *land authority*. Land authority—in particular the questions of who has it, who is experiencing an enhancement of it, and who is experiencing a loss of it—raises vexing questions in the context of increasing competition for land, urbanity, and telecoupling.

Taking into account current trends and challenges related to global land-use change, we should expect that the configuration of actors who can exercise such authority is undergoing significant change. Such reconfigurations are likely to produce differentiated biophysical and socioeconomic impacts across scales.[1] Given the increasing competition for land, urbanity, and telecoupling, it is imperative for land-use science to study this changing distribution of land authority and its implications. For further discussion on emergent global governance, see Margulis (this volume).

---

[1] Sikor and Lund (2009) make a similar argument for authority over access to resources at the national level.

Global Revalorizations of Land

Is land defined by its biophysical features or by the values attributed to it by different actors? While clearly it can and should be both, strong sentiments were expressed during the Forum that the key attributes or values associated with particular pieces of land depend primarily on the interests of the actors involved and how these change over time.

Another new aspect of this equation appears to be the efforts to move from seeing land as having one primary value (e.g., commodity production of maize or the provision of shelter) to land having multiple values (e.g., such as through the variety of ecosystem and other services that may be provided). Not only is land essential for producing crops, livestock, trees, wood, fiber, fruits, renewable bioenergy, and animal feed, it also plays a number of other key roles at different spatial scales (from local to global) by providing ecosystem "functions" or "services." Production functions have long been recognized. Other functions have been recognized quite recently, but are no less essential for the life support system of the planet and its human societies, and include:

- the carbon cycle and other nutrient cycling (soil fertility),
- the water cycle, the regulation of erosion, and flood control,
- biodiversity at the level of genes, populations, ecosystems, and global (the planet),
- pollination (and thereby agricultural production),
- waste recycling,
- aesthetic attributes (and values),
- recreational attributes (and values), and
- religious attributes (and values).

These more recently recognized services all contribute to the resilience and sustainability of natural and agro-ecosystems. One poorly understood issue is that of the existence (or nonexistence) of tipping points in these properties (through overuse or damage through ecological feedback mechanisms), beyond which a function is irreversibly lost, thereby decreasing the resilience of ecosystems.

Accompanying this discussion of the multiple values from land is a reassessment of the extent to which land should be seen as "just" a private good. Since land generates benefits—or costs when there is a breakdown in a function—which affect large numbers of people, land is being increasingly recognized as a global resource or a global public good. While individual interests in land are emphasized in many cultures, there are many formulations of these collective/common/shared interests in and affecting land, such as those from the following perspectives:

- biophysical: global biogeochemical cycling,
- economic: public goods,

- legal: nuisance laws,
- political: collective rule-setting processes,
- joint control of land: shared decision making on land management and disposition,
- indigenous peoples' traditions: stewardship, not ownership.

Some of the land grabbing and other efforts by global actors to gain control of land around the world are being driven by governments, NGOs, or the private sector in response to concerns that are shared by many at the national or global level. These include ensuring:

- food security for one's own domestic populations (e.g., investments by countries in productive agricultural land elsewhere),
- protection of global biodiversity (e.g., acquisition of large tracks of tropical rainforests by NGOs), and
- reduction or mitigation of greenhouse gas emissions (e.g., acquisition of agricultural lands made by countries, private investors, and/or NGOs for biofuels or forestland for the carbon stored therein).

Many of these common or public interests in the services provided by land have informed UN efforts to promote a "green economy," such as at the 2012 Earth Summit in Rio de Janeiro.[2] During our discussion of the potential impact of these efforts on land use, several different views of the "green economy" were suggested and a variety of opinions were offered as to whether any of these approaches would have positive, negative, or any impacts on land-use decisions or governance.

One approach viewed the green economy as being primarily about payments for ecosystem services (PES), frequently through individual offset or mitigation projects. To the extent that these projects involved "developed" country parties investing in carbon credits recognized under global rules from forestry projects in "developing" countries, many participants were highly critical of them—from their high transaction costs to the frequently low benefits provided to local communities. Support was voiced, however, when these payments were locally negotiated (e.g., by downstream water users to the managers of lands upstream). As such, our discussion of PES echoed many of the questions raised about the legitimacy of global rule-setting processes that affect land, particularly those which involve vulnerable peoples in developing countries. Support was also expressed when local communities or host country governments participate in designing the terms on which ecosystem services would be offered in exchange for payments from international actors.

A second approach was to consider the green economy as calling for an even broader incorporation of environmental externalities—both costs and

---

[2] See the resulting document adopted at the summit, "The Future We Want," at http://bit.ly/M18xLE

benefits—into markets using a wide variety of policy and other tools. This approach involves use of a wide range of economic incentives (such as taxes) to bring environmental considerations more directly into global markets. One appeal of this approach is that it is aimed directly at today's economic realities.

This focus on market incentives also raised a third approach, involving a broader critique of the continuing penetration of markets into environmental protection. Critics argue that such an economic approach to solving environmental problems will have adverse social and economic effects on poor and marginalized groups. According to Unmüssig et al. (2012), neither UNEP nor OECD address the issue of power and distribution of resources in their reports on the green economy; democratic control, human rights, and social participation also remain blind spots. The fear is that the current mainstream economic approach to the green economy will lead to further privatization of land and resources already used by indigenous peoples and other local groups, thus resulting in what has been called "green grabbing" (Fairhead et al. 2012; see also discussion below in the section, Conceptualizing Large-Scale Land Acquisitions).

Changes in Knowledge

The role of knowledge as a driver of change in global governance is multifaceted, as is the question of how the land-use change research community might best contribute new knowledge to those efforts.

Different Types of Knowledge about Land and Its Use

Efforts to assemble a "whole" picture of the values of land and its uses raises multiple framing and methodological issues. The two major ones covered in our discussions were those about linking the biophysical and social sciences, as well as global economic and indigenous/traditional ways of viewing land use.

Biophysical data about land lends itself to being quantified and modeled. Human behaviors which affect land are harder to quantify and predict. Since the purpose of land-change science is to track and understand the drivers of land use and ultimately to predict the direction of human-induced changes, both types of data need to be included, and ways to connect this data need to be found. Fortunately, research on land use in urban areas is developing new ways to bring these two areas together, even in terms of modeling.

Seen through the lens of our current global economic system, land use largely concerns the economic valuation of production areas for food, shelter, energy, etc. Like biophysical data on land, monetary valuations are quantified and can be modeled in a variety of ways. At the same time, other systems of knowledge of land place less value on its productive capacity and more on humankind's responsibility for its stewardship across generations, such as those of the indigenous cultures in some of the most rural areas on the planet.

As the global competition for land grows and questions of how land might best be managed in pursuit of a common good come to the fore, the attention being given to these other knowledge traditions is increasing. Concerns about the social impacts of land grabbing (see discussion below) feature prominently. When large-scale purchases of land by foreign investors leads to the displacement of local populations, teleconnections enable networks of concerned actors to amass political support for a range of responses: from direct efforts to stop a transaction, to those seeking to increase the benefits received by affected local parties or to develop/enforce new, global guidelines for more equitable transactions in the future.

Some responses to the outrage over land grabs involve an even more fundamental critique of the treatment of land as a mere commodity in global capitalism. Renewed discussions of fundamentally different ways of thinking about land are underway, including explorations of concepts held by indigenous people—that land is not a thing to be owned, but rather a gift to be stewarded for future generations.

This raises an interesting issue about the density of human populations and the balance between individual and collective interests concerning land: If the tradition of sharing land, as practiced by many indigenous peoples, arose when human populations were relatively small and densities were very low, and individual land rights emerged to encourage investment in more efficient production as populations grew, why are we now seeing a resurgence of common interest in land, given that human populations are now larger and more dense than ever before? Presumably, at the core of this lie concerns over the increasing scarcity of land and the welfare of future generations as well as access to the multiple values that are provided by land.

Knowledge As a Form of Power

How should knowledge of land and land change be used and to which ends? Given that knowledge is a form of power, we feel that the land-change scientific community should use its knowledge to influence the political economy of land use and help create equitable solutions. This means offering a neutrally presented menu of possible answers and their likely effects as part of a rational process of policy optimization.

Land-change scientists help to shape public awareness of the issues, as well as to support both the means to and the ends of change. This includes recognizing ethical perspectives (Boone et al., this volume) and actively coproducing science and knowledge with other actors.

With respect to ethics and knowledge, certain indigenous and otherwise "alternative" approaches to land and nature suggest that it is difficult to separate ethical and spiritual visions of land and society from the various types of knowledge of land and society that are produced. Frequently (perhaps even always?), these ethical positions carry with them particular conceptualizations, for example, of nature and society being unified or of nature having rights. Such conceptualizations are not purely abstract; in some societies (e.g., Bolivia, Ecuador), they have been inscribed in national constitutions (even if it remains to be seen how far laws, policies, and practices can or will be changed in such societies).

In principle, these conceptualizations would have direct implications on what would be accepted as legitimate categories and variables in a research project. They might also influence what is deemed to be a legitimate and illegitimate research process. *Ipso facto,* they would affect the types of knowledge produced by research.

In much formal scholarly work on land, ethics of land and society remain implicit. As a result, knowledge production is (implicitly at least) treated as an ethically neutral activity. However, if conceptual categories derive from ethical positions, such a position may not be tenable. This is not the same as saying that such scholarly work is based on ethics that are somehow distasteful. It merely means that it may be appropriate to make ethical positions clearer: Does the research process view land as something that humans should dominate and use for social purposes? Does it regard society as being separate from nature in principle? Does it believe that nature has rights? This, in turn, may ultimately foster communication across different communities, in particular, between science and stakeholders who are land users if not formal "owners." Initially, such conversations are likely to be difficult, but with time they are likely to be more fruitful.

In terms of the "coproduction" of science with other actors, many actors need to be considered, such as indigenous peoples, urban residents, and multinational companies. Coproduction efforts must also include the explicit recognition of the impacts of such collaborations on the framing of research questions, the collection of data (e.g., crowd sourcing, citizen science), and the intended use of research results.

Surprising results can be obtained. For example, when scientists in an international agricultural research center wanted to understand how local knowledge about ecological processes differs across locations, they worked with local farmers in Thailand, the Philippines, Kenya, and Tanzania, focusing on the farmers' understanding of how individual trees in agricultural fields and rainfall interact, as well as how this results in changes in soil fertility and soil water content. A simple model was designed by a group of anthropologists to capture the farmers' understanding in a systematic and formalized manner. The original hypothesis was that local knowledge would differ vastly across the four countries, because socioeconomic, cultural, and biophysical conditions varied greatly. This hypothesis was disproved, though, by the results: farmers across the four countries displayed remarkably similar knowledge and understanding of ecological interactions and, furthermore, their understanding was very detailed. Local farmers' knowledge was surprisingly very close to that of the scientists. In addition, because it was so detailed, scientists—working with the farmers—were able to revisit their research questions and hypotheses to make them more relevant (from the perspective of the farmers) and more insightful (from the perspective of the scientists) (Sinclair and Joshi 2001).

While many unresolved questions accompany the pursuit of these efforts to link different forms of knowledge, we identified three areas where some progress is being made:

1. Studies of land change in urban areas: it is essential to find effective ways of gathering data and modeling human preferences and actions so as to increase our understanding of the systems in which land-use changes occur.
2. Cultural ecosystem services are not easily quantified by monetized values used in provisioning and regulating systems, but they are clearly of "value" to all people, those who live in urban as well as rural areas, in traditional societies as well as in the global economy.
3. Knowledge networks are being used in agent-based modeling to capture different kinds of knowledge.

Work being done in these areas should provide a useful base from which we can continue to move forward on making new connections across different areas of knowledge.

Conceptualizing Large-Scale Land Acquisitions

Expanding efforts to protect or secure flows of both private and national or global "common" interests through the control of land, and the multiple values produced by it are generating a growing range of efforts (from local to global scales) to combat such acquisitions. They are also placing some national governments in internally inconsistent positions; for example, when a national governments acquire interests in other countries' lands, while restricting the ability of foreign parties to invest in theirs.

Many of these opposition efforts are framed around the concept of land grabbing, which emphasizes the inequity of land transactions for the vulnerable people affected and the illegitimacy of the methods used to acquire ownership. This strategic message is being delivered by advocacy groups as they track, publicize, and oppose such efforts by international investors to secure the flows of values from large parcels of land in other parts of the world. Frequently, these acquisitions involve conflicts between the "formal" interests in land held by domestic governments and the "informal" interests held by the people who actually live on and use the land. Since the economic and food crises of 2008, greater attention has been paid to this topic.

Scholars and policy makers have developed several definitions of large-scale land acquisitions to distinguish this practice from other forms of land purchases and/or appropriation of natural resources. Zoomers (2010) defines them as recent large-scale, cross-border land acquisitions that involve leases of over thirty years or which are direct purchases of a size greater than 5,000 hectares carried out by private firms or initiated by foreign governments. Although this definition can be applied beyond food and agriculture, it is useful because it (a) brings into focus several important features (such as acquisitions' transnational/trans-border quality), (b) identifies many of the modalities that make the acquisition of land possible, (c) provides a quantitative-based criteria for distinguishing "large" from "small," and (d) incorporates the key actors involved in making these deals into the analysis. A recent comprehensive study of large-scale land acquisitions at the global scale (Anseeuw et al. 2012a:18) employed a similar definition that emphasized the following dimensions in identifying land acquisitions: (a) transfer of rights to use, control, or own land through sale, lease, or concessions; (b) a conversion from land used by smallholders, or for important environmental functions, to large-scale commercial use; and (c) deals involving 200 hectares or more, which were not concluded before the year 2000.

While acknowledging that definitions such as these are important for establishing criteria for measuring the scale of land acquisitions, other land scholars have sought to develop more critical definitions. In a recent article, Franco et al. (2013) argue that there are severe limits to defining large-scale land acquisitions according to size and type, because doing so can obscure what is unique to contemporary land deals, most notably, potential long-term implications for the global agri-food system as a whole and trajectories of agrarian change in particular. To move the debate beyond criteria-based definitions, they build on more recent scholarship to suggest that definitions of large-scale land acquisitions should incorporate three key dynamics of social relations: the assumption of control over resources by one party over another and how benefits/costs are divided; the modalities of acquisitions and the source(s) of capital; and consideration of the contemporary global political economy context in reconfiguring power relations across and between the North and global South, as well as the political dynamics at play in transnational efforts to legitimate or challenge land deals (Borras et al. 2013).

The two sets of definitions discussed above have different implications. The first set seeks to establish criteria to distinguish land acquisitions from other forms of appropriation as well as to provide a basis for an empirical methodology. The second, itself a composite of other definitions, seeks to identify and understand the asymmetrical power relationships implicit in such land acquisitions.

It is worth noting that defining "land grabs" is also the subject of contemporary political processes. Each of these new transnational and international governance instruments are directly or indirectly seeking to establish an agreed definition of land grabs for policy-making purposes. Scholarship can play a significant role in helping to shape these negotiations and the resulting outcomes.

Links to National and Local Dynamics

Each of the components described above must link to national and local conditions in order to affect the use of individual parcels of land. Much of our discussion of these links took place around the illustrative examples described in the next section. Topics noted as possible areas for further research included:

- The links between global efforts needed to articulate and advocate procedures for land transfers—such as free, prior, and informed consent under the Tirana Declaration by the International Land Coalition—and the processes actually used for land transfers in different locations.
- The opportunities for using more traditional models of global standards, ratified and implemented by nation-states, with claims procedures at both the national and global levels, to link global concerns to local action.

We emphasize, however, that the interactions among these different drivers of changes in global land governance are themselves constantly changing and influencing each other. How best to track, much less predict, these changes across the system is a huge question facing land-use change researchers.

Illustrative Examples

In addition to these descriptions of the components of our conceptual framework on the drivers of changes in global land governance, we thought it useful to provide some brief illustrations of how these drivers have manifested themselves around particular values from land, including: food, minerals and hydrocarbons, biofuels, forest carbon, landscape amenities, and sacredness.

While all of these examples involve land located in rural areas, the competition or conflicts over the use of those lands are driven by flows both to and from urban areas: demand for food, provision of finance, digital dissemination of knowledge about the conflicts and possible routes to resolution, as well as many others. This is not to say that there is not competition for land in or closer to cities—witness the growth of peri-urban, informal settlements, as well as the conversion of surrounding farmland to built environments. Rather, the examples below are ones that have recently involved widely publicized clashes between global and local interests over changes in land use. As such, they provide a rich data set for exploring the connections from global to local governance of land use.

The examples also offer a range of perspectives on how one might approach the analysis of such conflicts; that is, what values or ways of knowing inform both the participants as well as the researchers. We have included a wide range of perspectives in terms of value, such as whose interests inherent in land issues should count and in which way. Each example described below offers perspectives on different governance conflicts over values from land and different ways of examining these issues.

Food

Recent changes in perceptions about food production and in the global politics of food have triggered a revalorization process and a scalar shift in global governance. Key drivers affecting how land for food production is perceived and valued include:

- An increasing global demand for food due to population pressure and income growth, including the "meatification" of diets (i.e., the tendency to consume more meat as incomes increase). At the same time, as the human population grows, the amount of arable land is constantly decreasing on a per capita basis (leaving aside the expected impacts of climate change). In many parts of the world, increased food production can now only take place on marginal lands with a low agricultural potential (West et al. 2010). These combined trends have resulted in an emerging concept of food security as a global issue, rather than simply the national and local issue it has traditionally been considered. Tied up with these developments is the related recognition that providing food security is an ever more important value from land.
- Increasing food prices, most commonly associated with the 2008 global food crisis that resulted in food riots, pushed an additional 600 million people into food insecurity (FAO 2009b), a state which continues still today (FAO 2011c). These increases have highlighted a number of food governance and policy failures at various levels. Rising food prices are linked to the search for arable lands outside of some countries' own national boundaries by states with little or no available additional arable lands, as well as by transnational corporations and financial investors. This increases food insecurity still further through land grabbing (HLPE 2011; see also Margulis as well as Hunsberger et al., both this volume).
- Vertical integration of agricultural commodity chains and the realization by the transnational agri-food sector that it shares an interest in having a reliable and increasing supply of agricultural produce of a given quality. This need has led to industry-led efforts to introduce private forms of governance, such as production and sustainability standards. Key examples include the Sustainable Agriculture Initiative (SAIP 2012), which covers coffee, dairy, fruits, and vegetables, as well as the guidelines from the sustainability "roundtables" for palm oil, soy, and ethanol (Bailis and Baka 2011; Schouten and Glasbergen 2011) that suppliers must meet regardless of where they are located in the world. This is, in turn, leading to a global "homogenization" of agricultural

practices by the crop grown, regardless of local conditions, as well as to the recognition that sustainable agricultural practices often have value. Demand by consumers for greater "social responsibility" of the agri-food sector and the increasing demand for food based on its qualitative dimensions have also reinforced the trend toward the notion that food production must preserve land and environmental integrity, for both current and future generations.

These drivers are propelled by leading producers, consumers, transnational agri-food corporations, states, and financial actors seeking to ascribe greater value to agricultural land than in the past. This is partly because high-quality arable land is now much scarcer because of (a) increasing human population, (b) past unsustainable agricultural practices (e.g., soil erosion, fertility depletion), and (c) present and future climate-change events that are expected to affect negatively the productive capacity of land in many locations (Robertson and Swinton 2005). This is confirmed empirically by globally rising farmland prices, which are an average of 400% higher worldwide since 2002 (Savills PLC 2012), and the expansion of a global market associated with the land grabbing phenomenon (Margulis et al. 2013).

Increasing interest in land for food production involves a revaluation of adjacent, moderately distant, and extremely distant lands. This revalorization has two fundamental dimensions: (a) land as the nexus from which food security and sustainable food supplies derive, including the wild crop relatives that are expected to breed crops better adapted to climate change (Maxted et al. 2012); and (b) land as the destination for the massive volumes of water (Mehta et al. 2012), petro-chemical inputs, organic matter, and genetic material required for modern food production. A case in point is the global livestock complex that involves significant transborder flows of animal feed (FAO 2011b), requires an ever greater area of land (currently estimated at 30% of the earth's terrestrial surface; Steinfeld et al. 2006), competes with other food crops (Thornton and Thornton 2010), and is widely recognized as a major challenge to sustainability (Steinfeld et al. 2006; Weis 2010).

Valuations of land for food production also reflect other social and cultural values, reminding us that multiple valuations are always at play over one piece of land (Overton and Heitger 2008; Pratt 2007). Examples are valuation in relation to well-defined geographical locations, as in the case of products which can only be named "Champagne" or "Parma ham" if they are grown in delineated territorial areas, or valuation defined by the type of production methods utilized, as in the case of organic foods. These more intangible cultural and historical attributes of land (i.e., enabling registration as "Champagne" or as "organically produced") result in significant increases in the material and social value of the lands concerned.

Through actions such as national food, agricultural subsidy and tariff policies, international trade and investment agreements, and the more recent global policy consensus statements on food by the Group of Eight (G8)/Group of Twenty (G20) (Margulis 2012), governments in the global North and South are actively influencing what is produced, where, how, and by whom at the national and global scales. This has a direct effect on the revalorization of the lands concerned. New agents, such as sovereign wealth funds, institutional investors, and commodities traders, increasingly make distal decisions about land use for food that play out through financial markets and global trade flows (see Hunsberger et al., this volume).

The combination of the above changes in perceptions, politics, and knowledge about the different attributes of land over and beyond its capacity to produce food have led to the emergence of a proliferation of governance mechanisms at the global level. These mechanisms reflect the tension that exists between land conceived as just an input into large-scale food production processes and land conceived as the locus of an array of economic, cultural, historical, and environmental attributes. Global governance mechanisms that have reinforced the concept of land as a mere input into food production by facilitating the development of a market-driven world agricultural trade include those under the WTO, most notably the *Agreement on Agriculture*, *Agreement on the Application of Sanitary and Phytosanitary Measures*, and *Agreement on Trade-Related Intellectual Property Rights* (Rosset 2006; Margulis 2011). At the same time, other global governance mechanisms have emerged to respond to the perceptions of land as more than a substitutable input into food production. The *International Treaty on Plant Genetic Resources for Food and Agriculture* established global rules to protect farmers' rights to access and benefit from the sharing of seeds and germplasm (Esquinas-Alcázar 2005).

Efforts to institutionalize food as a human right at the international and national levels have also gained significant political momentum, most notably since the year 2000 following clarification of the legal concept of the right to food (UN Commitee on Economic, Social, and Cultural Rights 1999) and the establishment of the Special Rapporteur on the Right to Food by the United Nations human rights' system (UN Committee on Human Rights 2000). These developments have been influential in bringing greater attention to the linkages among human rights, sustainable food production practices, and equitable land use and access (Kent 2008; De Schutter 2010a, b, 2011; FAO 2004, 2011a).

Minerals and Hydrocarbons

Different factors drive increased interest in land as a potential source of minerals and hydrocarbons (Bridge 2004; Beddington 2012):

- Prices of many minerals have risen rapidly for different reasons. For example, global urbanization and construction have driven increased copper prices; increased affluence as well as uncertainties regarding currency values have driven increased gold prices; technological

innovations require supplies of rare earths; interest in cleaner fuels has driven the search for natural gas.
- Technological changes have made it possible to extract subsurface resources from once inaccessible locations and lower grade deposits.
- Growth projections for particular economies may have driven interest in securing access to mineral deposits for some future use.
- Increasingly stringent governance instruments in a number of OECD countries have encouraged companies to seek operating environments with weaker regulations.

These factors have meant that different areas of land have come to be valued by international private and state-owned companies as potential sources of minerals and hydrocarbons. This interest drives investment in exploration (an exercise in trying to identify new roles for and values in land), as well as initiatives by national governments to make blocks of land available for exploration. In addition to revaluing land where mineral deposits are found, this process also leads to a revaluation of adjacent and moderately distant lands as sources of the large volumes of water that are necessary for mineral extraction.

This extension of the "extractive frontier" has been facilitated by a combination of factors: stock markets specializing in investments in exploration and extraction (e.g., Toronto, London-AIM), the lending policies of the World Bank Group, and changes in national policy frameworks to ease investment and information availability. It has also been contested by general activist movements, as well as by actors who (re)value the same lands in different ways. Of particular importance in this regard have been revaluations of land as sacred territories for indigenous peoples (Scurrah 2006), which will be discussed further below. Actors revaluing land for biodiversity have, in some cases, found ways to collaborate and coexist with extractive industry claims on land, such as through mitigation activities involving the protection or restoration of critical habitats.

This contestation over the valuation of land has driven several changes in global governance. The private sector has created instruments of self-governance and voluntary regulation. In the mining sector, for instance, the International Council on Mining and Metals (ICMM) was created in 2001 as a club of 22 of the world's largest companies. ICMM now produces guidelines for its members' operations, many of which affect company level codes of corporate responsibility. In the early 2000s, the World Bank sponsored the "extractive industries review"—itself a contested process that resulted in guidelines to orient bank group support to extraction (World Bank 2003). Indigenous organizations and their allies have sought to make use of and give greater visibility to the International Labor Organization Convention 169 (ILO 169), which addresses indigenous and tribal peoples in independent countries, viewing this as a governance instrument through which they can exercise greater influence over efforts to expand mineral extraction into lands they value as territory and sacred space.

Competing valuations of land coupled with global governance changes affect national governance arrangements. The extractive industry and its sponsors have encouraged governments to create new governance arrangements that will facilitate investment (e.g., arrangements regarding taxes, royalties, environmental impact assessment), though it is important to note that these pressures are not all the same. There is, in fact, some evidence of global companies encouraging arrangements that will align with the global reputations that these companies seek as "responsible corporations." Indigenous peoples' organizations and allies have also pressured national governments to recognize ILO 169 in national law so that it can become a domestic instrument that can be used to resist the presence of extractive industry on lands they view as their territory. In addition, these and other citizens' organizations have sought to influence legislation and land-use planning processes in ways that will prevent water resources from being used for the purposes of extraction.

Which governance arrangements are ultimately more powerful, and which forms of land valuation and land use will ultimately prevail will be determined by national political processes involving contestation and negotiation among multiple actors (governments, companies, consultants, citizens' organizations, NGOs, constitutional courts). In these national conflicts, actors are supported by different global allies and networks; they draw on different global governance principles and instruments in the process of arguing for the legitimacy and legality of their particular positions regarding how land should be used and by whom.

Biofuels

The production of liquid biofuels increased more than fivefold between 2000 and 2010 (Worldwatch Institute 2011). Rising interest in biofuels has produced new or strengthened connections between agriculture, energy, and the politics of climate change with strong implications for land use. Several factors have driven this trend:

- Fluctuating world oil prices, uncertainty over future fossil fuel supplies, and a desire for import substitution have driven interest in biofuels as an *energy security strategy*.
- Efforts to reduce greenhouse gas emissions, together with the perception that biofuels are low-carbon fuels, raised interest in biofuels as a *climate-change mitigation strategy*.
- A desire for economic growth, particularly in the agriculture sector, supported investment in biofuels as an *agricultural development strategy*.

Viewing land that could be used to produce biofuels as a source of energy security, climate-change mitigation, and new economic opportunities has increased its value to actors in all of these domains. The same objectives have motivated governments to create policies that encourage the production and use of biofuels (Howarth et al. 2009), solidifying this revalorization of land through mandates and incentives. These policies have encouraged international trade in biofuel crops, as states with fuel-blending mandates have been largely unable to meet their targets through domestic production.

Changes in knowledge and an emerging politics of resistance have influenced the governance and deployment of biofuels. Research on indirect land-use change has called into question the extent to which biofuels reduce greenhouse gas emissions, showing that where biofuel production leads to the conversion of landscapes that store large amounts of carbon, a "carbon debt" is incurred that can take decades or centuries to overcome (Fargione et al. 2008; Searchinger et al. 2008). As a result of this new understanding, the European Union has introduced measures to include indirect land-use change when evaluating the greenhouse gas outcomes of biofuels (EC 2012a). During the food price crisis of 2007–2008, "food versus fuel" became a strong theme in civil society resistance to the large-scale production of biofuels. Reports of social exclusion (Hall et al. 2009) and large-scale land deals (Cotula 2012) linked to biofuel production have also influenced the politics of biofuels.

Together, these shifts in the knowledge and politics surrounding biofuels have led to new developments in global governance. Certification schemes such as the Roundtable on Sustainable Biofuels have been created to encourage producers to meet a series of social and environmental criteria. While not exclusively aimed at biofuels, measures to address social concerns related to large-scale land acquisitions are also relevant. The "Principles for Responsible Agricultural Investment" (FAO et al. 2010) and the "Voluntary Guidelines on Responsible Governance of Tenure of Land, Fish, and Forests" (UN Committee on Food Security 2012) aspire to help protect land rights and food security in the context of large-scale investments in agriculture, which often characterize contemporary biofuel production.

At the national level, several countries have revised their biofuel policies in response to social and environmental concerns. Because of the international trade relationships involved, policy changes in one country can directly affect land use and social outcomes in other countries. The EU's sustainability criteria (European Union 2010) provide one example of policy reform on the part of a biofuel importer, whereas Tanzania's suspension of biofuel investments in 2009 (Browne 2009) reflects a policy shift in a potentially important exporter of biofuel crops.

The governance of biofuels involves actors across sectors and scales—intergovernmental bodies, states, local governments, the private sector, farmers, social movements, and NGOs with a wide variety of mandates—and, frequently, unusual alliances among them. While some of the motivations for pursuing biofuels are global in character, it is at the national and local levels that their effects are felt through direct and indirect land-use changes, competition and potential for conflict over land and water, crop substitution and shifts in food production, as well as changes in land tenure and social relations.

Forest Carbon

Struggles over the way forests are valued have long historic roots. To secure a supply of ship masts for the British Naval fleet, the United Kingdom sought control over forested lands in New England in the late 1600s and early 1700s (Perlin 2005:278). In France, the 1699 Colbert Ordinance enacted a vision of the forest as a "precious and noble resource, whose management needed to be rationalized for the nation's common good, even if at the expense of local subsistence-related uses" (Finger-Stitch and Finger 2003:14).

Contemporary debates over the value of forests have escalated as a consequence of the following drivers:

- technological change facilitating the use of lower-grade, smaller timber for structural applications (engineered wood products); and
- shifting production to areas of the planet where high growth rates support short rotation plantations, but that are also areas of high cultural and biological significance.

Contestation over an attempt to revalorize forests as a global commons first took form in the late 1970s and early 1980s. Assessments in the 1970s by the FAO were among the first to provide a worldwide perspective on the annual losses of forests in tropical regions (FAO 1981, 1980). This way of knowing forests informed the negotiations of the first International Tropical Timber Agreement, work of the FAO on its Tropical Forestry Action Plans, as well as the parallel work of the World Bank and the World Resources Institute on similar issues. The significance of forests—as critical global ecosystems, as a foundation for the culture and livelihoods of indigenous and local peoples, and as inputs for economic development—was an area of great debate.

The ultimate failure of the 1992 UN Conference on Environment and Development (UNCED or Earth Summit) to adopt an international forest convention helped set in motion parallel tracks of global governance within and outside the UN system. Private certification programs formed to promote responsible management of global forests outside the UN system (Cashore et al. 2004; see also Auld, this volume), and intergovernmental processes established regional and country-level criteria and indicator processes for defining sustainable forest management (Humphreys 2006). Both tracks have served as venues in which political struggles over the values of forests have continued.

The most recent revalorization of forests as global commons found legs with increased attention to the carbon storage provided by forests (above and below ground). A basis for this revalorization was established by the 1992 UN Framework Convention on Climate Change (UNFCCC), which in Article 4.1(d) noted that parties shall "promote sustainable management, and promote and cooperate in the conservation and enhancement, as appropriate, of sinks and reservoirs of all greenhouse gases not controlled by the Montreal Protocol, including biomass, forests, and oceans as well as other terrestrial, coastal, and marine ecosystems" (UNFCCC 1992). Ways of knowing forests were also quickly advancing an understanding of living and growing forests as a key carbon sink and, when cut down, an important source of carbon emissions (see e.g., Dixon et al. 1994).

Not only were these new values made visible, they were also cast as a comparatively inexpensive mitigation option under the UNFCCC (Stern 2006; Eliasch 2008). In a particularly powerful visual display, the consulting company McKinsey produced a cost curve that displayed the relative costs for greenhouse gas abatement of various activities, including preventing deforestation and (re)planting forests. These forestry measures were among the low-cost options: the analysis suggested that at 40 Euros a ton emission permit price, 6.7 gigatons of abatement could be achieved through forestry measures, which accounted for 25% of the potential abatements at this permit price (Enkvist et al. 2007). In other words, forests as a global commons were one of the less expensive means of mitigating societal releases of greenhouse gas emissions.

Nevertheless, for much of the 1990s and early 2000s, forest carbon was marginalized within climate-change governance processes. Other than through afforestation and reforestation under the Clean Development Mechanism, forest carbon remained off the agenda of the UNFCCC until the Bali Action Plan, which was adopted at COP13 in 2007. After a two-year process of lobbying and discussions led by the Coalition of Rainforest Nations, a mechanism referred to as REDD (Reducing Emissions from Deforestation and forest Degradation) became a central element of the post-2012 negotiations launched in Bali.[3] The core idea was that developing countries with tropical forests should be financially compensated (fund, market based, or both) for keeping forests intact.

REDD—or now REDD+ with a broader scope of activities to be included—has progressed rapidly both inside and outside the UNFCCC. Key issues being addressed include social and environmental safeguards, as well as monitoring, reporting, and verification (MRV) requirements. Capacity building in REDD+ countries and pilot projects on the ground are being carried out by a multitude of actors, including the United Nations, the World Bank, and various international NGOs. In parallel, private organizations have also experimented with forest carbon offsets (Angelsen et al. 2012). In the annual reports on the voluntary carbon offset market, forest projects have accounted for a high of 46% of the quantify of offsets (or 26 $MtCO_2e$) in 2010 and have been a consistently larger proportion of offsets over the period in which such data has

---

[3] For further information, see see unfccc.int/methods_science/redd/items/4531.php

been collected (Peters-Stanley et al. 2011; Peters-Stanley and Hamilton 2012; Hamilton et al. 2007, 2008, 2009, 2010).

These processes of revalorization have also been contested at the national and local levels. Key concerns involve the underestimation of the costs of forest carbon due to a focus on opportunity costs (e.g., the lost revenue for a landowner not using the forest for another use such as timber production or agriculture), as opposed to a broader focus on administrative, monitoring, and implementation costs (Dyer and Counsell 2010). National and local administrative structures, power relations, and contested tenure regimes all feature as key challenges which, when accounted for, have made the "cheap" global commons of forests more complicated to make legible and to govern.

Equally, there has been contestation over the re-centralization of decision making that a focus on forest carbon as a global good may encourage. The high administrative costs associated with REDD+ mechanisms mean that there will be strong incentives for centralization to take advantage of economies of scale. Such economies of scale in forest carbon governance have serious implications for smaller-scale initiatives and communities, and may unravel forest decentralization processes that have been heralded as important successes (Phelps et al. 2010). Still, funds allocated to forest carbon projects are substantial—reaching an estimate of USD 4.2 billion by 2011 (Nakhooda et al. 2011).[4] The potential to gain from these funds has led to place-based struggles over the value of land as a forest carbon reserve versus as a source of livelihoods from multiple goods and services for local peoples.

These processes of revalorization have also played out in the realm of knowledge and knowledge creation, with stark differences across levels of governance. REDD+ illustrates such tensions vividly: On one hand, local forest users tend to understand REDD+ to be principally about forest protection, offering opportunities for both income improvement and risk of livelihood deterioration (Angelsen et al. 2012). International climate negotiators and national government representatives, on the other, tend to understand REDD+ as a way of enhancing carbon stocks and mitigating climate change cheaply (developed, donor countries) or as an opportunity to generate international revenues (developing, host countries).

Driven in particular by international NGOs, justice and equity considerations are currently taking the form of social safeguards, including a "respect for the knowledge and rights of indigenous peoples and members of local communities" (UNFCCC 2011). While coming to the fore in Western scientific understandings, traditional practices of land management and traditional meanings of land are often ill-translated into Western schools of thought, and vice versa. For example, while much social science scholarship frames the human-environment relationship as two separate forces, traditional societies

---

[4] http://www.climatefundsupdate.org/themes/redd

view the world as nondualistic; that is, human-environment interactions are interwoven (Bamford 1998).

Biodiversity

Concerns about a global biodiversity crisis have led to the rise of new scientific fields (biological conservation, conservation planning, etc.) that have specified the amount and quality of land needed in various parts of the world to maintain viable animal and plant populations (Adams 2004). Having been taken up in the Convention on Biological Diversity and other governance processes, this has translated into large-scale demands for land (10–20% of global land surface) specifically devoted to biodiversity conservation (Rodrigues et al. 2004; Chape et al. 2005).

There are competing visions about how to achieve conservation through the exclusion or integration of people and land uses (Hutton and Leader-Williams 2003). The sustainable development agenda that emerged in the 1980s changed the dominant thinking on global biodiversity governance from a "fortress" to a community-based win-win discourse. Interestingly, the fortress conservation model applied more to former colonized territories (Brockington et al. 2008). It has, for instance, not been followed in Europe, where local communities continue to access natural resources in national parks to a much greater extent than in postcolonial settings.

The community-based biodiversity discourse consists of two main elements. First, it is necessary to let people who live in and around protected areas participate in the management of these areas. Second, local communities must benefit from conservation efforts. As in forest conservation, the primary concern in the win-win discourse is to conserve biodiversity. However, it promotes the integration of the interests of local communities as a means to achieve this conservation. Thus, the setup involves aspects of benefit sharing, compensation, and participation, and the partnerships are argued to constitute win-win situations, implying both environmental conservation and local development.

Four main factors have made the win-win discourse on conservation the approach of choice in post-colonial settings as well (Benjaminsen and Svarstad 2010):

1. It draws upon the basic ideas that became influential through the Brundtland Report (World Commission on Environment and Development 1987) of a win-win relationship between environment and development.
2. Since the 1980s, there has been increased pressure from development and human rights activists and indigenous groups to change conservation practices in a more "people-friendly" direction.

3. The shift to the win-win discourse on protected areas also forms part of a more general shift within developmental policy, from a top-down approach to a focus on participation.
4. Neoliberal economic policies have, in general, had great influence on conservation policy during the last couple of decades. These policies focus on involving civil society more at the expense of the state, as well as emphasizing how markets for tourism are needed to help finance conservation.

Critics argue, however, that there is a gap between discourse and practice in this community-based model as applied in a development context, and that conservation labeled "community-based" tends, in practice, to lead to dispossession of indigenous and local use resulting in forms of accumulation by dispossession (as defined by Harvey 2003) or "green grabs" (Fairhead et al. 2012; Benjaminsen and Bryceson 2012). This process is furthermore driven by a revalorization involving the opening up of spaces and landscapes previously controlled by villagers to ecotourism as part of biodiversity conservation efforts. Hence, noncapitalist spaces and resources are opened up for accumulation through the combination of tourism and conservation.

Landscape Amenities

In some parts of the world, particularly in the global North, land is increasingly being used for amenity purposes (Smith et al. 2010) and appreciated for the provision of "cultural ecosystem services." Defined as "nonmaterial benefits people obtain from ecosystems through spiritual enrichment, cognitive development, reflection, recreation, and aesthetic experiences" (MEA 2005:40), the cultural services of rural landscapes play an increasingly important role in peoples' quality of life. The growth of regional parks, private and public nature reserves, golf courses, tourism facilities, second homes, hobby farms, "ranchettes," and residential homes in the countryside are all expressions of amenity-based land uses.

Amenity uses are related to notions of a "postproductivist" countryside that includes shifts from pure food and fiber production to inclusion of biodiversity conservation, landscape esthetics, and rural quality of life; from quantity to quality in food production, growth of on-farm diversification and off-farm employment; and the promotion of sustainable farming through agri-environmental policy (Mather et al. 2006). Empirical studies, for example among ranch owners in California, have shown that "income maximization" in the conventional sense is often not an important landownership goal any more. Rather, amenity interests (e.g., having a good place to raise a family; using the land for private recreation, hunting, and fishing; leaving legacy values; welcoming friends and visitors to one's place; and enjoying a "country way of life") are of great importance to these landowners (Huntsinger et al. 2010; Plieninger et al. 2012). The fact that rural real-estate prices are frequently above those that can be justified by agricultural production value alone in North America and Europe confirms the substantial amenity benefits that landowners consume (Torell et al. 2005).

Two main factors have been responsible for this shift from "productive" toward "amenity" land uses in these regions over the past decades (Walker 2003; Primdahl and Swaffield 2010):

1. the decline of "old" primary products economies, which has resulted from the globalization of agricultural commodity markets, changing agricultural policies, and rising taxes and opportunity costs of farmland and forest land, and
2. the rising valuation and gentrification of these lands due to their development potentials, driven by the migration of early retired pensioners to less urban areas and long-distance commuters as well as by changes in technology and alternative job opportunities.

Both factors link directly to the impacts of an urbanizing world on rural land uses.

Such "estheticization" of rural land uses may, on one hand, contribute to sustainable land management under certain conditions. Amenity benefits and cultural services are, in contrast to most other ecosystem services, directly experienced and intuitively appreciated by people (Daniel et al. 2012), and they are therefore important motivators for responsible land stewardship (Chan et al. 2012). In addition, most cultural services are enjoyed in "bundles" and can thus guide land management toward multifunctionality, which is a frequently expressed, but rarely achieved desideratum in land-use science and policy (Cowie et al. 2007).

On the other hand, increasing valuations of land for its amenities can generate conflicts, especially when it is accompanied by restrictions of traditionally "productive" land uses, such as livestock ranching or timber harvesting. When such traditional land uses are set aside, competition for land increases (Smith et al. 2010). Amenity uses are particularly subject to dispute as their values are attached to a totality of many individual properties (the landscape qualities, whose "ownership" remains mostly undefined), rather than to individual natural resources (Walker and Fortmann 2003). In consequence, conflicts can emerge over the question of who will "possess" or "control" the landscape.

Sacredness and Territory

Revalorization of land, through the lens of sacredness and territory, reflects the wider recognition and new visibility of preexisting values and attributes previously confined to marginalized actors. Revalorization thus brings into focus attributes of land as sacred sites, spaces, and territories that were previously obscured by the primacy of extractive and/or productive uses. At a time of increasing concerns with global food security and mineral exploitation, this particular aspect of revalorization reflects the confluence of distinct yet interrelated agendas, value, and knowledge flows.

Human rights agendas and claims to rights-based development are integral here, as are global agendas emerging from the UN Permanent Forum on Indigenous Issues (UNPFII) and, more recently, from the 2012 Rio + 20 outcome document, "The Future We Want." Environmental justice agendas, which have recently gained more traction at a global scale, help to frame and shape the emergence of indigenous or locally valued attributes of land. Indigenous land claims thus draw on pressure for justice as well as for the recognition of diverse, primarily communal tenure systems, typically linked to extensive and/or subsistence-based land uses as well as to multiple dimensions of well-being and value. Recent revalorizations around biodiversity and conservation have been less concerned with new land-use values per se, but rather with recognition of the efficacy of alternative models of governance, knowledge, and stewardship in the production and maintenance of biodiversity. The explicit recognition of Indigenous and Community Conserved Areas (ICCAs), within mainstream conservation, is a case in point (Stevens 2010).

These manifestations of global revalorization are closely linked to changes in ways of knowing or, in the language of environmental justice framings, to cognitive and by extension to procedural justice. Recognition of changes in and multiple ways of knowing have been facilitated through global mechanisms and institutions such as the UNPFII (established 2002) and the International Labor Organization Convention (ILO) 169 (1989). These instruments and institutions have adopted rights-based approaches, through advocating for recognition of diverse tenure systems and, critically, by extension of alternative land relations, ways of knowing, or land ethics. The landmark UN Declaration on the Rights of Indigenous Peoples (2007) requires the recognition of indigenous peoples' rights to "their lands, territories and resources" and to the maintenance of their "distinct spiritual relationship...with land (and) territories." In a comparable vein, the Rio + 20 declaration argues for recognition of an alternative land ethic, encompassing rights of nature (Article 39) and for continued recognition of the rights of indigenous people, including recognition of their values and knowledge in decision making (Article 49).

Contemporary changes in ways of knowing also encompass the emergent ecosystem services (ES) paradigm, widely credited with the discursive transformation of the environment into an assemblage of services, and thus with transforming the ways in which we see and know land and the wider environment. Conceptually, the ES framing is characterized by an "anthropocentric, utilitarian perspective," but it may also open up new possibilities for recognition and revalorization (Potschin and Haines-Young 2011:575). This discursive transformation and associated practices of commodification have recently been contested through growing demands for attention to cultural ES and to nonmonetary valuation (Chan et al. 2012). However, the practical expression of the ES paradigm, through PES schemes, continues to emphasize utilitarian, extractive environmental perspectives.

These changes in ways of knowing or the recognition of diverse knowledge forms are integrally linked to changes in politics and to telecoupling, through knowledge flows, representations, and multiscalar connections between spatially remote and differentiated actors. Key aspects of changes in politics include:

- Contestations and resistance through social or indigenous peoples' movements. Facilitated through telecoupling, such movements are increasingly able to forge alliances with nonlocal NGOs, actors, institutions, and pressure groups to contest displacement from lands and eradication of alternative ways of knowing, not least through discursive representations around rights and culture.
- Increasingly powerful politics of responsibility, ethics, and (environmental) justice between the global North and South as enacted in multiple (including global) spaces, facilitate the recognition of and response to these claims. Again, telecoupling through knowledge flows is a powerful factor.
- Renewed definitions and politics of sustainability, in the context of Rio + 20, and the imminent deadline for realization of the Millennium Development Goals highlight aspects of intragenerational equity and broader nonmaterial aspects of well-being.

At the same time, the conjunction of global food crises, climate change, and biodiversity loss amplify powerful political currents around scarcity, productivity, and the commodification of nature, amplified through ES framings.

Global land-use competition is thus intensified through these telecoupling processes and the contested admission of esthetic, spiritual, and sacred values to the already fragmented sphere of visible land attributes. Emergent global governance arrangements to date provide somewhat limited institutional arrangements for mediation and management of the resulting competition.

Such recent and emergent global arrangements in land governance pertinent to issues of sacredness and territory may be summarized as follows:

- *Indigenous land governance and rights*: through UNPFII and ILO, Rio + 20 (see above). These encompass both formal organizations as well as rules, norms, and collective decision-making procedures. Telecoupling and knowledge flows increasingly enable local actors to appeal to these forums, norms, and procedures in alliance with nonlocal NGOs and activists, and in response to "land/green grabs."
- *Biodiversity conservation:* linked to the above instruments and procedures, but also specifically through Conferences of Parties to the Convention on Biological Diversity (2002:Article 8j); IUCN World Conservation Congresses and Theme on Indigenous Peoples, Local

Communities, Equity, and Protected Areas. Long-running tensions and conflicts between indigenous peoples and mainstream conservation are by no means resolved, however, as manifested in the continued phenomenon of "green grabbing" (Fairhead et al. 2012).
- *FAO voluntary guidelines on the responsible governance of tenure of land, fisheries, and forests in the context of national food security* (UN Committee on Food Security 2012). These guidelines emphasize the need for recognition of multiple (including customary and communal) tenure forms as well as indigenous/local rights and values, framed within the overall goal of national food security. Their adoption and efficacy remains to be seen, but they underscore an important principle in terms of the multiscalar dimensions of food security and links to rights, access, and values.

The emergent global governance arrangements outlined above are fragmented, with their efficacy likely dependent on the ability of diverse actors to access both the political and discursive spaces they provide. These abilities are also critically shaped by national contexts, whereby land-use trade-offs are typically negotiated and determined according to the degree of influence of various actors and local power dynamics. Global governance arrangements may shape these national policy contexts, for example, through enactment and recognition of ILO 169, UN Declaration on the Rights of Indigenous Peoples in domestic policy contexts. They may also act as a corrective, by enabling the contestation of perceived injustices at national and global scales.

Areas for Future Research

To the extent that land-change science seeks both to understand what is driving new uses of land as well as how to help put those changes onto a more sustainable path, the rules governing changes in land use are an essential area of study. Embedded in the conceptual framework and the examples offered in this chapter is a host of different topics on which more research is warranted:

- Differentiated actors: What attributes of actors and of land (both social and biophysical) are most likely to produce layers of contested values and/or conflicts over land use?
- Changes in global governance: What are the ethical, political, and administrative implications of a governance shift, including both processes and places in the urban era?
- Challenges for land governance: Do we need to revisit or develop a new land ethic? How should land governance be constructed beyond administrative boundaries? How can governance mechanisms be redesigned to go beyond a single territorial institution?

- Revalorizations of land: What attributes and/or values of land are enhanced by using different analytical lenses: (a) global change, (b) teleconnections, and (c) new actor and agencies.
- Changes in knowledge and ways of knowing: What strategies exist to address incompatibilities between different values or knowledge systems within particular frameworks, such as ecosystem services? Are there ways of co-producing new knowledge across different forms of knowledge without subsuming one type into another or having one assume a dominant position?
- Changes in politics: How do emerging attention to and politics about land affect the global discourse on land-use ethics and decision-making processes. Links to national/local dynamics: What are the options for legitimately and effectively linking global standards and local impacts on land use?

These examples provide opportunities to analyze the links between urban areas and the competition for land in rural places. The resulting knowledge would benefit all actors involved.

Urbanization and Land Use

14

Range of Contemporary Urban Patterns and Processes

Hilda Blanco

Abstract

Urbanization has historically been characterized by population density, durable built environments, governance, specialized economic activities, urban infrastructures, and their rural spheres of influence. This chapter highlights major contemporary patterns, trends, processes, and theories related to these dimensions, with special attention to the relation of central places to surrounding rural areas. It begins with a discussion of definitional issues related to the different dimensions of urban settlements and contemporary urban patterns. Theories and policies corresponding to these major characteristics of urban patterns and urbanization processes are presented, beginning with a brief overview of economic spatial theories. Focus is given to central place theory, where cities are conceptualized as central market places providing goods and services to lower-order cities and their rural hinterlands in exchange for food and materials. The impact of advances in technology and infrastructures on global trade connections is discussed, and insights from Castells' network society are highlighted. Empirical evidence of two urban policies—the compact city model and urban growth management—are reviewed for their connections to central place theory.

Introduction

To provide a theoretical background to the issues of global land use, this chapter reviews and reflects on major contemporary urban patterns and processes. Historically, urbanization has been represented by six characteristics: density of population, durable built environments, governance, specialized economic activities, urban infrastructures, and their rural spheres of influence. Recognizing these different dimensions or enablers of urban settlements, the chapter first discusses definitional issues and contemporary urban patterns. It then proceeds to various theories and policies that correspond to these major characteristics, beginning with a brief overview of economic spatial theories. Special focus is given to central place theory, a theory that explained the spatial pattern of industrial economies in developed countries in the early part of the twentieth century. This influential theory conceptualized cities as central market places that provide goods and services to lower-order cities and their rural hinterlands in exchange for food and materials. Proximate connections between cities and their regional spheres of influence have increasingly given way to distal relations between international locations for materials, food, manufactured goods, and services, as a result of globalizing trends, which have accelerated since the 1990s.

Advances in technology and infrastructures, in particular, transportation infrastructures and information communications technology and logistics have shaped our global trade connections. In addition, changes in governance and economic structural issues have contributed to the globalization of production and consumption. Consideration is given to how these impact urbanization. Thereafter, a brief review follows of the main insights related to the major socio-political-economic theory of globalization that focuses on urbanization: Castells' network society (Castells 1996). Two major urban policies (the compact city model and urban growth management), empirical evidence supporting them, and their connections to central place theory are discussed. The final section summarizes and reflects on the findings.

Features of Urbanization

Urbanization can be traced back in history to about 10,000 years ago. Çatal Hüyük, often credited as the first city in what is now modern Turkey, dates back to 6,500 BCE (Mellaart 1965). Since the earliest records, several essential features of cities can be identified as shown in Figure 14.1.

The features of urbanization (Figure 14.1) most often noted are (a) higher densities and (b) compact, durable environments. Whether agriculture preceded cities and provided the agricultural surplus for urban dwellers or, as Jane Jacobs argued (1969; see also Soja 2010), cities and agriculture coevolved, cities (from early on) traded special goods or services for food and other materials with their hinterlands and other cities. Many large ancient cities, such as Uruk or Ancient Athens, were city-states, exerting secular or religious power over their rural peripheries. For internal purposes, large urban populations also required formal governance to establish social customs, settle conflicts among strangers, as well as manage urban growth and provide infrastructures. Early cities also relied on road and water infrastructures. Roads facilitated trade between rural areas and cities and between cities. Water infrastructures, such as aqueducts, wells, and public fountains, were necessary to make local water supplies available to city residents. Urban water supply infrastructure, drainage and even sewers as well as irrigation canals in rural peripheries date back to 7,000 BCE. Ancient Greek cities, for example, were laid out in block patterns

Figure 14.1 Six features of urbanization. Some are associated with the character of urban settlements (density, durable built environments, and urban infrastructures), others refer to processes that enable urbanization (specialized economies and governance), while rural hinterlands or spheres of influence provide goods or services not available in urbanized areas.

with roads, and public fountains provided a public water supply (Wycherly 1962). Greek city-states—the *poleis*—were composed of cities and their rural hinterlands and gave the Western world the word for politics. *Polis* meant both the city, as a place, and the people, as a political entity. Cities have always depended on rural areas for food, fuel, materials for construction, as well as water for drinking and as sinks for their wastewater and solid waste. The rural hinterlands of urban areas have traditionally provided many benefits that we now understand as ecosystem services (Daily 1997).

Review of New Urban Patterns

In 2007, for the first time in human history, half of the world's population lived in urban areas. In the United States, Western Europe, and South America, by 2010, over 80% of the population was urbanized. China is undergoing the most rapid rate of urbanization. In 1990, its urban population was 26.4%; by 2010, 49.2% of its population was urban; and, by 2050, China's urban population is projected to reach 77.3% (UN 2012b). But, how do we identify urban areas? There is no recognized, worldwide definition of urban land or urban areas. The United Nations, in their periodic reports on global urban and rural populations, use the statistics that countries report as their urban areas or cities, and definitions vary across countries. Urbanized land is mainly defined in the following ways:

- as land in state-recognized cities such as municipalities or local authorities, as in the Dominican Republic,
- as land in agglomerations with threshold populations ranging widely from 1,000 persons in Australia to 10,000 persons in Italy, and,
- in terms of density per unit area, ranging from 386 persons per square kilometers (in the United States), to 1,500 persons/ km^2 (in the People's Republic of China).

The extent of the durable built environment of urban settlements or contiguous built environment remains implicit in many definitions. Definitions of urban places in a few countries, such as Costa Rica or Panama, also include the availability of urban services, such as electricity, water and sewerage systems, and other municipal services (UN 2012b).

In the twentieth century, many urban areas exceeded their official political boundaries and became metropolitan areas. The U.S. concept of metropolitan areas, first identified for the 1950 census, defines areas that contain a large population nucleus and adjacent areas that have a high degree of integration, typically through an integrated labor market and travel patterns. Metropolitan areas typically include at least one central city, suburban areas, and other urban areas (such as towns or villages) as well as surrounding rural land. The concept of functional urban regions in Europe (Nordregio et al. 2005) is analogous to the concept of U.S. metropolitan areas. Beyond metropolitanization, two other interconnected urbanizing trends complicate the study of urbanization: the growth of mega-cities (i.e., cities over 10 million) and the convergence of metropolitan areas into mega-scale urban regions, referred to by various terms: *megalopolis* (Gottmann 1961), *megapolitan areas* (Lang and Dhavale 2005),[1] or more recently *mega-city regions* (Hall 2009). Gottmann (1961) used the merging of the metropolitan areas of Boston, New York, Philadelphia, Baltimore, and Washington, D.C., or the Boswash urban corridor as a first example. This type of convergence of metropolitan areas is now understood as a functional, rather than a physical, or administrative concept, drawing on Castells' concept of urban space as a "space of flows" of people (Castells 1996), goods and information. Mega-city regions are polycentric, incorporating multiple cities and towns and their surrounding rural areas.[2] The term mega-city region was first applied to urbanized regions of eastern Asia, areas with populations of ten million or more, such as the Pearl River Delta, the Yangtze River Delta, the Tokaido (Tokyo-Osaka) corridor, and Greater Jakarta (Hall 2009). These urban regions are currently found throughout most parts of the world. Some examples include the Greater La Plata-Buenos Aires Metropolitan Region in Argentina, Mumbai-Pune mega-region in India, and the Randstad in the western portion of the Netherlands (for European examples, see Hall and Pain 2006). In Europe, a mega-city region is made up of several functional urban regions (Hall and Green 2005). These new urban patterns are typically represented by multiple units of government. This mismatch between function and political jurisdiction complicates the planning and governance of these urban regions (McCarney and Stren 2008).

The traditional urban-rural distinction is often used to define urbanization. As the urban-rural distinction has blurred, a new concern with peri-urban areas

---

[1] U.S. studies have identified ten megapolitan areas that house 197 million people (Lang and Knox 2009).

[2] In the U.S. census, consolidated metropolitan statistical areas, composed of several metropolitan areas, are similar to the concept of mega-city regions.

(transitional areas between rural and urban) has emerged in Europe over the past decade (Piorr et al. 2011; for European examples, see Hall and Pain 2006).[3] The physical expansion of urban areas at multiple times the rate of population growth fuels concern over peri-urban areas in Europe and urban sprawl in the United States. In Europe, for example, while population increased by approximately 7% from 1990–2006, urban areas grew by 37% during the same time period (Fertner 2012). In a meta-analysis of global urban land-conversion studies from 1970–2000, Seto et al. (2011:1) concluded that "across all regions and for all three decades, urban land expansion rates are higher than or equal to urban population growth rates, suggesting that urban growth is becoming more expansive than compact." According to another recent study of urban population and urban land-cover estimates that was based on the 1990–2000 growth rates of 160 cities around the world, the growth rates for urban land cover will more than double the rate of urban population growth (Angel et al. 2011). The ongoing expansion of urban areas diminishes the potential of rural areas to provide certain ecosystem services for urban agglomerations.

Understanding Urban Spatial Structures: Economic Spatial Theories

How do we make sense of the spatial configuration of urban areas? What factors or processes determine the size of cities with respect to each other and their spacing?

Economic theories concerned with spatial agglomeration focus on the role of economic agents and the factors that drive their decisions to concentrate or agglomerate in urban areas. These theories assume that economic agents agglomerate for two main reasons: (a) because a place provides a comparative advantage, such as proximity to a port or cheap labor or materials, and (b) because the concentration of activities in a place provides advantages for the producing or consuming agents. Several branches of economic theory have made contributions to spatial analysis: location theory with its focus on the location decisions of firms, urban economics with an emphasis on the location of households and firms within a city, and central place theory (Mulligan 1984).

Location and Spatial Equilibrium Theories

In the United States, Walter Isard's work was instrumental in the 1950s and 1960s in developing the fields of regional science and urban economics. His seminal work, *Location and Space Economy*, introduced the classic works of

---

[3] In the context of developing countries, the term peri-urban has been used to refer to informal settlements without adequate urban infrastructure that extend for miles beyond central cities (Adell 1999).

J. H. von Thünen, Lösch, Weber, Christaller, and others, made a strong argument for the centrality of space as a factor in economic analysis, and demonstrated how production theory can incorporate location factors, in particular transport inputs and rates (Isard 1956). Later work provided the basic analytic methods in regional science, including regional and interregional input-output analysis, econometrics, and spatial interaction models, among others.

The historical roots of location theory can be traced to the work of von Thünen (1826) who sought to determine optimal agricultural land uses based on transport costs to a central market. Von Thünen assumed that production methods and costs for any crop were independent of location. Alonso, a student of Isard, developed von Thünen's theory of agricultural land use into an intra-city theory of land uses. Alonso assumed a monocentric city as a featureless plain, with transport availability in all directions, and with employment and all goods and services available only in the central business district. In this model, households have to decide how much land and where to locate to maximize their utility, and firms decide their location to maximize their profits. Alonso extended the land price function from an agricultural to the urban context, in what is now the main feature of all formal residential location and urban housing market models, the "bid price curve" of a resident; that is, the "set of prices for land that the individual could pay at various distances" from the central business district that permits individuals to attain a constant level of satisfaction.

Going beyond the land models of von Thünen and Alonso, the work of Muth (1969) and Mills (1972) further developed a formal model of the housing market, in which producers combine land and capital to produce housing. The work of Mills (1967) focused on the question of why cities exist, and his response emphasized scale economies in production, people's utilities, and the cost of interregional trade. His work, more grounded on empirical research, also examined trends toward the decentralization of jobs and residences and foresaw the increasing importance of subcenters. Today, with increasing trends toward polycentricity, as evident in the discussion above on metropolitan and mega-city region urban patterns, the monocentric assumption underlying Alonso's urban location model have made the application of this theory less useful. However, recent research is beginning to explore how spatial agglomeration economic principles can be used to understand polycentric urban regions (Agarwal et al. 2012).

Spatial equilibrium theories focus on the issue of why cities exist, or as Glaeser and Gottlieb (2009:984) put it, "why dense areas are so much more productive." The theories get their name from the assumption that workers can migrate freely, which creates a spatial equilibrium where utility levels are equalized. This assumption is supported, at least for the United States,[4] by the

---

[4] Empirical evidence indicates that migration flows in Europe during the 1980s were not as large as in the United States (Decressin and Fatas 1995).

high mobility of workers: 40% of Americans change homes and 20% change counties every five years (Glaeser and Gottlieb 2009). Rosen (1979), Roback (1982), and Blomquist et al. (1988) have applied spatial equilibrium models that incorporate labor, land factors, and amenities in determining urban location decisions. Their research has identified prices for amenities and disamenities that quantify the quality of life across cities in the United States. In highlighting the role of amenities in urban location choices, this work is directly relevant for the reconceptualization of urban phenomena envisioned by Boone et al. (this volume), which emphasizes livelihoods and lifestyles, since lifestyles are often associated with a bundle of amenities.

Renewed interest in spatial agglomerations has generated empirical research on agglomeration economies. In his recent review, Puga (2010) discusses the research challenges and empirical evidence on the clustering of production beyond what can be explained by chance or natural advantage, including research that compares wages and rents across spatial patterns, such as the work of Combes et al. (2010), which analyzes data on the wages of French workers; the work of Dekle and Eaton (1999), which examines data on rents; the earlier work of Rosen and Roback, which employed both wages and rents; and research that uses data on outputs and inputs to estimate how productivity varies across space (Rosenthal and Strange 2004).

In addition to access to markets, since the mid-1990s, spatial equilibrium theories have been focusing on the role that access to ideas or human capital play in agglomeration economies (Glaeser et al. 1992). Duranton and Puga (2001), for example, argue that cities are "nurseries" for new ideas. However, the great strides in information technology (IT) over the past few decades challenge the role that urban agglomeration plays in the exchange of ideas and innovation. Glaeser and Gottlieb (2009) dispute this, pointing out the example of Silicon Valley, both a geographic cluster and IT center. Glaeser and Ponzetto (2007) further argue that changes in IT over the past few decades provide increasing returns to new ideas and make cities even more important. This line of theory and research will be important to the reconceptualization of urban phenomena and suggests that while the urban-rural distinction may be blurring in polycentric mega-regions that are greatly aided by IT and transportation technology, the physical proximity of dense urban areas may still hold idea-generating and innovation advantages.

Central Place Theory

Central place theory (CPT), developed by Christaller (1933/1966) and Lösch (1940/1954), integrates three key concepts in regional studies: consumer choice, firm agglomeration, and functional hierarchy (Mulligan et al. 2012). It focuses on trade and service activities, how they are distributed among settlements or central places, and, most important for understanding the spatial patterns of urban agglomerations, how central places are located across the landscape. The theory assumes a flat landscape and that people will buy goods from the closest place. Cities are conceived as central market places that provide goods and services to a surrounding population. In addition, the concepts of threshold and range are central to Christaller's theory, threshold meaning the minimum number of customers needed for a business to remain viable, and range referring to the average maximum distance that people will travel to buy goods or services. In effect, according to CPT, centrally located settlements have more purchasing power and offer more goods and services. Each type of settlement has a particular location in the hierarchy, and relative centrality determines both the type and variety of goods. Applying the theory to urban settlements in Southern Germany, Christaller arrived at a hierarchical spatial arrangement (depicted in Figure 14.2), where large cities are at the core of constellations of smaller towns and villages, with each central place providing

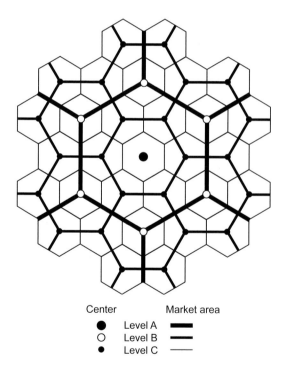

Figure 14.2 Hierarchy of hexagonal market areas after Christaller. Cities, towns, and villages are central places that provide goods and services to their residents, their rural hinterlands, and lower-order central places. At the center of this hierarchy are large cities, which serve as market places for their constellation of towns. Towns, in turn, provide goods and services to their constellation of villages and their rural hinterlands. Figure redrawn from Berry and Parr (1988).

some goods and services to their residents and their rural hinterlands, and where the size, spacing of, and the goods provided by the central places are determined by the size of their populations or consumer markets. An important element of CPT is that central places provide goods and services to both their urban residents as well as their rural hinterlands.

In the early industrial urban-rural landscape that Christaller's model sought to explain, there were multiple links of influence and effects between central places and their hinterlands. Central places provided goods and services to their rural hinterlands, and the hinterlands provided food and materials to central places. Due to their proximity, we can hypothesize that central places provided employment opportunities for rural areas, leading to an ongoing influx from rural to central places. Rural lands often had urban landlords. Multiple ties of language, culture, and travel linked urban places and their rural hinterlands. Finally, both urban and rural areas were under the same regional or national governance. Underlying the central place model and the early twentieth century economic spatial structure was a set of reciprocal relations between urban and rural areas. The proximate connections between cities and their regional spheres of influence has given way today to distal relations with international locations for materials, food, manufactured goods, and services as a result of globalizing trends, which have accelerated since the 1990s (Seto et al. 2012b; Seto and Reenberg, this volume).

Central place theory was further developed and applied after Lösch's modifications (Mulligan 1984), although its lack of a microeconomics foundation left it outside the dominant economic paradigm. The new economic geography (NEG) developed by Krugman (1991) and others (Fujita et al. 1999) sought to legitimatize concern with space and location decisions in mainstream economic theory by providing a rational choice-general equilibrium model of urban agglomerations. Krugman's NEG model changed one of the assumptions of classic general equilibrium models: instead of decreasing returns, he posited increasing returns. In his model, spatial structure is determined by the costs of transactions (e.g., shipping costs, transport structure, and geography) and types of increasing returns to scale (e.g., dense labor markets providing much better matching for workers). The model focused on industrial production. The spatial structure that results from NEG models, however, has been criticized as simplistic and at best capturing the industrial urban landscape of the early twentieth century (Mulligan et al. 2012). Even in the mid-1990s, NEG could not account for deindustrialization trends (Martin and Sunley 1996), and it is not clear how such models would address the international urban landscape in the wake of global supply chains.

Glaeser and Kohlhase (2004) pose more fundamental objections to both economic geography and urban economic models, including NEG models. They point out that such models are driven by transportation costs and that with the decline of transportation costs for goods movement, the models may provide good accounts of Western industrial cities in the 1900s but fail to capture the dynamics of urban patterns in the twenty-first century. They point out that in 1890, the average cost of moving a ton of goods one mile in U.S. cities was 18.5 cents (in 2001 USD), and that in 2004, the price dropped to 2.3 cents, arguing that theories of spatial development could almost assume that the transportation cost of goods movement is costless (Glaeser and Kohlhase 2004:199). Instead, they argue that agglomeration effects come from access to other people and not from transportation savings, that "natural resources in production are becoming increasingly irrelevant," and that factors that determine agglomeration are consumption-related natural amenities (such as warmth/climate) and regulatory policies of states or cities (Glaeser and Kohlhase 2004:200).[5]

Central place theory is garnering renewed interest today for several major reasons. First, empirical studies are providing evidence of the hierarchy of central places and its relation to the power law (Chen and Zhou 2006; Hsu 2012). This makes CPT an early precursor of scale-free network theory (Barabasi and Albert 1999), a branch of complexity theory, providing CPT with new theoretical pedigree and a claim to "feasible optimality" (Chen and Zhou 2006). Second, CPT is being recognized for its policy value in regional economic development and planning strategies (Mulligan et al. 2012). In addition, a recent study of the hinterlands and small urban centers in metropolitan areas in the United States (Partridge et al. 2008) finds that distance from higher-tiered urban areas was associated with lower population growth during the 1950–2000 period. The study concludes that given this result, the NEG model needs to take into account that distance has discontinuous effects, as well as the market pull of hierarchical central places. Furthermore, the study indicates that the costs of distance appear to be increasing more rapidly over the past few decades, despite the advances in IT and decline in transportation costs during this period. This research has important implications for the development of a new concept of urbanity envisioned by Boone et al. (this volume), since it suggests that distance is not a uniform measure, but rather weighted by the urban node's location in the urban hierarchy.

Motivated by interest in the environmental footprint or imprint of cities, recent scholarship is turning attention to rural hinterlands, a crucial aspect of CPT. In particular, recent empirical work provides evidence of the relationships between several Western cities and their hinterlands (Billen et al. 2012b). For example, Keene (2012) examines the food, drink, and fuel demands of Medieval London in 1300 and shows how the city drew supplies from its hinterland. Other studies focus on the actual extent of the hinterlands; for example,

---

[5] Responding to these criticisms, Krugman (2011) admits that since NEG was mainly focused on manufacturing, it really was designed to explain the 1900 spatial economic landscape in the United States. He argues, however, that while the theory may not be adequate to explain the current urbanization patterns in advanced economies, it can explain the spatial geography of emerging manufacturing powers, especially China.

New York City's food came from distant places as early as 1800s (Swaney et al. 2012), but 70% of Paris's food supply is still local (Billen et al. 2012a).

Contemporary Urban Patterns, Infrastructures, and Globalization

This section reviews the role that transportation and other infrastructures played in metropolitanization, before turning to a discussion of factors that led to globalization. Most important for understanding urban agglomerations and the distal supply hinterlands that have replaced local hinterlands are advances in technology, logistics, and the removal of financial barriers to trade.

Technology and Infrastructures Enabling Metropolitanization

The transportation infrastructures that facilitated the dense and compact early industrial city in the United States and other Western countries in the twentieth century were railroad and water transportation. Through these infrastructures, cities obtained food and raw materials from the countryside, and manufactured goods were transported from cities to other settlements and rural areas. The efficiency of these infrastructures depended on few hubs for shipping and, thus, railroad stations and ports reinforced central places, both in cities as well as in rural towns and villages. According to the 1952 U.S. Bureau of the Census, the mass production of automobiles (which began in the 1910s and made automobiles widely affordable) led to widespread car ownership after WWII, with the rate of car registrations per person surpassing 26% by 1950. Cheap automobiles, extensive road building by states and local governments, and the Federal Interstate Highway System of the 1950s were major factors contributing to the decentralization of the U.S. population into suburbs, leading to the first phase of metropolitanization. Communications media, such as the telephone and television, were well-established in the United States by the end of the 1940s and facilitated suburbanization. Rural electrification, which began during the New Deal of the 1930s, also contributed to opening up rural areas to suburbanization. In the 1960s and 1970s, retail in the form of shopping malls followed housing to the suburbs. With the shift in the U.S. economy from manufacturing to services in the 1960s and 1970s, office stock doubled, and then doubled again from 1970–1990. By 1986, over 57% of the U.S. office stock was located in suburban locations (Pivo 1990). Lang (2003) noted that by the late 1990s, more than 40% of total office space was located in outer suburban and exurban areas, in office clusters referred to as "edgeless cities." In the early 1990s, Garreau (1991) had identified "edge cities" as a new urban phenomenon in suburban America. Edge cities were primarily new office and retail centers, typically located at or near highway intersections. Tysons Corner, Virginia, was an early example of an edge city. A new metropolitan spatial model developed by Lang and Knox (2009) depicts metropolitan areas as polycentric with two or more central cities and their suburban areas, several edge cities near highway intersections, and isolated office clusters or the edgeless cities within the urban centers' spheres of influence, and micropolitan areas (self-standing towns of from 10,000 up to 50,000 people) outside the sphere of influence of the major urban centers.

Widespread automobile ownership and ubiquitous road networks in developed countries have facilitated the deconcentration of population in urban regions. The clear spillover of population, built environment, and economic activities into what used to be the rural hinterlands of central cities have either reduced the role of rural hinterlands in providing food and materials to their traditional central places or extended these areas much farther.

Technology and Infrastructures Enabling Globalization

The reshaping of the United States and European landscapes into polynucleated metropolitan areas occurring over the past twenty years encroached into the traditional rural hinterlands of central cities. At the same time, the globalization of production and services and international trade was accelerating. These complementary economic-spatial patterns led to both the diminishing importance of rural hinterlands for supplying food and raw materials to their central cities, and of central cities for the production of goods and services for their spheres of influence. We now turn to a discussion of how freight movement transportation advances, information communications technology, and logistics have been major factors in the quickening of the global trade experienced in the last few decades.

Containerization of Maritime Shipping

The containerization of cargo, its standardization, and the development of container ships and container port facilities were design innovations that revolutionized maritime shipping. It has made possible the globalization of trade that we are experiencing today. Before containerization, specialized goods, such as oil, were carried across the oceans in bulk, in ships designed to transport such goods, referred to as break-bulk shipping. Break-bulk ships carried goods in boxes, bags, barrels, or loose. Such shipping required individual handling and the potential for damage and theft was high. Before containerization of break-bulk cargo, longshoremen could spend as much time loading and unloading ships as the vessels spent in transit. Containerization, in effect, automated the handling of freight in port facilities. Malcom McLean developed the first commercial containers and container ship in 1956 (Cudahy 2006), but the widespread adoption of this innovation was slow. It was not until the late 1970s that container shipping reached developing countries (Hummels 2007). The container revolution included standardization of truckbeds and railcars for transporting containers over land. This enabled the secure transshipment of goods from the producer to its final destination.

By cutting out a large part of the labor costs to move cargo from port to vessel, increasing the speed of loading and unloading cargo (reported to have dropped from weeks to hours; Levinson 2008), and increasing the capacity of container vessels[6] and their speed, containerization reduced total time in transit, and shipping costs. The World Trade Organization estimates that maritime freight rates decreased by 65% between the 1950s and 1990s (WTO 2011); however, in a study of cargo and air shipping, Hummels (2007) concludes that containerization reduced transportation costs at best by 3–13%. The shortening of total time spent in transit and reduced transport costs enabled factories in emerging and developing countries and, given their lower labor costs, to compete with United States and European factories. As a result, United States and European imports diversified. According to Levinson, for example, the United States imported four times as many types of goods in 2002 as it did in 1972. In developed countries, this diminished the importance of urban centers as sources of manufactured products for their urban regions.

Growth in Air Freight

The widespread adoption of jet engines between 1957–1972, which were "faster, more fuel-efficient, and reliable and required much less maintenance compared to the piston engines they replaced" (Hummels 2007:137), was instrumental in lowering the price of aircraft and opening up opportunities for air freight. By the mid-2000s, air shipments represented less than 1% of total tons and ton-miles shipped, but were growing rapidly. A good illustration of the increasing importance of air shipping is the growth in the value of air imports and exports in the United States. Over the period 1965–2004, the share of trade value of air imports more than tripled, and the share of trade value of exports more than quadrupled. The total amount of air freight carried increased from 20 billion freight tons/km in 1975 to 160 billion freight tons/km in 2007 (Kupfer et al. 2011). This rapid growth of air imports occurred because of the sharp drop in the relative cost of air shipping: from 1955 to 2004, worldwide air revenue per ton-kilometer fell more than ten times (Hummels 2007).

Both maritime and air freight are urban-serving infrastructures. They are network infrastructures in which metropolitan ports and airports are nodes in global supply chains increasingly dominated by powerful shipping alliances and large carriers (Ducruet and Notteboom 2012). However, ports and airports are typically financed and maintained by local or regional governments.

---

[6] At the beginning, 20-ft containers (twenty feet long and eight feet wide) were the norm. Today the capacity of container ships and trade from container ports is measured in trailer-equivalent units (TEUs) or 20-ft equivalent units. A 2,000 TEU container ship can accommodate 2,000 20-ft containers or 1,000 40-ft containers. By 2010, new vessels could handle over 10,000 TEUs.

Information Technology, Logistics, and Supply Chains

The transformation of global trade was brought about by both advances in transportation and information technology as well as by techniques that optimized their use in global trade (i.e., logistics and supply and demand chain management). Logistics and supply chain management (often used interchangeably) refer to the planning and control of the flow of materials, supplies, and finished products for a firm (Stenger 2011). Logistics services, a growing subsector in the advanced services sector, rely on computing and digital communication technology to utilize programming and scheduling algorithms, spreadsheets, inventory optimization, electronic data interchange among businesses, on-line shipment, billing and payment, etc. (Schwarz 2006). The increasing speed of IT technology and sophistication of logistics services have made possible process innovations such as lean production and just-in-time manufacturing.

Governance and Economic Factors Enabling Globalization

Reductions in trade barriers among nations, the export-oriented growth strategies of emerging markets and other developing countries, and the increasing dominance of international financial institutions and multinational corporations were essential conditions for the globalization of trade. This has also contributed to the erosion in importance of the rural hinterlands of cities.

Reductions in Trade Barriers

Reductions in trade barriers in emerging markets,[7] especially in China and India—outcomes of either unilateral or multilateral agreements—have increased the opportunities for trade. For example, according to the World Trade Organization, by 2009 "China, India and Vietnam, countries which lowered overall tariffs the most, relative to 2001, had also seen the highest annual rates of trade growth" (WTO 2011:40). Another indicator of the decline of trade barriers over the past twenty years is the change in the trade openness ratio (i.e., ratio of total trade to gross domestic product) of these emerging economies. A Brookings Study (Kose and Prasad 2010:43) estimates that from 1985 to 2010, the trade openness ratio of emerging economies increased from less than 30% to around 80%, while advanced countries saw an increase from 26% to 46%. Moreover, the average growth rate of exports from emerging markets during this period was two times the rate than for advanced economies (Ahearn 2011:13).

---

[7] Emerging markets include China, India, Brazil, Indonesia, Mexico, Russia, Turkey, and Vietnam, of which China, India, and Brazil are the most dynamic.

Financial Flows and Foreign Direct Investment

As a result of the privatization of state-owned banks and removal of restrictions on the acquisition of assets by foreigners, private capital flows to developing countries have substantially increased over the past 25 years almost quadrupling their share of GDP, a much faster rate of growth than that of trade flows. During this period, foreign direct investment has been instrumental in incorporating developing countries into the global economy. By 2009, China was the second largest recipient of foreign direct investment, behind only the United States (Ahearn 2011:20–21).

The Transformation of the Industrial Production Model and the Rise of Multinational Enterprises

The industrial production model for most of the twentieth century revolved around the breakdown and automation of tasks on the factory floor or assembly line, *and* the vertical integration of production under one company beginning with the acquisition or processing of raw materials, design of products, to the manufacture of parts, their assembly, and, in some cases, even the delivery and marketing of the finished product. In the United States, this model of vertical integration was eroding by the 1980s, with outsourcing of parts, even in the automobile industry, becoming common. With advances in freight transportation and information communications technology, firms in emerging economies, and developing countries in general, sought opportunities to export goods and services to developed and other countries. In a similar way, companies in developed countries saw opportunities for foreign direct investment by offshoring segments of their production or by investing in new factories offshore, where the entire production process could take place to reduce overall cost (WTO 2011). Vertical fragmentation of production, or "slicing and dicing" the production process and distributing the segments around the world, enabled firms to seek out the cheapest and most appropriate labor for different aspects of production; it also provided the opportunity to open up multiple markets for their merchandise. In effect, the new model of vertical fragmentation internationalizes the assembly line. An example of a distributed global production chain is found in Table 14.1, which lists the distribution of production for Boeing's 787 Dreamliner. The new wave of multinational enterprises ushered in a network model of global production and distribution managed by IT-enabled logistics and global freight shipping infrastructures. In the early stages of globalization, multinational enterprises were typically large corporations from advanced economies that belonged to oligopolistic industries (Kogut 2001). Today, emerging economies are the source of an increasing number of multinational enterprises. By 2010, almost one-fifth of

Table 14.1 Fragmentation of production for Boeing's 787 Dreamliner: components are listed by country in which they were produced. Data from WTO (2011:95).

Components	Countries
Doors and windows	United States
Forward fuselage	Japan, United States
Escape slides	United States
Flight deck seats	United Kingdom
Flight deck controls	United States
Engines	United States, United Kingdom
Engine nacelles	United States
Center wing box	Japan
Landing gear	France
Electric brakes	France
Tires	Japan
Prepreg composites	Japan
Cargo doors	Sweden
Passenger doors	France
Auxiliary power unit	United States
Horizontal stabilizer	Italy
Raked wing tips	Korea
Vertical stabilizer	United States
Wing box	Japan
Wing ice protection	United Kingdom
Rear fuselage	United States
Lavatories	Japan
Center fuselage	Italy
Tools and software	France
Navigation	United States
Pilot control system	United States
Wiring	France
Final assembly of the airplane	United States

the Fortune Global 500 companies were emerging market multinational enterprises (Ahearn 2011:23).

Trade globalization has resulted in a multiplicity of urban global linkages over which cities have little direct control or influence. As a result, urban places have no meaningful links to or influence over the global location of their suppliers (their global hinterlands). In addition, because of the globalization of production and services, urban places have fewer ties with their traditional, surrounding hinterlands.

Castells' Theory of the Network Society

Castells, who recognized early the new social reality made possible by information communications technology and transportation advances, argues that our current stage of world development is not properly characterized as the information society, but rather as the network society (Castells 1989, 1996). As he explains, network organizations, such as families or friends, are not new, but before digital IT, communication and coordination across a network was slow and difficult, and so major social processes were carried out by hierarchical organizations (e.g., government, military, vertically integrated business firms, religions). In contrast, digital information and communication technologies enable networks that make possible communication and coordination among large numbers of people in real time across the world, diminishing the importance of space (Castells 2010). Networks are multiple, fluid, and are characterized by different nodes and linkages of shifting strengths. What others call global cities, due to their prominence in financial services or technological innovation, Castells insists are nodes in networks of global finance or technological innovation. It is the networks and not the places which are formative, and networks are always in flux. In his conception of the network society, Castells conceives urban space as a space of flows of people, communication, and goods, and, as noted above, he provides the rationale for the definitions of metropolitan and mega-city regions. He recognizes the role of transportation and communication in urban agglomerations and the lack of government or institutional cohesion in these vast regions.

Castells' insistence on the importance of networks, and not places, raises the question of why the hubs of global networks are located in metropolitan regions. In an age of instantaneous and multiple means of communication, why do networks still concentrate in urban regions? Part of his answer is that urban agglomerations of "services, finance, technology, markets, and people" generate economies of scale. More intriguingly, he points to another type of economy that drives urbanization: spatial economies of synergy. By this, he means that sharing a space with a valuable partner in a particular network increases the "possibility of adding value as a result of the innovation generated by this interaction" (Castells 1996). The network society still requires metropolitan spaces where interpersonal interaction can occur "because communication operates on a much broader bandwidth than digital communication at a distance" (Castells 1996). From his perspective, these vast metropolitan areas are likely nodes for various networks, which act as attractors for capital, talent, or other valued things. However they lack, except in a few cases, such as Toronto, the type of government or institutions that match their vast territories, and thus remain powerless to plan or govern these vast regions. As a result, Castells concludes, these new urban patterns display both dynamism and marginality, and face a fundamental contradiction between the wealth and power attached to network flows and the meaning that places hold for people.

Changing Urban Values and Urban Policy: The Compact City Model

While urban development has become increasingly decentralized into vast metropolitan regions, the worldwide social movement toward sustainable development has embraced a compact city strategy as key to sustainability and revived the concept of place-based communities. The compact city model has become a near-paradigm for urban design and planning (Kenworthy 2006; OECD 2012). At its theoretical core, the compact city approach argues for higher densities, clear boundaries between urban and nonurban places, a mix of land uses, transit accessibility, as well as a pedestrian- and bicycle-friendly street environment. The central values that underlie the contemporary compact city model are related to transit accessibility, to the reduction of the number and length of household automobile trips, their associated energy savings, and the consequent reduction of air pollution and greenhouse gases. In Europe, the United States, and other countries around the world, the exemplars for the compact city model are the older, compact, pedestrian-oriented, and transit-accessible European cities such as Paris, London, or Barcelona. In the United States, prompted by and reacting to ongoing suburbanization, the compact city model is aligned with urban growth management efforts and the New Urbanism, especially transit-oriented development (TOD).

The early phases of urban growth management policies were primarily concerned with protecting rural and resource areas from urban encroachment, in effect from suburbanization, which led to various urban containment mechanisms, in particular, the urban growth boundary pioneered by the state of Oregon. New Jersey, which was the most suburbanized state in the country in the 1980s, used the central place concept as the major regional organizing concept for managing growth in its first state-wide urban growth management plan (New Jersey State Planning Commission 1991). The plan's general strategy was to guide future growth into compact forms of development and redevelopment. The plan identified compact forms of development as "centers," ranging from urban areas to towns, villages, and hamlets, and emphasized the market function of these centers and their potential for public transit and walkability.

The New Urbanism: Transit-Oriented Development

The New Urbanism is an influential urban design movement that emerged in the United States in the mid-1980s focused mainly on reforming suburban development (Duany et al. 2001). It sought to replace suburban subdivisions with more traditional town plans incorporating interconnected street grids, better defined lot/block and block/street patterns, and higher densities than existing suburbs. In addition, many new urbanist designers emphasize vernacular architectural styles. Peter Calthorpe, an urban designer, initiated the concept of TOD in the late 1980s. This concept has been incorporated into the new urbanism agenda and is a major tenet of the compact city approach. As with New Urbanism in general, the concept was very much aimed at new suburban development, but it is also applicable to the redevelopment of existing neighborhoods. Calthorpe (1993) argues that reduction in automobile use and increase in public transit should be pursued as a regional urban form strategy that links TODs. The TOD hierarchy is similar to the central place hierarchy. The TOD strategy,[8] however, is part of a broader sustainability agenda that includes the following elements (Dittmarr and Ohland 2004):

- organizing growth at a regional scale to be compact and transit supportive,
- locating retail, jobs, housing, parks, and civic uses within walking distance of transit stops (mixture of land uses),
- creating pedestrian-friendly street networks that connect local destinations,
- providing a mix of housing types at different densities and prices,
- preserving sensitive habitat, riparian zones, and high-quality open space, and
- making public spaces the focus of building orientation and neighborhood activity.

A variation of the TOD concept is Seattle's urban village strategy,[9] which has been used to densify existing low-density residential neighborhoods in Seattle. The organizing concept of urban villages (also similar to the central place concept) was applied within the city itself to develop hubs of neighborhood-oriented commercial areas mixed with higher-density residential uses. While the compact city model is a general internal urban form model that specifies broad features of a city, such as a mix of uses or density, the urban village concept and TODs are hierarchical organizing concepts for the internal structure of cities or urban regions.

More recently, sustainable urbanism has promoted local food and materials, in effect, calling for greater reliance on local food suppliers. This is an understandable attempt to bring back Christaller's world. Reviving a city's connection to its rural hinterland would provide greater ties of influence between urban consumers and proximate hinterland producers and avoid the unknown and hard to influence multinational supply chains. These calls for "buying local" are often motivated by concern for a city's ecological footprint, and, more specifically, for the reduction of energy used in transporting food from distant regions and the resultant emission of greenhouse gases. However, life-cycle analyses draw a more complex account of the energy and greenhouse gas costs

---

[8] To view the "Urban Strategy: A New Framework for Growth" see http://www.calthorpe.com/files/Urban%20Network%20Paper.pdf

[9] For more information, see http://www.seattle.gov/dpd/static/Urban%20Village%20Element_LatestReleased_DPDP_021118.pdf

of distal versus local food production (Sim et al. 2007), and theorists warn against assuming that local supplies are more just or sustainable than nonlocal ones (Born and Purcell 2006).

Empirical Research on Compact Cities

Is the compact city model supported by the findings of empirical research? Much research has been conducted on this topic. A recent meta-analysis of research on travel and the built environment (Ewing and Cervero 2001) gives a good indication of the amount of research on the topic. Ewing and Cervero reviewed 200 studies that quantified the relation between travel and the built environment and included fifty of these in their meta-analysis. In addition, a comprehensive study conducted by the U.S. National Research Council examined the empirical evidence on the effects of compact development on motorized travel, energy use, and CO_2 emissions (NRC 2009). Based on the few studies that used data from individual households and applied the best available statistical methods, the National Research Council found that doubling residential density across a metropolitan area might lower the number of vehicle miles traveled (VMT) per household by about 5–12%, and perhaps as much as 25%, if coupled with higher employment concentrations, significant public transit improvements, mixed uses, and other supportive demand management measures. This means that the elasticity of VMT, with respect to density, ranged from –0.05 to –0.12.[10] Ewing and Cervero's meta-analysis of research on travel and the built environment analyzed, for the most part, studies that used aggregate data. Using aggregate data rather than individual household data makes it difficult to claim causality. However, the meta-analysis untangled the effects of different aspects of the compact city model and found that the elasticities of VMT, with respect to different aspects of the built environment, varied from land-use mix at –0.09, intensity/street density at –0.12, destination accessibility –0.20, distance to downtown –0.22, and density –0.04. Density in the meta-analysis had a very weak correlation with automobile travel, at the low range of the National Research Council results. Both of these major research syntheses conclude that combining several compact city strategies (e.g., density, land-use mix, and roadway connectivity) can have a synergistic effect and significantly reduce VMT. In effect, these studies support a mix of compact city strategies, such as increasing density and mixing appropriate land uses around areas of high public transit accessibility, centralizing jobs in city centers and surrounding transit nodes or TODs, as well as urban containment strategies. Research also indicates that higher densities than typical suburban densities are more cost-effective from the perspective of infrastructure and public facilities provision (Burchell et al. 2005).

---

[10] Elasticity is a measure of the change in density to the associated change in VMT. In this case, as the density increased 100% or doubled, VMT decreased from about 5% to 12%.

Summary and Conclusion

Defining the Urban

When we turn to definitions of urbanization, it is clear that what we mean by urban settlements varies around the world. Traditional definitions revolve around total population, population density, and authorized local units of national or state governments. More recently, some advanced nations have shifted their definitions from state-authorized units (such as municipalities or contiguous dense physical development) to flows of population (as in commuting patterns) and flows of goods, services, and communication. This has led to definitions of new urban patterns, such as metropolitan areas or mega-city regions that incorporate some rural lands. It also reflects the blurring of the traditional urban-rural dichotomy. Around the world, urban regions have outgrown the local government jurisdictions of the past century, and most urban agglomerations today are faced with the lack of adequate government power to manage their growth, raise revenue, and provide public goods and services. This makes the urban governance issue a crucial challenge for the urbanizing world.

Urban Economic Dynamics and Spatial Form

Economic geography theories attempt to explain urban spatial patterns through the interaction of economic drivers; as such, they are the main way through which we make sense of the economic aspects of urbanization. Urban spatial theories offer insights on why firms and people aggregate in cities, and on the location of firms and households within cities. Central place theory, drawn from observations of the pattern of urban settlements in Germany's industrial economy in the early twentieth century, developed a hierarchical scheme of central market places serving lower-order central places and their hinterlands, with hinterlands providing food and materials to their central places. The 1990s brought a new interest in the theory from mainstream economics and the NEG approach provided a microeconomic foundation for it, although the spatial result of NEG was simplistic. Renewed interest in CPT from regional scientists has been fueled by confirmation that in Japan and the United States, settlement size and numbers follow a power law, analogous to the hierarchical network of central places posited by Christaller and Lösch. Central place theory has also become particularly relevant in urban policy, since compact city strategies at the regional and intracity scales incorporate policies to develop hierarchies of central places. In addition, recent research finds that not all central places have the same centripetal pull on their rural spheres of influence, providing evidence of a hierarchy of urban places. Most recently, empirical studies of rural hinterlands, which are of theoretical interest because of the increasing substitution of the economic exchanges involved in central places and rural hinterlands with global, distal exchanges, are beginning to be published. To a large extent, these studies are motivated by the policy interest in more sustainable urban footprints.

Globalization Theory

Castells' influential theory of the network society incorporates a sophisticated understanding of our global economy and has special relevance to urban areas. Important insights include: the role of information communications technology in making possible global networks of production; the transformation of hierarchical, vertically integrated production processes to networked distributed global production; how economic networks as formative drivers are changing patterns of urbanization from contiguous urban places to the fluid spaces of mega-city regions; the continuing importance of urban agglomerations and their central places as attractors of capital, innovation, research, etc.; the evident contradictions in urban regions between wealth and marginality; and the fundamental contradiction for urbanism between the flows associated with the wealth and power of networks, and the meaning and attachment people have for places. Both Castells and Glaeser see the potential for human interaction in shared places as a major driver for urban agglomerations in our increasingly placeless global societies—the city as a setting for encounters. Both underestimate the value of infrastructure systems and built environment as economic attractors in themselves.

The Role of Infrastructures

Reflecting on the features of urbanization discussed at the beginning, definitions have emphasized density and, more recently, flows of population, goods, and information. Our review of factors and features of global trade highlighted the formative role of transportation infrastructures, information communications technology, and logistics (which essentially refers to strategic and implementation planning) in transforming the world economy into a global economy. The brief review of U.S. metropolitanization also emphasized the role of transportation and other infrastructures. In both cases, of course, other factors (e.g., government actions) were also necessary. For example, as discussed above, in the case of globalization, trade liberalization served as an essential driver; in the case of suburbanization, the role of government in insuring home mortgages, which led to the 30-year mortgage, opened up home ownership to millions. Infrastructures make possible the functioning of an urban region: they are the road and rail networks that channel the flows of people, the communication infrastructure that facilitates communication, water infrastructures which are essential for life and health, etc. Infrastructures are sources of continuing value for urban agglomerations. Unlike the assumption of CPT, the urban landscape is not a featureless plain; it is endowed with durable infrastructure networks that make urban life possible in vast urban agglomerations, and on which both people's lives and economic activities depend. Economic activities, especially in network societies, are fluid. An urban region's fortunes may ebb or flow, but infrastructures are long-lived endowments of urban civilizations. The persistence of primate cities in the world have much to do with the increasing returns of their urban infrastructures. Infrastructures represent the shared durable capital of urban regions and require government institutions, commensurate with their scale, to plan, establish, maintain, and upgrade them.

Urban Policy: The Compact City Model

While the economic and political-economic theories reviewed here seek to explain the economic aspects of urbanization, urban policy reflects normative values as well as an understanding of explanatory theories and relevant empirical research. The compact city model, although it flies in the face of trends toward deconcentration, has become a normative model for sustainable urbanism with its goals of reducing automobile travel, energy use, and greenhouse gas emissions. Within a regional or even within the city context, CPT can offer an organizing framework for mega-city regions. The NEG and complexity theory provide frameworks and models that support CPT as an organizing schema. Economic geography, on the basis of rational choice theory and general equilibrium models, and complexity theory, on the basis of self-organizing complexity models, provide theoretical support for CPT-based regional policies. Also, the hierarchy of central places aspect of CPT, with its focus on trade or retail centers, retains more relevance in the face of global production chains than NEG models, with their focus on industrial production. One of the major motivations behind support for the compact city model is to reduce automobile travel, although other motivations are also at work, such as providing a more livable built environment and reducing land conversion to urban uses. Reducing automobile travel has become very important because of the high energy consumption of automobile travel when compared to other modes of travel, and the associated air pollution, especially, greenhouse gas emissions and their impact on climate change. Empirical research supports a mix of compact city strategies for reducing vehicle miles traveled, although density alone is weakly correlated with travel behavior.

Rural Hinterlands and Ecosystem Services

Both economic geography, based on microeconomic theory, and Castells' sweeping account of the network society within a global economy fail to incorporate the reliance of cities on ecosystem services, what CPT partially captured through its recognition of the rural hinterlands of central places. Neither confronts the ecological challenges that we face (in particular, the challenge of climate change), which have significant implications for the future of urban agglomerations. For example, urban regions in many parts of the world already face water scarcity, a problem which will be compounded in this century under climate change. Doubtless, many of these metropolitan regions will have the institutional and financial means to forge sustainable solutions, but in other cases, water supply issues may reduce the growth and viability of urban regions.

The Governance Challenge of Metropolitan Regions

Around the world, urban regions have outgrown the local government jurisdictions of the past century, and most urban agglomerations today are faced with the lack of adequate government power to manage their growth, raise revenue, and provide public goods and services. This makes urban governance a crucial issue.

The concept of the ecological footprint has emphasized the extent of goods and services that urban areas consume. In a roundabout way, it has focused attention on the extent to which many of the ecosystem services, which local hinterlands provided to cities in an earlier age, are now being provided through the distal connections of our global economy. Correspondingly, the clients of the specialized goods or services that urban regions provide are not necessarily regional but global networks. An element in the movement toward sustainable urbanism is the "buy local" program, intended to reduce our ecological footprint. Assessing the value of the policy of buying local can be determined on the basis of transportation costs, energy savings, reduction of greenhouse gas emissions, or alternatively on the basis of supporting sustainable livelihoods in the local or in a distant region. Multiple criteria can and should be applied in this calculus. A neglected criterion that should be included is a concern for the future governance of urban regions. In a world where national governments over the past quarter century have increasingly ceded power over finance and economic activities to international financial institutions and multinational enterprises, fostering a common regional identity and solidarity may be important first steps in developing effective metropolitan governance regimes to confront climate change and other environmental challenges, as well as to temper the volatility of financial and trade markets in the twenty-first century.

15

How Is Urban Land Use Unique?

Dagmar Haase

Abstract

This chapter discusses current urban land use (in terms of form, size, and shape of cities and urban areas) against a global background. The specifics of urban land use (surface characteristics, dynamics of change, and impacts on the environment) are examined using different conceptual approaches (e.g., ecosystem services, risk, and governance aspects). Although urban land use is a special case (i.e., small in scale, yet dominant in influence), a range of commonalities exist between urban and nonurban land use. A discussion on shrinking cities underlines that there are more pathways to urban land (-use) development than growth. The current extent and rates of urbanization force us to rethink land connectivity, competition, and decision making, and the resulting knowledge can be used to generate a new concept of land use. The connections and implications of urban land-use patterns need to be examined on a global scale, as local-scale patterns may be affected by global-scale outcomes and vice versa.

Current Importance

There is no consensus as to what defines a city or an urban area, but a commonly accepted definition stipulates that cities are large and permanent settlements characterized by high population densities and complex supply systems for housing, business, transportation, sanitation, and utilities (Jenks and Dempsey 2005). A key characteristic of cities is that they facilitate strong and multiple interactions between people (Satterthwaite 2007). Cities vary in size and form: large "core" cities, for example, may be surrounded by peri-urban settlements that have lower population densities. Whereas a city describes a place that can be demarcated by administrative boundaries, an urban area is spatially less defined. Worldwide, cities are the primary type of settlements, and the lack of consensus on a definition does not lessen their importance in terms of economic or human well-being.

Briefly viewed from a historical perspective, only 3% of the world's population lived in cities in 1800. By the end of the twentieth century this figure rose, however, to 47%. In 1950, there were 83 cities with populations that exceeded one million. By 2010, this number had risen to more than 460 cities. If this trend were to continue, the world's urban population would double every 38 years.[1] Presently, more than 50% of the world's population live in cities and urban areas[2] and future population growth, whether natural or migratory, is projected to occur in urban areas. Globally, cities and urban land use have gained increasing importance. Today, approximately 95% of the global gross domestic product (GDP) is produced in cities (Seto et al. 2012b). Many universities and research centers are located in cities (although this applies only partially in the United States), and almost all financial and insurance services, and their resultant profit flows, are found in urban areas (predominantly in the United States and Western Europe) (Angel 2012; Angel et al. 2010).

Over the last two centuries, cities have developed rapidly in terms of scale and rate of change. Urban population projections would have us assume that this trend may abate in the Western industrialized world while continuing in the mega-cities of the Southern Hemisphere (Beauregard 2009). Although the majority of the world's population lives in urban areas and final goods and services are produced in urban areas, less than 5% of Earth's surface is urban. A recent forecast of urban land expansion predicts that urban areas will triple by 2030 (Seto et al. 2011) but still cover less than 10% of Earth's surface, further underscoring the concentration of people and activities within a small part of the available land-surface area. Regions like Europe, North America, and Oceania have high levels of urban populations (i.e., more than 75% of the population lives in urban areas) and relatively low—and partially shrinking—population growth rates. In contrast, Asia and Africa have relatively low levels of urbanization and of the world's urban population, but they exhibit high growth rates and have the largest number of mega-cities overall. Central and South America have a high share of urban population and high growth rates yet few mega-cities.[3] Thus, most of the world's future urbanization is projected to occur in the Global South (Seto et al. 2011).

Not all urban areas are expanding or expected to grow in the future. In regions where the bulk of the urbanization and industrialization processes occurred more than a century ago, there is evidence that urbanization will abate if not decline. Declining urban populations have been observed in some North American and European cities. In addition, many experts predict a re-urbanization from suburbia or peri-urban areas to city centers (Kabisch et al. 2010; Buzar et al. 2007). While the process of re-urbanization will result in an increase in population in the city centers, it will also involve a

---

[1] http://www.citypopulation.de
[2] http://esa.un.org/unup
[3] http://esa.un.org/unup

depopulation of the already less dense peri-urban areas. This relatively new phenomena of urban shrinkage, decline, and related land-use perforation (Haase 2012b) pose new challenges to sustainable land use and urbanization. Globally, more than 370 cities are shrinking or in decline, with their net populations decreasing at $\geq 1\%$ per year. The phenomenon of shrinking cities is found predominantly in Europe, Russia, Japan, and in the U.S. Rust Belt (Rieniets 2009; Kabisch and Haase 2011). Urban shrinkage is characterized by a set of socioeconomic and land-use characteristics: population decline, an aging population, and the accompanying processes of building and infrastructure underutilization, land abandonment, and land-use perforation (Haase et al. 2012). Land-use perforation illustrates the uniqueness of urban land use: unlike abandoned agricultural lands, which may become pasture or some other type of "productive" unfarmed land, abandoned urban land does not regenerate in the same way, although lots can "recover" in the form of secondary plant succession (Haase 2008). The transformation of abandoned urban land to green spaces (e.g., parks, allotment gardens, urban forests) is, however, costly and requires continuous maintenance. In addition to abandoned lands, lots, and buildings, shrinking cities also suffer from inequalities of income, poverty, and quality of life (Haase 2012b), which can pose serious socioenvironmental problems for a city, such as the windblown dispersal of contaminated substrates of brownfields (i.e., land previously used for industrial or commercial purposes) or the establishment of run-down areas (e.g., illegal dump sites).

Along with population growth, inequalities of wealth and income distribution and energy consumption are likely to increase further (Sahakian and Steinberger 2011). Moreover, the resulting impacts from increased population growth and consumption will stress cities, urban spaces, and urban ecosystems (Satterthwaite 2009).

Although urban areas occupy a relatively small percentage of the global land area, they exact an enormous impact on global land use through the flows of people, materials, ideas, finance, and supply and demand. Not only do the resource demands of urban populations (i.e., diet, travel behavior, demand for goods) impact nearby rural areas, they link urban activity to land use throughout the entire world. Urban land "teleconnections" (i.e., the distal flows and connections between people, economic goods and services) and land-use change processes, which drive and respond to urbanization, provide a conceptual framework with which to examine these linkages (Seto et al. 2012b). Two examples of this are when changing diets of populations in one area of the world (e.g., Europe or the United States) force an increase in shrimp farming which results in greater coastal water pollution, or when the aging population of northern European cities spurs increased demand for the construction of retirement residences in southern European countries, accompanied by the problems associated with soil sealing in the Mediterranean coastal areas.

Path Dependency of Urban Development and Importance for Land

Where do we find the "urban sphere" and where do we find urban land use? More importantly, how does this affect our understanding of land-use patterns in and of cities?

In terms of the global urban land-use distribution across ecological zones, the biome of the temperate zone shows the highest share of urban land use, followed by the Mediterranean and coastal biomes. Remote sensing techniques, such as the Defense Meteorological Satellite Program (DMSP) satellite system, show the extent of urbanization and location of the world's cities in the form of night lights. In the DMSP images, the brightest areas of the planet are the most urbanized, except for regions in India and parts of China, where population density in rural areas is considerable and produces significant amounts of light. Cities tend to grow along coastlines and large transportation networks. The U.S. interstate highway system and the motorway triangle between the Netherlands, Belgium, and Germany appear as lattices connecting the brighter dots of cities and urban centers. Japan is also a highly urbanized region. Thus, a path dependency of urban land use is evidenced by the lattice-shaped transport systems around the world. There are large zones of Earth that still remain untouched by urbanization. The interior jungles of Africa and South America as well as the desert areas of Africa, Arabia, Australia, Mongolia, and the United States show up in DMSP images as poorly lit areas. The boreal forests of Canada and Russia and the great mountains of the Himalaya are also undeveloped (Cinzano et al. 2001; Lambin and Geist 2006). Thus we can conclude that cities tend to cluster. In fact, cities are more likely to develop in clusters over time than they are to emerge out of nowhere.

A recent survey of the world's largest cities (those with at least 1,000,000 inhabitants), which was based on findings of the City Mayors 2011 survey,[4] shows that mega-cities with more than 5 million inhabitants are growing, particularly in Asia, Africa, and Latin America. Clearly, the South has been replacing the West in terms of the largest urban agglomerations, but the West still has the highest share of urban land. With the exception of city-states such as Singapore, Monaco, or Vatican City, European countries are the most urbanized countries in the world. For example, Belgium is 97% urbanized, Denmark 87%, and Sweden 85%. The United States is 82% urbanized. It therefore appears that the size of urban areas and the population density in cities are becoming increasingly important for the image and quality of life of a city or urban region and that they also shape the teleconnections and impacts on distant land.

In considering the socioeconomic and demographic characteristics of cities, we must assume that the development of cities is creating polarized camps of

---

[4] http://www.citymayors.com/statistics/largest-cities-mayors-intro.html

"winners" and "losers" across the globe. In general, population and economic activities are being concentrated in large urban high-density areas, while the rural peripheries are depopulating. At the same time, we must expect that cities in the early industrialized Western democracies as well as in Japan, Russia, or parts of China will face a fundamental demographic change, including individualization, below-replacement fertility rates, and an aging population. It is not clear, however, whether stagnation or population shrinkage has purely negative effects on a city or the well-being of its population. Findings from European perception surveys in cities[5] suggest that cities which have passed the point of their highest population peak develop the highest levels of wealth and quality of life (e.g., Vienna, Pittsburgh, Newcastle, Manchester, and Leipzig). Thus, we can conclude that due to the path dependency of cities, urbanization can exacerbate existing inequalities in income or industrialization.

A Special Case of Urban Land Use: Urban Shrinkage and Decline

While much of this volume addresses urbanization, growth, and the increase in urban areas, it is important to note that not all urban areas are expanding. Haase et al. (2012) define urban shrinkage as a phenomenon of massive population loss in cities resulting from a specific interplay of the economic, financial, demographic or settlement systems, environmental hazards, and changes in the political or administrative systems. For example, the systemic changes experienced in Germany, Eastern Europe, and Russia after 1990, coupled with the introduction of a market economy, led to urban shrinkage (Rink 2009; Moss 2008). Urban shrinkage is a special case of urban land use partly because abandoned urban lots and buildings are subject in the future to different land-use trajectories and dynamics than other human-managed land systems that are abandoned. For example, abandoned agricultural systems revert to pasture or other types of vegetation; however, abandoned lots or buildings do not revert to their natural biophysical state.

Urban shrinkage and decline have been examined through the lens of uneven economic development (Harvey 2006) and the underlying dynamics of the territorial division of labor (Amin and Thrift 1994). Shrinkage can also result from tremendous environmental hazards, such as the devastation of New Orleans by Hurricane Katrina in 2005. Urban shrinkage can also occur through demographic change, namely low fertility rates and massive out-migration (Müller 2004). For example, in eastern Germany, current urban shrinkage has resulted from the decline in traditional industries (e.g., coal and petro-chemistry) after German Reunification. This decline induced a general economic crisis, unemployment, and out-migration to other prospering

---

[5] See the European Commission' Survey on Perceptions of Quality of Life in 75 European Cities
http://ec.europa.eu/regional_policy/sources/docgener/studies/pdf/urban/survey2009_en.pdf

regions. In turn, this caused a subsequent decline in fertility and an increase in the aging population (Haase et al. 2012). Furthermore, rampant suburbanization in peri-urban zones around shrinking cities leads residents to abandon the city. Both processes can result in a rapid increase in the age of the remaining population, which causes further demographic decline (Nuissl and Rink 2005). This pattern has been observed in many shrinking cities across Europe and the United States (Couch et al. 2005). Shrinkage is, without doubt, a socioeconomic process chain, but it can also have physical and spatial land-use components.

Over the past fifty years, approximately 370 cities with more than 100,000 inhabitants have experienced population losses of more than 10%. These cities are distributed across the world, but are located predominantly in early industrialized and developed regions. Europe has more than 70 (19% of the total) shrinking cities while the United States has more than 93 (25% of the total) (Kabisch et al. 2012). Urban shrinkage is already on the political agenda in countries such as Japan, Russia, and China.

The impact of shrinkage on urban land use is complex because it affects both the urban fabric (i.e., the physical aspects of urbanism) and open space in an uneven manner. In the United States, for example, cities experience a "doughnut effect": city centers are hollowed out by brownfields and unused plots, while the suburbs grow (Beauregard 2009). In eastern Germany, urban shrinkage has perforated the urban fabric: specific parts of a city face more drastic demolition, which results in a larger proportion of vacant land and thus an alteration of the built space (Haase et al. 2012; Schwarz et al. 2010). Despite the emergence of brownfields as a result of deindustrialization, such dramatic land-use changes have not yet been observed in eastern Germany. The impacts of shrinkage on land use can have a time lag component; for instance, the effects of land-use change after a vacant building is demolished or a new land use is finally established can take time to appear. Often, various types of interim land uses can be observed, representing the subtler processes of land-use change (Lorance Rall and Haase 2011). However, urban shrinkage should not only be associated with losses and negative connotations. Urban shrinkage also creates new spaces and affordable land, which is then available for alternative land-use options such as new public or green spaces (Haase 2008).

As outlined above, population decline in shrinking cities leads to a decrease in residential density. It also leads to an oversupply and underuse of urban land, namely of housing stock, infrastructure, and services (Haase et al. 2007). The underuse of building stock poses problems for public as well as private suppliers, specifically in regard to the dense urban fabric, industrial buildings, and storage depots. Underuse, in turn, leads to housing and commercial vacancies and to a more rapid dilapidation of unused buildings (Bernt 2009). In some places, buildings are demolished to balance the housing or real estate market (Couch et al. 2005), whereas in others, they simply become unusable after a period of disuse. Although a decreasing building stock density may lead to a relaxation of housing prices in a densely built city, vacant lots can lead to a disruption of the urban space in the form of dissolution of streets or block structure. (Haase et al. 2007). Moreover, ongoing construction activities in peri-urban areas and the rural hinterland of shrinking cities reinforce the decline of the inner city.

Land-use perforation itself poses challenges for superficial and subterranean urban infrastructure provision (Hasselmann et al. 2010). This is obvious for network-dependent infrastructures, such as water, sewage, or electricity. Vacant houses and derelict land no longer require a supply of water, electricity, or wastewater transport and thus the pipes and cables leading to a vacant house are no longer used. In an area with a larger proportion of vacant and derelict land, underutilization can pose severe maintenance problems for the service provision to the whole area (Moss 2008). Vacancy also affects the social urban infrastructure (e.g., schools, day-care centers, roads, and public transport). All of these infrastructure systems are optimized for a certain demand structure in an area, which is usually determined by population density and commercial or industrial activity. In the best case scenario, efficiency decreases in areas with higher rates of vacancy (Schiller and Siedentop 2005). In the worst case scenario, an area might enter a vicious cycle of declining population accompanied by an underutilized and then dismantled infrastructure and may, as a result, become less attractive. This, in turn, drives even more residents to relocate to another area in the city, thus perpetuating the cycle (Schwarz and Haase 2010).

Urban Land Use: What Is It and Is It a Special Case?

In contrast to nonurban land use, urban land use is characterized by a high degree of impervious cover, a high share of artificial surface material, and a high degree of built space (Haase and Nuissl 2010). Urban areas also have higher population densities and utilize land/space in multifunctional (often multistory) ways. To illustrate, the highest population densities are found in Indian, Pakistani, and Chinese cities. For example, Mumbai has 29,650 residents per square kilometer, Kolkata has 23,900 residents/km^2, Karachi has 18,900 residents/km^2, and Shenzhen has 17,150 residents/km^2. U.S. and European cities report average population densities of 2,000–4,000 residents/km^2. Rural areas, by contrast, often reach population densities of <500 persons/km^2.

One might argue that one aspect of the uniqueness of urban land use lies in its unidirectionality of change: once land becomes urban, it is highly unlikely to revert back to pre-urban land-cover conditions. However, this argument is incomplete, as a regeneration of nature does occur in shrinking cities and at abandoned industrial sites (Haase 2008; Breuste et al. 2013).

The degree of impervious cover differs among different urban land-use types and across cities in various regions of the world. Generally, there is an urban to rural gradient of impervious surfaces, where the highest degree of impervious cover is usually found in the center or central part of a city. In most cities, this is represented by the central business district. High rates of impervious cover are not limited to commercial or industrial areas but are found in high-density residential areas as well (Haase and Nuissl 2010). Building height and density at commercial sites are, however, generally lower compared to that found in the central business district. Lower soil sealing rates are often found in low-density residential areas (single and detached houses and villas) and along transport corridors where lawns, yards, and trees which line the streets increase the ratio of vegetation to impervious cover. As expected, leisure and recreation areas, including allotment gardens, sporting grounds, and cemeteries, have the lowest degree of impervious cover. The spatial patterns of impervious services vary substantially among different countries and continents. These variations reflect differences in local planning, historical land use, and stages of economic development. Particularly in emerging economies (e.g., India and Latin America), high-density residential areas with high-rise buildings are characterized by a high population density, and are often located in close vicinity, or in the same neighborhood, to lower income areas, including slums, favelas, or villas miserias. According to Fuller and Gaston (2009), the average proportion of green-blue (parks, urban forests, lakes, rivers) land use in European cities ranges from less than 2% to almost 50%; considerable variance exists, however, between countries and regions. Similarly, estimates of urban tree cover in the United States vary by more than two orders of magnitude (Nowak et al. 1996). In mega-cities, the amount of urban green space is often very low and limited to high-income housing areas. In summary, the patterns of impervious gradients observed in North America and Europe may not hold for other parts of the world, but impervious cover is nonetheless a differentiating aspect of urban land use.

When urban areas grow disorderedly and sprawl to the peri-urban area, this process can be referred to as "peri-urbanization." A peri-urban (or suburban) area refers to a transition, interaction zone, where urban and rural activities and patterns are juxtaposed and landscape features are subject to rapid modification induced by human activity (Douglas 2006). Peri-urban areas, which may include valuable protected areas, forested hills, preserved woodlands, prime agricultural lands and important wetlands, can provide essential life support in terms of ecosystem goods and services for urban residents. Moreover, along the urban rural gradient, this land consumption is often characterized by dispersed developments, mono-functional and low-density land uses, and reliance on private car ownership—all typical features of veritable urban sprawl (Nuissl et al. 2008; Squires 2002; Couch et al. 2005).

In addition to the impervious nature of urban land use, multifunctionality is a unique characteristic of land use: Its primary purpose is to provide economic wealth and quality of life for both society and urban dwellers. Housing, commercial land use, and transportation are equally crucial. At the same time, urban dwellings and courtyards offer places for urban vegetation and gardening. Trees planted along streets represent both nature and transportation infrastructure. Across vertical and horizontal dimensions, urban land use is highly complex. Multifunctionality may vary at different vertical levels within a single spatial response unit (patch, pixel). For example, a building can be commercial on the lower levels, residential in the middle and higher levels, while supporting gardens on the rooftop. Similarly, urban areas can vary highly in the horizontal domain: viewed from above, urban areas resemble mosaics or patchworks of mixed land use. The three-dimensional structure of the built space is an important indicator of the urban form, but unfortunately, this is often ignored in studies. Given the large number of mega-cities with extremely tall buildings (predominant, e.g., in Asian, African, and Latin American coastal cities), the importance, impact, and perception of building height in cities and urban areas need to be more carefully considered in the future.

Urban land use is also highly dynamic: it undergoes change rapidly. Urban land use involves the (partial) destruction and transformation of urban soils and can lead to enormous environmental pollution of soils and water with organic and inorganic compounds (Wessolek 2008). Although urban land use is unique in many ways, there are similarities and shared elements with other land uses (e.g., the degree of impervious surfaces, the presence of trees and green spaces, or the provision of ecosystem services). Impervious soils can also be found in rural and peri-urban areas (e.g., residential areas, greenhouses, roads). The type of impervious cover in cities does not differ from that found in nonurban areas; there is simply a higher share of impervious cover in cities. We find green space and water within cities. Furthermore, cities are biodiversity hot spots (Kühn et al. 2004).

Urban Form, Function, and Structure

The spatial form of an urban area or a city, such as its density or compactness, is an important factor for both quality of life and environmental impacts (Schwarz 2010). Urban form reveals the relationship between a city and its rural hinterland (Grimm et al. 2008). It also characterizes the impact of human actions on the environment within and the environment surrounding a city (Alberti 2005; Weng et al. 2007) and relates to transportation patterns (Dieleman and Wegener 2004). An ongoing debate differentiates two separate and contrasting examples of urban form: that which is often found in United States and Asian cities versus the European compact city (i.e., a city characterized by high population densities, relatively shorter commuting distances, and high accessibility) (Burton 2000; Dieleman and Wegener 2004; Frenkel and Ashkenazi 2008). Definitions of urban form vary significantly in the literature.

Some studies use only land use or land cover to measure urban form to characterize a city's physical structure (Herold et al. 2002; Huang et al. 2007a). Others include socioeconomic aspects, such as population number or density, in their characterization (Frenkel and Ashkenazi 2008; Kasanko et al. 2006; Tsai 2005). Studies of urban form have used concepts and indicators developed in the field of landscape ecology: the area of discontinuous urban fabric, edge density, mean patch size, number of patches, the compactness index of the largest patch, population number, and population density (Schwarz 2010). In contrast, from a planning perspective, urban form tends to focus on smaller grains, such as at the neighborhood or block level. Here, urban form could refer to building orientation, height, and size, combined with width and configuration of street networks.

Because urban areas exact such a large impact on other land uses, much effort has been put into examining whether there is an *optimal* urban form. Research in this area has examined whether the compact city is more sustainable than scattered and disconnected patterns of low-density urban development, sometimes pejoratively characterized as sprawl. Urban sprawl is defined by the large expansion of cities into surrounding areas through the creation of new low-density developments, such as with detached or semi-detached housing and large commercial strips (Dieleman and Wegener 2004; Schneider and Woodcock 2008). Much literature argues that patterns of low-density development are unsustainable due to their impact on resources (e.g., Jabareen 2006). Studies also show that compact cities may (a) be less resource intensive, (b) minimize the use of undeveloped land, and (c) lessen environmental impacts (Nuissl et al. 2008). However, compact cities may also produce higher noise levels and air pollution than less dense areas (Briggs et al. 2000).

In addition to the resource argument for compact urban form, there is evidence that human welfare is positively affected by the compact urban form. Studies of European cities with high population densities exhibit higher levels of welfare as measured by income and overall life satisfaction (Schwarz 2010). However, a negative correlation between per capita GDP and population density has also been found (Huang et al. 2007a), which implies that cities with very high population densities have low per capita GDP (Kraas 2007).

The large impact and long reach of urban areas raises a number of questions about sustainable global land use in an urbanized era. Optimal urban form becomes even more pressing given the large numbers of people that are expected to live in cities in the future. Moreover, other trends may affect urban form and function, such as teleworking, the aging of urban populations, and rising energy prices. Will global sprawl abate in favor of the compact city? Or will cities start to approach optimal medium and low-density patterns with high shares of urban green space? Which forms are more sustainable and livable? To what extent does our answer depend on our point of view? Do we judge the success of urban form from the perspective of nature conservation or from the perspective of the urban resident?

In terms of their functional form, urban regions are divided into a core city and a surrounding fringe (Kabisch and Haase 2011). These two parts depend on each other in terms of their demographic development, economic development, and urban dynamics (Bettencourt et al. 2006). Often, the core city and the suburban hinterland do not belong to the same administrative unit and are thus separated in terms of planning and land-use policy making. This pattern has been found globally and has been discussed as a worldwide phenomenon and problem (Antrop 2004). Given our current state of multi-actor governance, participatory planning schemes, globalization, and high-speed communication, do we still face the "core versus hinterland" problem that has been thought to plague urban regions? Does today's urban region constitute a single unit, or do financial, demographic, and economic forces still centrifuge the elements of an urban region into discrete parts? Has the urbanization of lifestyles in the rural hinterland and urban land teleconnections (Seto et al. 2012b) transformed "urban" realm into a matter without physical place?

Green and Blue Urban Land Use: Urban Ecosystem Services

Urban land use is unique for another reason: it is situated in close proximity (one could say overlap) to the built and ecosystem services provisioning area. In other words, regardless of the multiple impacts that humans may have on the natural environment of cities, urban spaces (including its land uses) also provide a range of benefits to sustain and to improve human life. These benefits are known as urban ecosystem services (TEEB 2011). Although urban ecosystem services have been classified differently, they are most commonly divided into four categories (Cowling et al. 2008): provisioning services, regulating services, habitat or supporting services, and cultural services. Provisioning services involve the material outputs from ecosystems, including food, water, medicinal, plants, and other resources. Regulating services act as regulators to control, for example, the quality of air and soil or provide flood, storm, and disease control. Habitat and supporting services are fundamental to almost all other urban ecosystem services, as they provide living spaces for organisms. Supporting services also maintain a diversity of different breeds of plants and animals. Cultural services include the nonmaterial benefits and enhancement of well-being that people obtain from their interactions with ecosystems, including aesthetic, spiritual, and psychological benefits as well as tourism benefits (TEEB 2011). Healthy ecosystems provide the basis for sustainable life in both cities and urban regions, undoubtedly determining, influencing, and affecting human well-being and most economic activities (Ash et al. 2010). However, the urban ecosystem is also affected by the city (Nuissl et al. 2008; Vitousek 1997). Since cities cannot fully satisfy their demand for ecosystem services (Cox and Searle 2009), additional ecosystem services must be obtained from the immediate peri-urban regions as well as from regions scattered across the globe.

From Hierarchical Urban Land-Use Planning to Multi-Actor Land Governance for Urban Areas

Across the world and regardless of income levels, both cities and urban regions suffer from weak or lack of governance and institutions needed to manage their growth or decline, land use, the generation of revenue or other funds, and provision of public goods (such as transportation infrastructure, health care, and educational facilities) (see Blanco, this volume). Cities depend on their hinterlands, national developments and programs as well as, increasingly, on international networks. Cities are also part of a global network and commodity chain (Sassen 2009, 2011). More than national governments, cities are able to foster proactively a sustainable urban and urban-regional land development through multiple policy tools. For example, they could encourage the compact city model as a normative following a comprehensive top-down planning model (cf. NLGN 2006) in combination with an agency or voluntary model comprising appointed agencies, interest groups, or joint bodies with both strategic planning responsibilities and adviser implementation functions (e.g., the Öresund Committee). This would require more actors and stakeholders to be involved in urban land governance—something that is currently not the norm.

Socioenvironmental concerns of climate change and quality of life closely related to land use as well as new integrative concepts of urban planning and urban governance are needed to address the diverse local agendas (Ravetz 2000; Breuste et al. 2013) that deal with urban ecosystem services (TEEB 2011; Haase 2012a), green and blue services (Westerink et al. 2012), the urban footprint (Grimm et al. 2008), and sustainable land-use governance (Westerink et al. 2012; Ravetz 2000). Integrative planning means the removal of sectoral thinking and actions in favor of more holistic concepts of urban regional and local regeneration, resilience and adaptation without ignoring the global context or network. Accordingly, expert-driven formal landscape and green-space planning will be enhanced through processes that involve many different actors ("multi-actor"). Not exclusively, but most notably in shrinking cities, concepts of green and blue services provide the potential to move from a comparatively simple "land-use view" of green, brown, and blue areas in cities toward a valuation of ecosystem goods and processes and spatial potentials for each piece of land (Haase 2012a; Bastian et al. 2012; Lorance Rall and Haase 2011).

Hypotheses for Our Planet's Urban Future

Using urban as a special case of land use suggests three provocative hypotheses for our planet's urban future: First, the "urban" is everywhere today. The rest of the world will soon function as a service area for a growing number of cities in which almost all of the world's population will be concentrated. We will live in a teleconnected urban world that no longer physically touches the surroundings or rural hinterland (Boone et al., this volume). This implies that areas of resource use, production (rural/wildlands), and consumption (urban) will be increasingly separated from each other. This hypothesis envisions, however, increased production and consumption loops within the same city, leading to a more closed system in which cities eventually incorporate the task of farming (Vogel 2008b). Thus, the uniqueness of urban land use or "the urban" could increase in the future. Whether it becomes more dependent on the hinterland or remote areas is an open question.

Second, the city of the future will be higher and more densely built, and will extend further underground, in terms of service areas and commercial and office buildings. Due to the high population density and short distances, the urban resident will live more sustainably than a rural resident. Future urban residents will "save open space" as they increasingly inhabit vertical and multistory spaces. Thus, space saving is another urban (ecosystem) service and needs to be valued as such. Moreover, the urban city will be more connected by wireless communication (Boone et al., this volume), which will save energy due to reduced physical transport. Simultaneously, urban green spaces will be spread over different vertical levels, starting at the ground level but also covering roofs and walls in diverse forms. Thus, higher population densities could be reached by offering an ever-increasing number of people a place to live. This new concept of "the urban" city would offer a place to live for more than 90% of all people on Earth. In turn, rural and urban land would be viewed from quite a different perspective.

Third, one persistent question regarding urban land-use development may not be answered by the city of the future: Is the optimal urban form characterized by sprawling green livelihoods for a growing number of people or by a compact, high-density, and short-distance space? Although contradictory at first glance, both ideals could be visions for the city of the future. Both could embody the sustainable or resilient city, depending on the overall target of urban development. The current state of aging and decreasing populations in cities could unwittingly support this vision of future cities.

First column (top to bottom): Christopher Boone, Hilda Blanco, Harini Nagendra, Steward Pickett, Chuck Redman, Makoto Yokohari and Stephan Pauleit, and Group discussion
Second column: Chuck Redman, Shuaib Lwasa, Group discussion, Makoto Yokohari, Karen Seto, Jennifer Koch, and Harini Nagendra
Third column: Karen Seto, Dagmar Haase, Christopher Boone, Shuaib Lwasa, Group discussion, Dagmar Haase and Steward Pickett, and Hilda Blanco

16

Reconceptualizing Land for Sustainable Urbanity

Christopher G. Boone, Charles L. Redman,
Hilda Blanco, Dagmar Haase, Jennifer Koch,
Shuaib Lwasa, Harini Nagendra, Stephan Pauleit,
Steward T. A. Pickett, Karen C. Seto, and Makoto Yokohari

Abstract

Current systems to classify land are insufficient, as is the delineation of Earth's surface into discrete categories of land covers and uses, because they ignore the multiple functions that land provides and the movement of people, materials, information, and energy they facilitate. To address sustainability challenges related to urban lifestyles, livelihoods, connectivity, and places, new conceptualizations are needed which have the potential to acknowledge and redefine the extent, intensity, and quality of urbanness on Earth. This chapter proposes a framework which focuses on people and institutions as agents of change and examines changes in urban lifestyles and livelihoods over larger regions, regardless of whether an area is delineated as "urban" or "rural." It views urbanization and the urban era to be an integrated system and provides a multivariable approach to urbanity. It discusses a new land ethic and highlights challenges that exist to facilitate a sustainability transition.

Introduction

Standard urban-rural land classification systems are insufficient for analytical or planning purposes. The delineation of Earth's surface into discrete categories of land covers (e.g., forest, rural) and uses (e.g., recreation, agriculture) ignores the multiple functions those areas may provide and the movement of people, materials, information, and energy they facilitate (Cadenasso et al. 2007; McHale et al. 2013). The Earth's surface is far from static, and new conceptualizations are needed which incorporate an understanding of the *processes* that shape, take place on, and are facilitated by land. This is especially urgent in an era of rapid urbanization and globalization, where the structure and function of lands may affect one another even at great distances and at near instantaneous speeds, a phenomenon known as teleconnections (Seto et al. 2012b). These teleconnections are amplified by urbanization, with the result that even seemingly remote areas may have urban characteristics. For these reasons, traditional concepts of urban and rural have become increasingly less useful in describing the function and structure of land as places of human activity.

In this chapter we propose a conceptualization of land that measures and analyzes *urbanity*, the urban-ness of places, and the economic activity and population characteristics of the land. We define urbanity as how people support themselves through various livelihoods, the material culture and patterns of consumption representing different lifestyles, their spatial connectivity, and how they identify with the places where they reside and upon which they rely. The magnitude and qualities of livelihoods, lifestyles, connectivity, and place create the degree of urban-ness of intertwined human experiences and land configurations. Key terms used in this chapter can be summarized as follows:

- Urbanity: urban-ness of land defined by the physical and functional characterstics that support and facilitate urban-like livelihoods, lifestyles, connectivity, and places.
- Livelihood: means of securing necessities for life, such as occupations, access to resources and information, reliance on social networks, or supporting institutions.
- Lifestyle: way of life that defines and reinforces self-identity. In the urban era it is often defined by and associated with occupation, socio-economic status, consumption, behaviors, and other activities that distinguishes individuals or groups from others.
- Connectivity: ability to connect between nodes in a network, and the magnitude, speed, direction, kinds, and infrastructure of those connections.
- Place: an area or location defined by physical or social characteristics that create meaning (sense of place) and distinguish it from other areas or locations.

It is possible to define a continuum of urbanity that is not defined by administrative boundaries of cities, but by the activities and functions that occur in places even far removed from what are traditionally understood as urban areas.

In this chapter we demonstrate how the concept and elements of urbanity can be used to assess and visualize the potential for sustainability of places. We explore the notion of a new land ethic in an urban era, one that includes the elements of urbanity as a potentially positive set of attributes, and how explicit attention to ethics informs our choice of human well-being, ecological integrity, and social equity as sustainability dimensions.

Conceiving Land as Places for and Defined by Human Activity, Relations, and Experience

Urban areas are often conceptualized in a discrete manner; for instance, statistics that extend over an area of 2–4% of the terrestrial world are used to describe urbanization, when in fact these numbers refer to areas of urban land cover (Seto et al. 2009). This delineation refers to the morphological face of urban land, the densely built-up areas with comparatively high population density. As urban areas grow, updates of the urban space are usually made by remote sensing methods to detect and classify land use. The problem is that these discrete and cover-based approaches are unable to capture the complexity of processes, outcomes, and impacts of urbanization, which are spatially related, scale dependent, and operate across large and interconnected regions through distal teleconnections (Seto et al. 2012b). We have attempted to develop a framework to conceptualize urban areas, focusing on people and institutions as agents of change, and examining changes in urban lifestyles and livelihoods over larger regions, regardless of whether an area is administratively or morphologically delineated as "urban" or "rural." This framework draws on approaches that perceive urbanization and the urban era as an integrated system, where humans, energy, and matter flows are involved with urban lifestyles, livelihoods, connectivity, and place. The conceptualization provides a multivariable approach for defining the degree of urbanity.

The standard morphological and land-cover driven delineations and the new urbanity-driven delineations of urban areas presented here differ considerably, both in terms of the spatial extent they cover and the degree of continuity and interlinkages between places. Consequently, within the new framework, global maps of urbanity would differ substantially from standard definitions of urban areas. Further, the multiple variables of the conceptualization—lifestyle, livelihood, connectivity, and place—permit the detection of areas of high urbanity within administratively defined rural areas as well as areas of low urbanity within morphologically or administratively determined urban areas, recognizing the reality of gradients and heterogeneity of such distributions in many parts of the world (Figure 16.1).

Urbanity: Elements and Processes

Our proposed conceptual framework of land change and urbanization is dynamic, relational, and spatially heterogeneous, as opposed to static, dichotomous, and gradient based. It focuses on people and institutions as agents of change, rather than on administrative boundaries as locations for dense demographic settlements. The framework aims to identify explicitly how these agents of land change and urbanization are connected at multiple scales. Furthermore, it allows users to diagnose the implications of urbanization for environmental

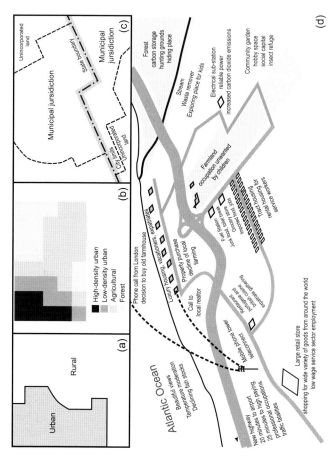

Figure 16.1 Ways of representing or conceiving land in an urban era. Urban-rural classifications and delineations (a) and pixels of urban and rural cover (b) are inadequate for representing human relationships (between individuals and groups but also with ecosystems) that influence land change and human behavior and ultimately opportunities for sustainability. Governance districts (c) can represent jurisdictional and decision-making authority, which is important in guiding the relationships described, but are still relatively static representations of those dynamics. Configurations of land (d), or landscapes, emphasize the *qualities* and *functions* of those configurations, rather than just the spatial arrangements or physical characteristics of the land, as influencing and representing the outcomes of human activities, relations, and experience.

integrity, social equity, and human well-being in ways that help us to identify pathways for sustainable urbanization. Improving understanding of the implications of an urbanizing world for global land use, including the detection of phase transitions, is another goal of the framework.

Urbanization can be characterized as simultaneous processes across three dimensions. First, urbanization is a process whereby livelihoods become less agrarian, less dependent on local production, and less directly linked to the use of natural resources through such activities as fishing, hunting, or forestry. Second, urbanization involves changes in lifestyle, including such features as degree and mode of mobility, social identity, behavior, personal values, consumption choices, and modes of action. Third, urbanization increases the number, diversity, distance, dynamism, and redundancy of connections of the locality, its inhabitants, its economies, and ecosystems to distant places. Although places and people have been connected throughout history by trade and migration, urbanization today is changing the nature, frequency, and intensity of these teleconnections (Seto et al. 2012b). In short, we suggest that:

- Urbanization results in increased teleconnections, and vice versa.
- Land change is increasingly driven by nonlocal actors through these teleconnections.
- Understanding land change through a new framework of global land use in an urban era can help us identify planetary responsibility and Earth stewardship.

The key point is that to understand contemporary urbanization and land change, we must first understand how these processes are continuously distributed and connected through space and across time, in contrast to prevailing frameworks that are largely place based and discrete in space and time. The approach will avoid using nominal, contrasting categories, such as rural versus urban, and it differs from other approaches that assess change along neatly defined urban-rural gradients.

We summarize the insights above as a continuum of urbanity, emphasizing the dimensions and factors around which actors work their effects. Actors influence urbanization through the creation and use of livelihoods, the generation and expression of lifestyles, and the creation and exploitation of built environments, infrastructures, and teleconnections. Their actions take place in ecological and cultural contexts embodied in specific places, and those places are in turn affected by the actions (Figure 16.2).

Actors, Livelihood, and Lifestyles

People and institutions (actors) create and depend on livelihoods and the institutions that support livelihoods, such as markets, governance structures, and laws. Urban areas typically have a great diversity of actors and livelihoods, historically related to industry but increasingly to service and information

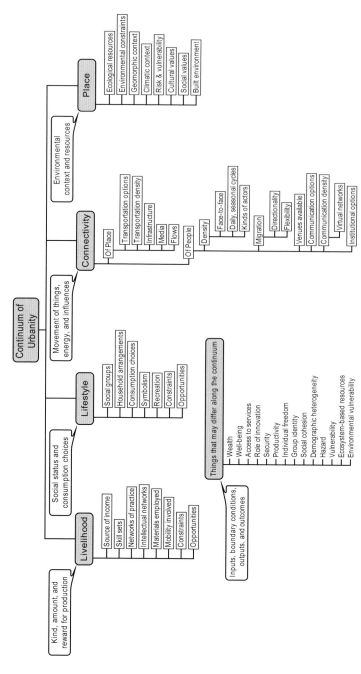

Figure 16.2 The continuum of urbanity, as defined by characteristics of livelihoods, lifestyle, connectivity, and place. Listed are sample phenomena to consider when building a functional model of urbanity. These items inform the addition of detail to the flow model (Figure 16.3) linking livelihood/lifestyle, comprising individual actors and institutional actors, and connections to other systems. These are the proposed specific elements of measuring access to the degree of urbanity.

economies. These contrast with livelihoods associated with natural resource extraction and management, forestry, and agriculture.

In pursuing different livelihoods, individuals influence, and are influenced by, the lifestyle groups with which they associate, the skills they possess, the networks of practice or support where they are embedded, and the mobility involved. Lifestyle is a way of life that expresses and reinforces self and group identity. Lifestyle is strongly tied to socioeconomic status, occupation, consumption patterns, behaviors, worldviews, and a variety of demographic factors. Urban lifestyles, often associated with high levels of consumption (typically nonlocally produced products) and nonagricultural occupations, are one dimension of urbanity.

Connectivity

To characterize the urbanity of a site, including the system it supports and the actors affecting it, the kinds, number, direction, and strength of connections between that system and other systems nearby or at a distance must be evaluated. Connections may include flows of energy, water, food, waste, pollution, financial investment, personal remittances, tourists, migrants, information, knowledge, communication, and construction materials. In addition to the entities moving between sites, the necessary infrastructure is also a component of connectivity.

Connectivity extends linkages from the city region to global scales. Connectivity can set in motion interdependencies and influence between places. The diversity of goods available in urban markets is one manifestation of connectivity. It is important to note that connectivity does not neccesarily imply the existence of linear gradients; it can encapsulate sharp changes and reversals in gradients, as well as encompass a significant degree of spatial heterogeneity. Connectivity can decrease and break down due to natural and social disturbances, such as natural disasters and political constraints. These changes in connectivity can affect intensity of urbanity and ultimately urban sustainability, an idea elaborated below.

Place

Place is the biophysical and social particulars of specific areas that give it meaning and distinguish it from other areas. Place incorporates social, built environment, and ecological capacities and constraints. Spaces where urbanity develops reflect social and biophysical-environmental capacities, flows, and constraints at any given time. These set the larger context for the development, maintenance, or decline in urbanity in specific sites or regions.

Actors driving urbanization set in motion trajectories of change, often without adequately taking into account the impacts of such change on social and ecological capacity of the urbanizing places. Change in social and biophysical-environmental context, as well as the type and degree of connectedness, can impact the sustainability of urban locations, with the expansion or collapse of one location giving rise to distributed, often unanticipated consequences for other locations due to their complex connectivity (Figure 16.3). Thus, such a framework should enable one to identify previously unconsidered "tipping points," thresholds, and trajectories that have implications for sustainability. Below we present three examples of how such a framework can be applied to understand land change in the urban era.

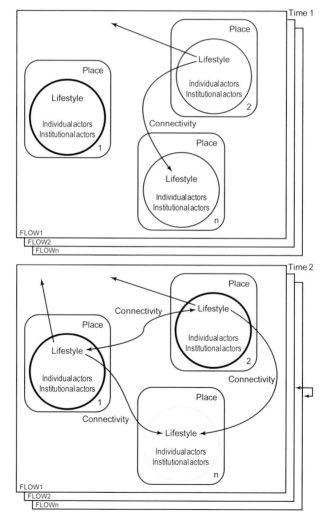

Figure 16.3 Stylized change in urbanity over time. Changes in lifestyle, livelihoods, or connectivity may change degree of urbanity directly or indirectly between places. Thicker dark lines indicate increased urbanity.

British Second Homes in Portugal

Our framework suggests that a high degree of urbanity connects traditional urban places in the United Kingdom with areas in Portugal that were traditionally labeled as rural. The shared urbanity of these seemingly different places is established by teleconnections between them.

Through the media or personal travel, wealthy individuals in the United Kingdom become aware of the pleasures of a sunny, warm place. This creates a desire, reflecting the lifestyle of these people, to have a second home in Portugal. The actors who originate this demand are distant urban dwellers with high incomes and livelihoods capable of satisfying the demand. Through modern communication networks, they contact developers and mediators, who often participate as actors outside the immediate vicinity of either the rural Portuguese or the city-dwelling U.K. residents, to connect buyers and sellers to satisfy the desire to possess a second residence in a sunny place along the Portuguese coastline. In addition, changing livelihoods and lifestyles on the Portuguese coast generate a stock of vacant houses that are associated with the out-migration of young generations, due to dwindling or no interest in rural life and livelihoods (farming). The houses are bought by British urbanites from local actors (farmer families), mediated by the developers, leading to further changes in livelihood along the Portuguese coast, as the lifestyle of the second homeowners requires infrastructure development for health, transport, and leisure. This, in turn, leads to increased consumption of energy, water, and land, higher levels of air pollution and noise, and it affects ecosystems and biodiversity. As the livelihoods of the local population change, unemployment in the agricultural sector rises as a result of land competition. In the service sector, by contrast, new jobs and sources of income are created through the presence and needs of new homeowners. This further impacts and changes the lifestyle of the local population. All in all, the new type of connectivity between the two places and their actors leads to a higher degree of urbanity along the Portuguese coastline and creates "urban patterns" that did not exist there before—patterns which nonetheless differ from a traditional "city" (Bell et al. 2010).

The change in urbanity along the Portuguese coast has causal teleconnections back to the United Kingdom. Flows of investment and people between the new vacation settlements and U.K. residences creates a temporary vacancy of houses in Britain, which opens the door to new influxes of students and tourists, providing income to the second homeowners and impacting livelihoods and lifestyles in British cities. Second homeowners return to Britain with new diets and habits from Portugal, which increase a demand for imported new foods to satisfy a Mediterranean-style diet.

The new patterns of actors at both places and the impacts on livelihoods can be detected and measured using prevailing, traditional spatial conceptualizations. However, to understand more fully the process of land change in an urban era, the multiple and interacting variables of urbanity are useful. The framework helps tease apart this complex example to show how urbanity extends over great distances based on teleconnections of different actors located far from each other, and following or adjusting their own livelihoods and lifestyles.

Meat Production and Dietary Preferences in Saudi Arabia and Uganda

Let us now consider the connectivity and resulting changes in urbanity between Uganda and the Middle Eastern States of Saudi Arabia, United Arab Emirates, Qatar, and Kuwait. The number of tourists and pilgrims to the Middle East has increased over the years, and this has increase the amount of urbanized land in the Middle East as well as the demand for meat and meat products, reflecting an important part of the lifestyle influenced by culture and religion. Actors are individuals and institutions. Demand for meat in the Middle East has triggered a relational connection to Uganda, one of the many meat-producing countries that provide meat to urbanizing Middle Eastern countries. The primary actors in Uganda are the farmers, particularly those in the cattle zones of Western Uganda as well as northeastern subregions, which produce 40% of the meat products in Uganda (UBOS 2012). Contractual arrangements between Ugandan farmers and Middle Eastern intermediaries have increased meat production in response to meet the quota of meat protein required in the consuming region. In the process of complete transactions and exportation, the production place not only experiences increased income, it is also influenced by Middle East connections. This results in changing Ugandan lifestyles to include greater meat protein consumption, changes in housing, changes in dietary composition, and connections with other parts of the world through mobile communication and television. These changes not only drive the urbanity in the production place but also increase confidence in consumption to expand and urbanize more. This relationship can trigger or enhance other regional relations between the production place and source areas at various scales along different dimensions, among which imports from the Middle East are significant (UBOS 2012). Import data shows increased goods and services flowing from the Middle East to Uganda and is further evidenced by cargo and passenger airliners plying the route.

Rurality within Cities in Bangalore, India

Rapid urbanization in the south Indian city of Bangalore has led to a huge expansion in the city's boundaries, with a more than tenfold increase in area since 1949 (Sudhira et al. 2007). Bangalore is located in a fertile agricultural catchment with a number of villages that have a history extending over several centuries. As the city expanded, it engulfed a number of these small village settlements within its boundaries, and they now constitute part of the city. Although administratively considered as part of the city, some of these areas are termed slums because of the obvious differences in house construction, livelihoods, and lifestyles (Gopal 2011). Other villages, even though not administratively defined as such, have retained a strong cultural identity through celebrations of iconic festivals, such as annual temple processions that celebrate the worship of local (typically female) village deities (Srinivas 2004).

Conventional urban classification approaches map the spatial boundaries of an entire city, within which statistical information is sometimes collected at the level of administrative subunits. Such a discrete approach runs the risk of homogenizing information across different types of land use and livelihoods within a city like Bangalore. For instance, a recent residential layout called HRBR (Hennur Road Banaswadi Road) in eastern Bangalore contains within its core a remnant of the village that was originally located here: Kacharakanahalli. Software engineers with highly urbanized lifestyles, who gain their livelihoods from the city's famous information technology industry, live here in high-rise apartment complexes adjacent to a small group of families of original inhabitants who rear pigs (Nair 2005). While the city attempts to impose its notion of standardized urban form on the landscape it engulfs, the villages located within the city are instantly recognizable, with features such as a central village square with a peepal (*Ficus religiosa*) tree on a platform that serves as a meeting place for the village (Srinivas 2004; Nair 2005).

As this example of Bangalore demonstrates, there is a need for more continuous approaches of urban representation in many parts of the world. Just as many areas classified as rural are often urban in lifestyle or livelihoods (as in the example of Portugese coastal villages), in many parts of the world, areas that are classified as urban exhibit many forms of rural life, especially in terms of their cultural character and lifestyle, but often even in terms of their livelihoods, rearing of livestock, or practice of other traditional rural occupations such as agriculture. The example provided here of Bangalore has resonance elsewhere, with similar observations of rurality in urban livelihoods being noted in areas as diverse as Kampala (Ishagi et al. 2003) and Mexico City (Losada et al. 1998). Similar to the case of Bangalore, an expansion of the city limits in Mexico City has engulfed a number of peripheral villages, with the result that traditional livelihoods, such as livestock rearing and agriculture, have intersected with Western forms of urbanization, such as pet rearing and hobby gardening. The result is new, hybrid lifestyles and livelihoods that do not fit neatly into discrete rural or urban categories (Losada et al. 1998).

Summary

These examples show how the continuum of urbanity can be applied. We do not claim that the empirical examples were discovered as a result of the new framework we propose. Indeed, components of these narratives have in some cases been well known for some time. Rather, these examples highlight that what has traditionally been considered to be discrete territories—the rural and the urban—are in fact intimately connected, even over very long distances, and through the decisions of actors who are initially engaged in vastly different lifestyles, livelihoods, and who have different places of primary residence or diverse perceptions of the place where they ultimately come to interact. The examples illustrate the need for and the value of a new framework (Figure 16.2) in which urbanity is not spatially restricted to traditionally recognized cities or dense, older suburbs. These narratives can be considered to be emerging models that call for a conceptual framework to unify and compare their details and similarities. They show that the characterization of the interaction between so-called rural and urban systems in the globalizing world involves more than a collision of lifestyles, contrast in livelihood, divergent senses of place, or even a matter of teleconnection. A complete understanding of the interactions and dynamics of these complex systems requires that all four components of the urbanity framework be exercised. Furthermore, the bidirectionality of influences between the traditionally urban and the traditionally rural is seen in several of the examples. The existence of this framework invites a search for new examples of the urban continuum and the comparison of cases throughout the world under different conditions, defined by the attributes of place.

Sustainable Urbanity Informed by Ethics and Values

The way a system is framed, investigated, analyzed, and interpreted is significantly affected by the values and attitudes of the investigators and stakeholders. We must explicitly recognize the implications of this situation for our proposed framework. Land-related analyses cannot solely reflect physical or spatial units. Rather, it is human activities, networks, and attitudes that give meaning to places. These sources of meaning may not be linked permanently to a specific physical space. Hence, our approach to measures of urbanity uses categories of measurement—related to livelihood, lifestyle, connectivity, and place—that reflect our own normative attitude toward land-use analysis. Measured attributes and resultant patterns—either as patches, networks, or gradients—may be manifest in specifiable locations, and these physical locations are likely to change over time and alter shape and boundaries as various attributes are considered. The ecological content of any place (i.e., the resources, fluxes of materials, energy, and biodiversity) and the action of environmental regulating factors are significant to the well-being of people living there.

The second domain where values must be explicitly integrated is in the analysis of how various inputs affect the system's formulation and its position on a pathway toward what we term "sustainable urbanity." The definition of sustainability and its measurement should be coproduced by stakeholders and investigators (Gibson 2006). As a starting position we believe that pathways toward sustainability involve indicators that can be conceptualized in three dimensions: increase of human well-being, social equity, and environmental integrity. We explicitly do not use the classic three pillars of sustainability (i.e., economy, environment, and society) because the directionality of sustainability dimensions presented here reflects our ethical standpoints, as well as those of others.[1] In addition, we chose these dimensions because they reflect shared normative goals of sustainability. Rather than using society as a dimension, for example, we chose social equity as a sustainability goal toward which we should strive, rather than just a metric alone (such as a Gini coefficient of income distribution).

Most decisions and processes involved in changing urbanity differentially affect indicators in each of these three dimensions. That is, some actions may enhance overall human well-being (e.g., improved health) but not be fairly distributed, thereby diminishing social equity. Other actions may enhance well-being or equity, but diminish environmental integrity. In an ideal world, every action would enhance all three dimensions of sustainability, but in many situations even the best intended actions will require trade-offs (McShane et al. 2011). The key is that as investigators we observe, measure, and analyze situations in terms that are sensitive to impacts in the various and normative aspects of sustainability.

A fundamental aspect of sustainability is concern for the viability of future generations (WCED 1987). How do we develop or conduct human activities in a way that our activities do not threaten but rather enhance environmental quality, social equity, and the well-being of future generations? The conditions of life for future generations is, to some extent, undetermined. With the exception of the near future, we do not know the extent and speed of technological innovation and social change that will influence future livelihoods and lifestyles, connectivity, places, and our use of natural resources. However, certain aspects of our world today will influence prospects for future generations. Although neoclassical economists argue that manufactured capital can be substituted for natural capital (Ayres et al. 2001), the precautionary principle combined with an ethical imperative in an age when humans dominate the planet requires us to err on the side of caution to ensure environmental quality for future generations (Costanza and Daly 1992; Foster et al. 2000). In addition, our concern for future generations leads us to emphasize the importance of social institutions. We are a social species and even more so now, as the world's population becomes increasingly urban. Our lives are mediated by social institutions; different political, economic, and social scales and qualities shape our livelihoods and lifestyles and influence connectivity and our interactions with the natural and built environments. Institutions are the bridge between present and future generations, and concern for the prospects of future generations requires

---

[1] For example, the Social, Technological and Environmental Pathways to Sustainability Centre, http://steps-centre.org

concern for the quality of institutions. Do current institutions advance opportunities for social equity, environmental quality, and human well-being?

Human livelihoods and lifestyles change across time as well as across societies, and the preservation of specific livelihoods and lifestyles is not necessarily an aspect of sustainability (although negative changes in livelihoods and lifestyles within short periods of time can have deleterious effects on populations). However, current livelihoods provide the social resources for near-future generations. From the standpoint of future generations and their sustainability, what is needed to ensure and enhance opportunities for livelihoods, especially for poor populations? Indicators of importance to gauge the prospects of future generations include educational attainment, investment in physical capital stock, percentage of labor force engaged in knowledge creation, as well as degree of environmental degradation (Llavador et al. 2011). The last three indicators are important because physical capital has a long life and is hence a multigenerational resource, whereas knowledge creation is crucial for innovation that bolsters well-being and environmental protection. In general, our ethical concern for future generations can be measured or monitored in terms of current trends, in particular through the directionality of such trends, or whether actions or institutions actually increase or reduce ecological integrity, social equity, and human well-being.

Visualizing Sustainable Urbanity

The three axis model of sustainability (Figure 16.4) can be used to visualize how the components of urbanity (Figure 16.2) relate to sustainability goals. Livelihood, lifestyle, connectivity, and place can be represented by specific features relevant to particular situations. Measurements or estimates of these four main features of urbanity can be assembled on the three axes. For example, livelihood may identify the kind of productive activity with which households engage and how much, if any, income is derived from that activity. The number of livelihoods and their remunerative capacity can be indexed primarily on the human well-being axis of the sustainability visualization, but will also have impacts measurable on the social justice and environmental integrity axes. Similarly, lifestyle can be represented by the diversity of social groups present and the degree of social cohesion within those groups. Equity of access to political power by groups with different lifestyles and decision-making processes can be used to index the axis of procedural social equity. The third axis of sustainability can be assessed by whether the collection of livelihoods and lifestyles in a specified area is associated with degradation of biodiversity or water quality or whether indicators of ecosystem integrity increase.

Our intention is not to provide a purely objective scheme for mapping characteristics of livelihood, lifestyle, connectivity, and place to the three axes of sustainable urbanity. Rather, we consider the proposed method for visualizing

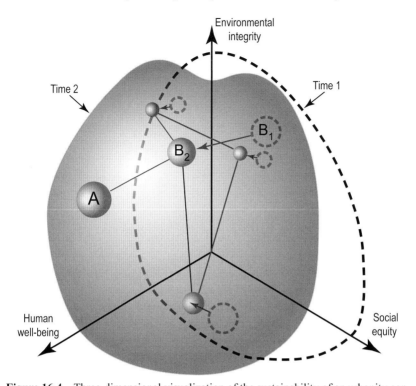

Figure 16.4 Three-dimensional visualization of the sustainability of an urbanity continuum. Each of the small spheres or circles represent a place or set of actors within a system that represents some bounded, connected unit of analysis (e.g., the sustainability outcomes of the consumption patterns of lifestyle groups within a specified region or connected network). This representation also needs to take into account the teleconnected outcomes along the three dimensions. Sphere A represents a new teleconnection, the purchase of land by foreign investors in a small community in Portugal, shown as B_1 (see text). Such a teleconnection can shift the shape of a sustainability bubble from Time 1 to Time 2. The small community in Portugal (B_1) is most greatly affected by the new investment, shifting it to a new space (B_2) that has higher human well-being due to income increases but reduced ecological integrity caused by development and reduced social equity due to increased wealth disparities. Nearby villages (represented by unmarked spheres) shift as well, but with less magnitude than B_1. Visualizations can change in direction and magnitude either through time or with modifications to the unit of analysis, such as the addition of new nodes and places in the network.

sustainable urbanity to be an approach that is flexible enough to be applicable under a wide range of spatial configurations and thematic settings, allowing the categorization and comparison of different measures and actions under those specific conditions. In addition to flexibility, another aim is to focus on directionalities and visualize trends in the sustainability of an analyzed system.

The challenge is that there are almost limitless ways to measure system characteristics. As in other areas of science, one must prioritize indicator selection based on hypothesized relevance to the issues and context being addressed. The key is to make the rationale for these priorities explicit and logical, given the underlying assumptions and objectives of the analysis. In a specific application, the selection of criteria representing the different axes of sustainability might be a very small set.

Building the Visualization

Measures of the different dimensions of sustainable urbanity can be composed of a set of weighted measurable and observable components, combined in a flexible manner and normalized. Each entity of analysis can then be translated to a three-dimensional shape located in the sustainability space. The kind of entity analyzed is directly related to the four characteristics of urbanity: livelihood, lifestyle, connectivity, and place. The shape of the entity in sustainability space is constituted by the subcomponents of the entity, which can vary in kind and number. Examples for subcomponents are actors or places. Temporal dynamics of the analyzed entity might result in a reallocation of the subcomponents in the sustainability space, inducing changes in volume, surface area, and form of the shape, representing pathways of the system. Based on the direction of the pathways, they can be characterized with regard to change in their sustainability. The flexibility inherent in the proposed concept allows application to different research questions as well as spatial and cultural settings. Whereas other visualization approaches such as spider diagrams allow an easy translation into one specific number, this approach considers the notion that there is not only one solution to problems; problems require interpretations within a variety of spatial and cultural contexts.

Visualization needs to focus on specific analyses and often on subcomponents of analyses. These visualizations would be most effective to represent comparisons such as between cities at a specific time point, differences within a city at a specific time point, or changes in a city or place over a time period. The three-dimensional diagrams and associated three-dimensional shapes (bubbles, spheres) are holistic representations which could be complemented with spider diagrams reflecting single points or units. Different viewers will favor different shapes as optimal pathways toward sustainability. For example, some will favor changes in overall well-being over the other two, whereas others may be most concerned about ecological integrity, and so forth. The figure may also take one of several characteristic shapes which may, in fact, reflect signatures of differing types of change.

The shapes traced out by the indices of sustainability in three-dimensional space in Figure 16.4 make it possible to compare how different degrees of urbanity affect pathways toward sustainability. For example, within an urban area, different neighborhoods or patches, identified by their joint biotic and built structure and the character of their connectivity, will have an index-derived shape in the visualization. This visualization could be used to assess the potential for each neighborhood or patch to contribute to citywide sustainability. Similarly, different urban areas may have contrasting volumes in the visualization. This allows the sustainability potential of different cities, or even of different metropolitan areas in a regional conurbation, to be compared.

Finally, the visualization can help make comparisons through time. This strategy can be applied to historical data or future projections. The three-dimensional visualization can also portray the pathways of how current degrees of sustainability have emerged. Proposed changes based on municipal or larger-scale sustainability plans can be evaluated for their impact on the sustainability bubble. Whether proposed policies, human interventions, or natural disturbances make an area of interest more or less sustainable can be understood using sustainability bubbles representing before and after conditions. The ability of the proposed visualization to change behavior and decision making toward our stated normative goals constitutes another important research question. Ultimately, this may lead to what we term a "new land ethic" for an urban era, to which we now turn.

A New Land Ethic in an Urban Era

Known for his contributions to the development of environmental ethics, the work of Aldo Leopold focused on conservation and the natural amenities of land. His concept of the "land ethic" (Leopold 1949) focused on humans as stewards of the land but not on the human experience, or on how land shapes the interactions among people as well as between humans and land. Leopold's land ethic, though broad, does not account for the experience and geographic context of a large and growing percentage of the population that is part of the urbanity continuum.

Thus, there is a need for a new land ethic: not an *urban* land ethic, but a land ethic that encompasses urbanity. A new land ethic for the urban era requires broadening the focus from the land to include two interactions: the interaction between people and places, and the interaction between people with each other. In other words, a new land ethic moves the focus from the role of humans in stewardship of the land (for the sake of the land) to stewardship of the land to enrich the human experience, by improving human well-being, social equity, and the ecological integrity on which all life depends. Under this new ethic, land is no longer conceived as a passive landscape within which the human experience occurs. This new land ethic recognizes that landscapes and the configuration of landscapes—including the built environment and natural landscapes—are fundamental for shaping the human experience. Human development challenges take place on landscapes and in places, and thus we must recognize the role that land plays in enabling, fostering, or constraining human development.

Pathways to Sustainable Urbanity

A reconceptualization of land based on the idea of an urbanity continuum can serve as a useful analytical framework. It has the potential to acknowledge and redefine the extent, intensity, and quality of urban-ness on the surface of the earth, a useful starting point when addressing sustainability challenges related to urban lifestyles, livelihoods, connectivity, and places (e.g., dietary changes, energy consumption, and resource demands). We have argued that an understanding of the elements of urbanity can also be tied to normative sustainability goals and perhaps a new land ethic that works to improve human well-being, ecological integrity, and social equity for all people and places. A formidable challenge that remains will be to develop or to use in new ways the institutions that could facilitate such a sustainability transition. How, for instance, do we plan for a sustainable urbanity continuum that transcends jurisdictions, cultures, or value systems? On the other hand, could the concept of the urbanity continuum circumvent or overcome entrenched institutional obstacles to sustainability? Since the idea of urbanity does not constrain human activity to static and discrete administrative boundaries, it may lead to a more distributed notion of stewardship rather than concentrating solely on local concerns. Teleconnections, embedded in the idea of urbanity, link local with distal actors and places. In doing so, they urge us to think about the stewardship of interconnected people and land. At the very least, the urbanity continuum conceptualization should enable novel pathways to sustainability.

Looking Forward

17

Ways Forward to Explore Sustainable Land Use in an Urbanizing World

Anette Reenberg and Karen C. Seto

Introduction

Our original ambition for this Ernst Strüngmann Forum was to refine existing and develop new perspectives to address and better understand the challenges and opportunities for sustainable land use in the 21st century. Whereas sustainable land use is often considered to be a global issue, *urban* land use is often a local issue. By jointly addressing the planetary and local dimensions of land use, this Forum sought to develop a new conceptual framework of land change and urbanization that would explicitly identify how these two processes are connected, delineate pathways for sustainable urbanization, and expose the implications of an urbanizing world for global land use.

The reflections and discussions that led to the ideas and suggestions presented in this volume all emerged from a shared concern: The magnitude and rate of change over recent years and the simultaneous trends in urbanization, economic integration, and emergence of new land agents require us to rethink sustainable global land use. This not only involves a reconsideration of how sustainable land use is conceptualized and achieved. It is equally critical to think about the implicit normative and fairness issues. These perspectives underlie decisions made about who has the right to use land, what types of data are considered relevant, and what types of analyses are considered appropriate.

To address these issues, it was critical to go beyond the expertise of the core land science and urban research communities so as to bring together a much broader set of skills and perspectives from relevant disciplines—areas which also consider land use, urbanization, and resource governance in their inquiries. Radical advancements of a more conceptual nature seem to be important if progress in land-related research is to be made beyond the traditional plea for a higher level of cross-disciplinary integration.

Once convened, we collectively posed a relatively simple question: Do existing frameworks capture the relevant connections and interactions that affect land use today? More specifically, the starting point for the Forum was to decompose the exploration of land use and look at it through four lenses that directly influence the ways in which land use can be explored with respect to sustainability (see Figure 1.1, Seto and Reenberg, this volume). While this decomposition was not meant to be the end goal or a proposition for the fundamental revision of the way we explore sustainable global land use, it spurred constructive dialog between the different disciplines, exposed novel perspectives, and identified crucial areas for future exploration.

Fresh Perspectives

One of the key outcomes from this Forum was a more comprehensive and nuanced contextualization of the growing competition for, access to, and use of land. This stressed the necessity of going beyond well-established, place-based inquiries of land to explore flows and process. Five issues emerged as critical to internalize in future global land-change and urbanization research: (a) time and space interdependencies, (b) distal land connections, (c) the move away from territorial definitions, (d) revalorization of land, and (e) equity and fairness.

Time and Space Interdependencies

One important common denominator present across the disciplinary expertise at the Forum was the concern about issues related to temporal and spatial scale. For example, land-use legacies are acknowledged to add explanatory power to our understanding of trajectories of change in land systems in specific places because past processes shape current conditions and may constrain future responses. Meanwhile, contemporary land-use processes and patterns will affect future land uses because of path dependencies. Land-related processes may, however, also result in lagged or contemporaneous effects in distant places. In other words, places may be enabled or constrained in complex ways by local legacies as well as by their connections to distant places.

Cascading effects and feedbacks that are triggered by changes in one place but appear in distant places also deserve attention. Divisions of space are becoming increasingly complex in the course of the land-use transformation, involving multiple jurisdictions (local to international), multiple services (production, biodiversity, protection), and multiple actors in unusual connections. Hence, to capture such complexities in our exploration of the transformation processes requires a new conceptual framework—one that goes beyond the current emphases in the scientific communities of land-change science and urbanization, respectively, and is both place-based as well as discrete in space and time. Approaches must be developed that are able to capture how processes are continuously distributed and connected through space and across time; that is, they must (a) be dynamic and continuous, as opposed to discrete/dichotomous and static, (b) focus on people and institutions as agents of change, rather than on administrative boundaries as locations for specific demographic settlements, and (c) explicitly identify how these processes of land change and urbanization are connected at multiple scales (Boone et al., this volume). At the Forum, there was general consensus that current conceptualizations do not permit nor account for these connections, although it was recognized that they are important and must be considered. Our understanding of global land-change processes must therefore embrace these two important dimensions of time and space.

Distal Connections

As humanity continues to urbanize, land-use transformation will result in significant ways. Some of these transformations are direct: urbanization requires the development of land. Others are connected through space: urbanization requires resources that are often produced or extracted in places far from the urban locus of consumption. Globalization of the economy, information flows, and knowledge have, equivalently, large implications for global land uses. The process of globalization and urbanization, and their inherent institutional changes, has accentuated the agency of some actors in the global system, and emerging limitations on land availability and land-use options imply fewer degrees of freedom in system response (Haberl et al., this volume).

Regardless of where it occurs, urbanization creates new demands on rural, often distal, places to provide land-based products. This, in turn, blurs the distinction between the urban and the rural. Long-distance transport of land-based products to distant consumers is not a phenomenon entirely connected to our contemporary world. However, the separation of land production from the location of resource consumption has grown in magnitude and reach, and importantly, the mediating social, institutional, and economic processes have been accelerated. As a result, the proportion of land used worldwide by large, capital-intensive agri-businesses that produce for distal, urban markets has risen at the expense of labor-intensive family farms, which respond to more local demands, and has been accompanied by positive and negative implications for resource use efficiency. Nearly all land systems are now affected to some extent by these forms of long-distance connectivity, addressed in this book using the notions of teleconnections and telecouplings. The current level of interactions between two or more independently coupled social-ecological systems, which constitute the telecoupling, has potentially far-reaching implications for system function, compared to previous points of history. The central concept of telecoupling captures not only "action at a distance" but also the feedback between social processes and land outcomes in multiple interacting systems (Eakin et al., this volume).

The increased significance of telecoupling implies a need to integrate diverse epistemological perspectives, methodology, and analytical approaches. Together these approaches complement the long-standing emphasis of land and urban science on place-based research but add a new focus on the networks and system interactions involved in land change. Many of the methods that could potentially be brought to bear to account for distal connections are well developed in other disciplines, in which geo-referenced space and place are not always prominent attributes of concern. There seem to be new entry points for the land and urban science communities to explore ways in which they can productively engage with other disciplines to improve knowledge about the nonmaterial dimensions of telecoupling, such as financial flows, social networks, norms and values, as well as information and symbolic relationships.

Moving Away from Territorial Definitions

As already stressed, the classical underpinnings of land change and urbanization are largely place-based and do not immediately lend themselves to inclusion of processes related to flows of information and products. Biophysical and economic data about land may be quantified and modeled in a territorial context. Some human behaviors, such as norm-based regulations that affect land, are more difficult to administer or limit within well-defined, continuous spatial entities. However, since one of the goals of land-change science is to track and understand the drivers of land use and, ultimately, to predict the direction of human-induced changes in land use, both types of data need to be included and ways to connect them must be found.

The apparent disregard of linkages in theoretical building blocks may be explained by the fact that well-established canons about land use and urban studies were theoretically and conceptually developed as isolated themes. Hence, defining issues of importance for global land uses and livelihood sustainability may be overlooked. For example, when we think of "urban" as a place, we miss nonterritorial issues that extend beyond the urban boundary.

Particularly important in a teleconnected world is how governance of land influences land-use trajectories. Although land governance is restricted to territorial boundaries, it is affected by the flows of material, people, and information. Classical territorial arrangements continue to have a significant importance in contemporary land governance. However, territorial governance faces increasing challenges from the emergence of flow-based institutional arrangements, which has a large impact on land-based production and the corresponding commodity flows, as exemplified by emerging transnational production or sustainability standards, certification schemes, etc. Such flow-based governance mechanisms related, for example, to trade have changed the relative role of the territory-based land governance mechanisms for land-use decisions, because new land users are favored over others.

The contemporary trend in land governance points away from the dominance by classical territorial forms, such as land-use regulations made by central governments, land-use planning made by local governments, and land management undertaken by local communities. Hence, it will be crucial to continue to explore how the spatial patterns of land-use changes are connected to geographical spread of the ongoing shift in emphasis from territorial to flow-centered land governance.

Revalorization of Land

The value of land resources has multiple meanings that influence how land changes and land competition will be assessed under various conditions. Thus the notion of revalorization is necessary for reconceptualizing sustainable land use, inasmuch as it means that qualitatively or quantitatively distinct values, which differ from those previously recognized, are given to specific lands, be it monetary, political, or cultural.

The perceived and actual competition for global land is strongly influenced by current and future revalorizations of land as compared to previous points in history (Gentry et al., this volume). Global revalorizations of land have fuelled global competition over land. Few, if any, major types of land competition are unique to today's world. There are, however, a number of ways in which competition for land has entered a new stage of significance. These include, for example, the competition between potential agricultural land and land allocated to conservation of biodiversity, or land allocated to biomass for energy production. Future land trajectories will depend on how competition for land continues to develop and how value to land is ascribed by emerging users. Climate change and biodiversity conservation are, for example, likely to stay on the agenda for the coming decades, thereby creating increasing funds, economic opportunities, and institutions to devote land for conservation activities worldwide, but particularly in developing countries.

A shift in global land governance has occurred as a consequence of underlying changes in the values attached to land, including the creation of new monetary values (e.g., through forest carbon markets) and the assertion of social values in the form of biodiversity habitats, aesthetic beauty, or recreational potential. Many actors manage and impact various local- to global-scale territories, and decisions are at play, often in complex and interacting constellations. This poses new challenges to land management as well as to a deconstruction of the complex dynamics for generic analyses of land-change processes.

The issue of potential future revalorization of land poses an immense challenge as forward-looking scenarios are constructed and sustainability considerations are contemplated. Values may change significantly. Moreover, contemporary societies appear to have shifted their view of land as having one primary value (e.g., production of maize) to that which has multiple values (e.g., a number of different ecosystem services).

Equity and Fairness

Urbanization has already changed livelihoods and land uses in distal places and will continue to do so. A common starting point for the discussion of sustainable land use is one where land uses remain static, and it is often assumed that rural areas have more intrinsic or environmental value to sustainability than urban areas. However, in an urbanizing world where few land uses or livelihoods are untouched by urban processes, it is necessary to rethink concepts of sustainability so that they are not bound to a location or sets of conditions at any one place. Both rural and urban places have value and must contribute to improved global sustainability.

Correspondingly, issues of fairness or justice are also not limited to a single location. Equity and justice developed for a specific location may be neither equitable nor just when the impacts on distant place are considered. Furthermore, the ways in which researchers frame, examine, and interpret a system are significantly affected by their values and attitudes (Boone et al., this volume). Because local norms, values, and customs give meaning to places and are not tied exclusively to a physical place, the growing interconnectedness of distant places and processes can set into motion unexpected trajectories of change in land use, livelihoods, and values. Moreover, since the concept of sustainability is socially constructed, the goals and underpinning values and norms are likely to vary between different places.

Therefore, in an increasingly connected and urban world, we need to strive toward concepts of sustainability whereby urban and land processes are intimately tied and jointly considered. Only by doing so will we be able to examine the implications of urbanization for global sustainability, social justice, and human well-being in ways that will enable us to identify pathways for sustainable urbanization and global land use.

Looking Ahead

Thus far, the land-change science community has been primarily comprised of geographers, who have also been influential in a large proportion of urbanization research. While geography itself is a diverse discipline, there are many other analytical lenses—political science, law, ecology, sociology, planning—yet these have been used less in land-change science and urbanization. These other perspectives have important contributions to offer, especially in view of contemporary trends. However, the relevant research communities do not have a well-established forum within which to interact and, consequently, potential important theoretical and conceptual innovations have so far not easily materialized. Although the need for this type of dialog has been expressed by research communities in the natural and social sciences as well as by policy communities, the obstacles for the co-creation of ideas across disciplines have not been overcome. Hence, this Forum provided a unique opportunity to discuss and explore new ideas and concepts that have been developed external to the land-change science and urbanization communities, and this proved to be important to a rethinking of how we analyze land and examine sustainable land use.

Reflecting a wider range of perspectives represented at the Forum, we have highlighted a few of the most compelling issues. Further insights are presented in the individual chapters and synthesis reports of the group discussions (see Haberl et al., Eakin et al., Gentry et al., and Boone et al., this volume). Together, they clearly reflect how the concept of sustainable land use must incorporate new processes and connections associated with global urbanization, large-scale foreign investments, international investments in agricultural production, and the new forms of global rule-making and land-related governance. To meet the challenges associated with the globalization of land resources and their implications for land sustainability, further dialog is required to bring together the perspectives from different spatial scales, goals, and geographic contexts. It is our hope that the results of this Forum will prompt further discussion and provide fresh perspectives with which to examine the challenges and opportunities for sustainable land use in the 21st century.

Bibliography

Abbott, K., and D. Snidal. 2009. The Governance Triangle: Regulatory Standards Institutions and the Shadow of the State. In: The Politics of Global Regulation, ed. W. Mattli and N. Wood, pp. 44–88. Princeton: Princeton Univ. Press. [12]

Abdullah, R. 2011. Labour Requirements in the Malaysian Oil Palm Industry in 2010. *Oil Palm Indus. Econ. J.* **2**:1–11. [9]

Adams, W. M. 2004. Against Extinction: The Story of Conservation. London: Earthscan. [13]

Adell, G. 1999. Theories and Models of the Periurban Interface: A Changing Conceptual Landscape, Strategic Environmental Planning and Management for the Peri-Urban Interface-Research Project. Development Planning Unit. University College, London. http://discovery.ucl.ac.uk/43/1/DPU_PUI_Adell_THEORIES_MODELS.pdf [14]

Adnan, S., and R. Dastidar. 2011. Alienation of the Lands of Indigenous Peoples in the Chittagong Hill Tracts of Bangladesh. Copenhagen: International Work Group for Indigenous Affairs. [1]

Aerts, J. C. J. H., E. Eisinger, G. B. M. Heuvelink, and T. J. Stewart. 2003. Using Linear Integer Programming for Multi-Site Land-Use Allocation. *Geograph. Anal.* **35**:148–169. [1]

Agarwal, A., G. Giuliano, and C. Redfearn. 2012. Strangers in Our Midst: The Usefulness of Exploring Polycentricity. *Ann. Reg. Sci.* **48**:433–450. [14]

Agence France-Presse. 2012. Laos "Halts New Investment, Land Concessions." http://www.google.com/hostednews/afp/article/ALeqM5gfck8obPN9pivAmcrHVIpVFTCpyA?docId=CNG.702337e7101d78bbd1e37945c35a064b.701. (accessed Nov. 24, 2013). [7]

Ahearn, R. J. 2011. Rising Economic Powers and the Global Economy: Trends and Issues for Congress. Congressional Research Service Report 7-5700. Washington, D.C.: Congressional Research Service. [14]

Alberti, M. 2005. The Effects of Urban Patterns on Ecosystem Function. *Intl. Reg. Sci. Rev.* **28**:168–192. [15]

Alden Wily, L. 2012. Looking Back to See Forward: The Legal Niceties of Land Theft in Land Rushes. *J. Peasant Stud.* **39**:751–775. [10]

Alig, R., G. Latta, D. Adams, and B. McCarl. 2010. Mitigating Greenhouse Gases: The Importance of Land Base Interactions between Forests, Agriculture, and Residential Development in the Face of Changes in Bioenergy and Carbon Prices. *Forest Policy Econ.* **12**:67–75. [4]

Allan, J. A., M. Keulertz, S. Sojamo, and J. Warner. 2012. Handbook of Land and Water Grabs in Africa: Foreign Direct Investment and Food and Water Security. New York: Routledge. [10]

Alonso-Fradejas, A., J. L. Caal Hub, and T. Chinchilla Miranda. 2011. Plantaciones Agroindustriales, Dominación y Despojo Indígena-Campesino en la Guatemala del Siglo XXI. http://www.congcoop.org.gt/images/stories/pdfs-congcoop/Plantaciones_y_despojo-Guatemala-sXXI.pdf. (accessed Dec. 8, 2013). [11]

Amin, A., and N. Thrift, eds. 1994. Globalization, Institutions and Regional Development in Europe. Oxford: Oxford Univ. Press. [15]

Andersen, T., J. Carstensen, E. Hernandez-Garcia, and C. M. Duarte. 2009. Ecological Thresholds and Regime Shifts: Approaches to Identification. *Trends Ecol. Evol.* **24**:49–57. [4]

Anderson, K. 2010. Globalization's Effect on World Agricultural Trade, 1960–2050. *Phil. Trans. R. Soc. B.* **365**:3007–3021. [2]

Andrews, R. N. L. 1998. Environmental Regulation and Business "Self-Regulation." *Policy Sci.* **31**:177–197. [12]

Angel, S. 2012. The Planet of Cities. Cambridge, MA: Lincoln Institute of Land Policy. [15]

Angel, S., J. Parent, D. L. Civco, and A. Blei. 2010. Atlas of Urban Expansion. Cambridge, MA: Lincoln Institute of Land Policy. http://www.lincolninst.edu/sub-centers/atlas-urban-expansion/ (accessed Oct. 23, 2013). [15]

Angel, S., J. Parent, D. L. Civco, A. Blei, and D. Potere. 2011. The Dimensions of Global Urban Expansion: Estimates and Projections for All Countries, 2000–2050. *Prog. Plan.* **75**:53–107. [1, 2, 6, 14]

Angelsen, A. 2008. Moving Ahead with REDD: Issues, Options and Implications. Bogor: Centre for Intl. Forestry Research (CIFOR). [4]

———. 2009. Realising REDD+: National Strategy and Policy Options. Situ Gede, Indonesia: Centre for Intl. Forestry Research (CIFOR). [4]

———. 2010. Policies for Reduced Deforestation and Their Impact on Agricultural Production. *PNAS* **107**:19639–19644. [2, 4]

Angelsen, A., M. Brockhaus, W. D. Sunderlin, and L. V. Verchot. 2012. Analyzing REDD+: Challenges and Choices. Bogor: CIFOR. [13]

Anseeuw, W., L. Alden-Wily, L. Cotula, and M. Taylor. 2012a. Land Rights and the Rush for Land: Findings of the Global Commercial Pressures on Land Research Project. Rome: ILC. [7, 10, 13]

Anseeuw, W., M. Boche, T. Breu, et al. 2012b. Transnational Land Deals for Agriculture in the Global South. Analytical Report Based on the Land Matrix Database. Bern/Montpellier/Hamburg: CDE/CIRAD/GIGA. [2, 6, 7, 10]

Antrop, M. 2004. Landscape Change and the Urbanization Process in Europe. *Landsc. Urban Plan* **67**:9–26. [15]

Araghi, F. 2009. The Invisible Hand and the Visible Foot: Peasants, Dispossession and Globalization. Peasants and Globalization: Political Economy, Rural Transformation and the Agrarian Question, A. H. Akram-Lodhi and C. Kay, series ed. London: Routledge. [11]

Armstrong, S. 1995. Rare Plants Protect Cape's Water Supplies. *New Scientist* **11**:8. [7]

Ascher, K. 2005. The Works, Anatomy of a City. New York: The Penguin Press. [6]

Ash, N., H. Blanco, K. Garcia, et al. 2010. Ecosystems and Human Well-Being: A Manual for Assessment Practitioners. Washington, D.C.: Island Press. [15]

Auld, G. 2009. Reversal of Fortune: How Early Choices Can Alter the Logic of Market-Based Authority. Ph.D. dissertation, Yale University, New Haven. [12]

———. 2010a. Assessing Certification as Governance: Effects and Broader Consequences for Coffee. *J. Environ. Dev.* **19**:215–241. [12]

———. 2010b. A Review of "Setting the Standard: Certification, Governance, and the Forest Stewardship Council." *J. Intl. Wildlife Law Policy* **13**:257–260. [12]

———. 2011. Certification as Governance: Current Impact and Future Prospects. In: The Politics of Fair Trade, ed. M. Warrier, pp. 68–86. London: Routledge. [12]

———. 2013. Confronting Trade-Offs and Interactive Effects in the Choice of Policy Focus: Specialized versus Comprehensive Private Governance. *Regul. Govern.* first published online, Aug. 26, 2013. http://onlinelibrary.wiley.com/doi/2010.1111/rego.12034/abstract. [12]

Auld, G., and G. Q. Bull. 2003. The Institutional Design of Forest Certification Standards Initiatives and Its Influence on the Role of Science: The Case of Forest Genetic Resources. *J. Environ. Manag.* **69**:47–62. [12]

Auld, G., B. Cashore, C. Balboa, L. Bozzi, and S. Renckens. 2010. Can Technological Innovations Improve Private Regulations in the Global Economy? *Business Polit.* **12**:1469–3569. [12]

Auld, G., L. H. Gulbrandsen, and C. McDermott. 2008. Certification Schemes and the Impact on Forests and Forestry. *Annu. Rev. Environ. Res.* **33**:187–211. [12]

Australian Bureau of Agricultural Resource Economics Sciences. 2012. Food Demand to 2050: Opportunities for Australia. http://data.daff.gov.au/data/warehouse/Outlook2012/fdi50d9abat001201203/Outlook2012FoodDemand2050.pdf. (accessed Nov. 30, 2013). [1]

Ayers, A. J. 2013. Beyond Myths, Lies and Stereotypes: The Political Economy of a New Scramble for Africa. *New Polit. Econ.* **18**:227–257. [10]

Ayres, R. U., J. van den Bergh, and J. M. Gowdy. 2001. Strong versus Weak Sustainability: Economics, Natural Sciences, and "Consilience." *Environ. Ethics* **23**:155–168. [16]

Bae, J.-S., R.-W. Joo, and Y.-S. Kim. 2012. Forest Transition in South Korea: Reality, Path and Drivers. *Land Use Policy* **29**:198–207. [6]

Bai, X., R. R. McAllister, R. M. Beaty, and B. Taylor. 2010. Urban Policy and Governance in a Global Environment: Complex Systems, Scale Mismatches and Public Participation. *Curr. Opin. Environ. Sustain.* **2**:129–135. [4]

Bailis, R., and J. Baka. 2011. Constructing Sustainable Biofuels: Governance of the Emerging Biofuel Economy. *Ann. Assoc. Am. Geograph.* **101**:827–838. [13]

Baird, I. G. 2010. Land, Rubber, and People: Rapid Agrarian Changes and Responses in Southern Laos. *J. Lao Stud.* **1**:1–47. [7]

———. 2011. Turning Land into Capital, Turning People into Labour: Primitive Accumulation and the Arrival of Large-Scale Economic Land Concessions in the Lao People's Democratic Republic. *New Proposals: J. Marxism and Interdis. Inq.* **5**:10–26. [7]

Balmford, A., R. E. Green, and J. P. W. Scharlemann. 2005. Sparing Land for Nature: Exploring the Potential Impacts of Changes in Agricultural Yields on the Area Needed for Crop Production. *Global Change Biol.* **11**:1594–1605. [4]

Bamford, S. 1998. Humanized Landscapes, Embodied Worlds: Land and the Construction of Intergenerational Continuity among the Kamea of Papua New Guinea. *Soc. Analysis* **42**:28–54. [13]

Barabasi, A.-L., and R. Albert. 1999. Emergence of Scaling in Random Networks. *Science* **286**:509–512. [14]

Bartley, T. 2003. Certifying Forests and Factories: States, Social Movements, and the Rise of Private Regulation in the Apparel and Forest Products Fields. *Politics & Society* **31**:433–464. [12]

———. 2007a. How Foundations Shape Social Movements: The Construction of an Organizational Field and the Rise of Forest Certification. *Soc. Prob.* **54**:229–255. [12]

Bartley, T. 2007b. Institutional Emergence in an Era of Globalization: The Rise of Transnational Private Regulation of Labor and Environmental Conditions. *Am. J. Sociol.* **113**:297–351. [12]

Bartley, T., and S. N. Smith. 2010. Communities of Practice as Cause and Consequence of Transnational Governance. In: Transnational Communities: Shaping Global Economic Governance, ed. M. L. Djelic and S. Quack, pp. 334–373. Cambridge: Cambridge Univ. Press. [12]

Barton, M. 2005. The Possible Implications of a Nitrogen Cap on Farming Businesses within the Lake Taupo Catchment. Masters Thesis, Southern Cross Univ., Lismore, Australia. http://www.motu.org.nz/publications/detail/the_possible_implications_of_a_nitrogen_cap_on_farming_businesses_within_th (accessed Nov. 25, 2013). [4]

Bastian, O., D. Haase, and K. Grunewald. 2012. Ecosystem Properties, Potentials and Services: The EPPS Conceptual Framework and an Urban Application Example. *Ecol. Ind.* **21**:7–16. [15]

Bationo, A., B. Waswa, A. Adamou, et al. 2012a. Overview of Long-Term Experiments in Africa. In: Lessons Learned from Long-Term Fertility Management Experiments in Africa, ed. A. Bationo et al., pp. 1–26. Dordrecht: Springer. [3]

Bationo, A., B. Waswa, J. Kihara, et al. 2012b. Lessons Learned from Long-Term Fertility Management Experiments in Africa. Dordrecht: Springer. [3]

Batisse, M. 1982. The Biosphere Reserve: A Tool for Environmental Conservation and Management. *Environ. Conserv.* **9**:101–111. [4]

Beauregard, R. A. 2009. Urban Population Loss in Historical Perspective: United States, 1820–2000. *Environ. Plann. A* **41**:514–528. [15]

Bebbington, A. J. 2005. Donor-NGO Relations and Representations of Livelihood in Nongovernmental Aid Chains. *World Devel.* **33**:937–950. [8]

Beddington, A. 2012. Extractive Industries, Social Conflict and Economic Development: Evidence from South America. London: Routledge. [13]

Bell, S., S. Alves, E. S. d. Oliveira, and A. Zuin. 2010. Migration and Land Use Change in Europe: A Review. *Living Rev. Landsc. Res.* 4, http://www.livingreviews.org/lrlr-2010-2. (accessed Nov. 25, 2013). [8, 16]

Bendell, J. 1999. Civil Regulation: A New Form of Democratic Governance for the Global Economy? In: Terms of Endearment: Business, NGOs, and Sustainable Development, ed. J. Bendell, pp. 239–254. Sheffield: Greenleaf Publishing. [12]

Benjaminsen, T. A., and I. Bryceson. 2012. Conservation, Green/Blue Grabbing and Accumulation by Dispossession in Tanzania. *J. Peasant Stud.* **39**:335–355. [13]

Benjaminsen, T. A., and H. Svarstad. 2010. The Death of an Elephant: Conservation Discourses versus Practices in Africa. In: Forum for Development Studies, vol. 37, pp. 385–408. New York: Taylor & Francis. [13]

Beringer, T., W. Lucht, and S. Schaphoff. 2011. Bioenergy Production Potential of Global Biomass Plantations under Environmental and Agricultural Constraints. *GCB Bioenergy* **3**:299–312. [4]

Bernstein, S. 2000. Ideas, Social Structure and the Compromise of Liberal Environmentalism. *Eur. J. Intl. Relat.* **6**:464–512. [10]

Bernstein, S., and B. Cashore. 2007. Can Non-State Global Governance Be Legitimate? An Analytical Framework. *Regul. Govern.* **1**:347–371. [12]

Bernt, M. 2009. Partnerships for Demolition: The Governance of Urban Renewal in East Germany's Shrinking Cities. *Intl. J. Urban Reg. Res.* **33**:754–769. [15]

Berry, B. J. L. 1991. Long-Wave Rhythms in Economic Development and Political Behavior. Baltimore: Johns Hopkins Press. [6]

Berry, B. J. L., and J. B. Parr. 1988. Market Centers and Retail Location. Englewood Cliffs: Prentice-Hall. [14]

Besky, S. 2011. Colonial Pasts and Fair Trade Futures: Changing Modes of Production and Regulation on Darjeeling Tea Plantations. In: Fair Trade and Social Justice: Global Ethnographies, ed. S. Lyon and M. Moberg, pp. 97–122. New York: NYU Press. [12]

Bettencourt, L. M. A., J. Lobo, D. Helbing, C. Kühnert, and G. B. West. 2006. Growth, Innovation, Scaling, and the Pace of Life in Cities. *PNAS* **104**:7301–7306. [15]

Billen, G., P. Chatzimpiros, S. Barles, and J. Garnier. 2012a. Grain and Meat to Feed Paris: The Changing Balance between Cereal Cultivation and Animal Farming in the Paris Basin (18th–21st Centuries). *Reg. Environ. Change* **12**:325–335. [14]

Billen, G., J. Garnier, and S. Barles. 2012b. History of the Urban Environmental Imprint: Introduction to a Multidisciplinary Approach to the Long-Term Relationships between Western Cities and Their Hinterland. *Reg. Environ. Change* **12**:249–253. [14]

Binswanger, H. P., K. Deininger, and G. Feder. 1995. Power, Distortions, Revolt and Reform in Agricultural Land Relations. In: Handbook of Development Economics, ed. J. Srinivasan and T. N. Behrman, vol. 3. Amsterdam: Elsevier. [4]

BIS. 1996. Bank for International Settlements. *Quartery Review*, Statistical Annex. Basil: BIS. [6]

———. 2012. Bank for International Settlements. *Quartery Review*, Statistical Annex. Basil: BIS. [6]

Blanke, M., and B. Burdick. 2005. Food (Miles) for Thought: Energy Balance for Locally-Grown versus Imported Apple Fruit. *Environ. Sci. Pollut. Res. Int.* **12**:125–127. [4]

Blomquist, G. C., M. C. McBerger, and J. P. Hoehn. 1988. New Estimates of Quality of Life in Urban Areas. *Am. Econ. Rev.* **78**:89–107. [14]

Bordo, M. D., B. Eichengreen, and D. A. Irwin. 1999. Is Globalization Today Really Different Than Globalization a Hundred Years Ago? NBER Working Paper No. 7195. http://www.nber.org/papers/w7195.pdf. (accessed Dec. 8, 2013). [6]

Born, B., and M. Purcell. 2006. Avoiding the Local Trap: Scale and Food Systems in Planning Research. *J. Plan. Educ. Res.* **26**:195–207. [14]

Borras, S. M., Jr., D. Carranza, J. Franco, and M. L. Alano. 2011. Land Grabbing and the Contested Notion of Marginal Lands: Insights from the Philippines. International Conference on Land Grabbing. Sussex: Institute of Development Studies. [11]

Borras, S. M., Jr., and J. C. Franco. 2010. Towards a Broader View of the Politics of Global Land Grab: Rethinking Land Issues, Reframing Resistance. In: ICAS Working Paper Series No. 1. The Hague: Transnational Institute. [11]

Borras, S. M., Jr., J. C. Franco, S. Gómez, C. Kay, and M. Spoor. 2012a. Land Grabbing in Latin America and the Caribbean. *J. Peasant Stud.* **39**:845–872. [10]

Borras, S. M., Jr., J. C. Franco, and C. Wang. 2013. The Challenge of Global Governance of Land Grabbing: Changing International Agricultural Context and Competing Political Views and Strategies. *Globalizations* **10**:161–179. [10, 11, 13]

Borras, S. M., Jr., C. Kay, S. Gómez, and J. Wilkinson. 2012b. Land Grabbing and Global Capitalist Accumulation: Key Features in Latin America. *Can. J. Dev. Stud.* **33**:402–416. [10, 11]

Borras, S. M., Jr., P. D. McMichael, and I. Scoones. 2010. The Politics of Biofuels, Land and Agrarian Change: Editors' Introduction. *J. Peasant Stud.* **37**:575–592. [10]

Börzel, T. A., and T. Risse. 2005. Public-Private Partnerships: Effective and Legitimate Tools of Transnational Governance? In: Complex Sovereignty: On the Reconstitution of Political Authority in the 21st Century, ed. E. Grand and L. Pauly, pp. 195–216. Toronto: Univ. of Toronto Press. [12]

Botzem, S., and S. Quack. 2006. Contested Rules and Shifting Boundaries: International Standard-Setting in Accounting. In: Transnational Governance: Institutional Dynamics of Regulations, ed. M. L. Djelic and K. Sahlin-Andersson, pp. 267–286. Cambridge: Cambridge Univ. Press. [12]

Bowen, L., S. Ebrahim, B. De Stavola, et al. 2011. Dietary Intake and Rural-Urban Migration in India: A Cross-Sectional Study. *PLoS ONE* **6**:e14822. [1]

Bradley, B. A., A. Houghton, J. F. Mustard, and S. P. Hanburg. 2006. Invasive Grass Reduces Aboveground Carbon Stocks in Shrublands of the Western U.S. *Global Change Biol.* **12**:1815–1822. [7]

Breuste, J., D. Haase, and T. Elmquist. 2013. Urban Landscapes and Ecosystem Services. In: Ecosystem Services in Agricultural Amd Urban Landscapes, ed. H. Sandhu et al., pp. 83–104. Chichester: Wiley-Blackwell. [15]

Bridge, G. 2004. Mapping the Bonanza: Geographies of Mining Investment in an Era of Neoliberal Reform. *Prof. Geograph.* **56**:406–421. [13]

Briggs, D. J., C. de Hoogh, J. Gulliver, et al. 2000. A Regression-Based Method for Mapping Traffic-Related Air Pollution: Application and Testing in Four Contrasting Urban Environments. *Sci. Total Environ.* **253**:151–167. [15]

Brink, A. B., and H. D. Eva. 2009. Monitoring 25 Years of Land Cover Change Dynamics in Africa: A Sample-Based Remote Sensing Approach. *Appl. Geog.* **29**:501–512. [2]

Brockington, D., R. Duffy, and J. Igoe. 2008. Nature Unbound: Conservation, Capitalism and the Future of Protected Areas. London: Earthscan. [13]

Brown, I., L. Poggio, A. Gimona, and M. Castellazzi. 2011. Climate Change, Drought Risk and Land Capability for Agriculture: Implications for Land Use in Scotland. *Reg. Environ. Change* **11**:503–518. [3]

Brown, I., W. Towers, M. Rivington, and H. I. J. Black. 2008. Influence of Climate Change on Agricultural Land-Use Potential: Adapting and Updating the Land Capability System for Scotland. *Climate Res.* **37**:43–57. [3]

Brown, L. R. 1995. Who Will Feed China? Wake-up Call for a Small Planet. New York: W.W. Norton for the Worldwatch Institute. [5]

Browne, P. 2009. Tanzania Suspends Biofuel Investments. *New York Times*, 14 Oct. [13]

Bruinsma, J. 2003. World Agriculture: Towards 2015/2030, an FAO Perspective. London: Earthscan Publications. [3]

———. 2009. The Resource Outlook to 2050. Expert Meeting on How to Feed the World in 2050. Rome: FAO. [1]

Burchell, R. W., A. Downs, B. McCann, and S. Mukherji. 2005. Sprawl Costs: Economic Impacts of Unchecked Development. Washington, D.C.: Island Press. [14]

Buresh, R. J., P. A. Sanchez, and F. Calhoun. 1997. Replenishing Soil Fertility in Africa. Wisconsin: Soil Science Society of America. [3]

Burney, J. A., S. J. Davis, and D. B. Lobell. 2010. Greenhouse Gas Mitigation by Agricultural Intensification. *PNAS* **107**:12052–12057. [4]

Burton, E. 2000. The Compact City: Just or Just Compact? A Preliminary Analysis. *Urban Stud.* **37**:1969–2001. [15]

Bustamante, M. M. C., C. A. Nobre, R. Smeraldi, et al. 2012. Estimating Recent Greenhouse Gas Emissions from Cattle Raising in Brazil. *Clim. Change* **115**:559–577. [2]

Büthe, T., and W. Mattli. 2011. New Global Rulers: The Privatization of Regulation in the World Economy. Princeton: Princeton Univ. Press. [12]

Buzar, S., P. E. Ogden, R. Hall, et al. 2007. Splintering Urban Populations: Emergent Landscapes of Reurbanisation in Four European Cities. *Urban Stud.* **44**:651–677. [15]

Cadenasso, M. L., S. T. A. Pickett, and K. Schwarz. 2007. Spatial Heterogeneity in Urban Ecosystems: Reconceptualizing Land Cover and a Framework for Classification. *Front. Ecol. Environ.* **5**:80–88. [16]

Cadman, T. 2011. Quality and Legitimacy of Global Governance: Case Lessons from Forestry. Basingstoke: Palgrave McMillan. [10]

Cai, Y. M., W. X. Zhang, and Y. S. Liu. 2007. Forecasting and Analyzing the Cultivated Land Demand Based on Multi-Objectives in China. *Resources Sci.* **29**:134–138. [5]

Calthorpe, P. 1993. The Next American Metropolis: Ecology, Community and the American Dream. Princeton: Princeton Univ. Press. [14]

Campbell, D. J., H. Gichohi, A. Mwangi, and L. Chege. 2000. Land Use Conflict in Kajiado District, Kenya. *Land Use Policy* **17**:337–348. [4]

Canadell, J. G., C. L. Quéré, M. R. Raupach, et al. 2007. Contributions to Accelerating Atmospheric CO_2 Growth from Economic Activity, Carbon Intensity, and Efficiency of Natural Sinks. *PNAS* **104**:18866–18870. [4]

Carlson, K. M., L. M. Curran, G. P. Asner, et al. 2013. Carbon Emissions from Forest Conversion by Kalimantan Oil Palm Plantations. *Nature Clim. Change* **3**:283–287. [4]

Caroko, W., H. Komarudin, K. Obidzinski, and P. Gunarso. 2011. Policy and Institutional Frameworks for the Development of Palm Oil-Based Biodiesel in Indonesia. CIFOR Working Paper 62. http://www.cifor.org/publications/pdf_files/WPapers/WP62Komarudin.pdf. (accessed Dec. 8, 2013). [9]

Carruthers, J. I., and A. C. Vias. 2005. Urban, Suburban, and Exurban Sprawl in the Rocky Mountain West: Evidence from Regional Adjustment Models. *J. Reg. Sci.* **45**:25–48. [4]

Cashore, B. 2002. Legitimacy and the Privatization of Environmental Governance: How Non-State Market-Driven (NSMD) Governance Systems Gain Rule-Making Authority. *Governance* **15**:503–529. [10, 12]

Cashore, B., G. Auld, and D. Newsom. 2004. Governing through Markets: Forest Certification and the Emergence of Non-State Authority. New Haven: Yale Univ. Press. [12, 13]

Cashore, B., F. P. Gale, E. E. Meidinger, and D. Newsom. 2006. Confronting Sustainability: Forest Certification in Developing and Transitioning Societies. New Haven: Yale School of Forestry and Environmental Studies Publication Series. [12]

Cassman, K. G. 1999. Ecological Intensification of Cereal Production Systems: Yield Potential, Soil Quality, and Precision Agriculture. *PNAS* **96**:5952–5959. [4]

Cassman, K. G., A. R. Dobermann, D. T. Walters, and H. Yang. 2003. Meeting Cereal Demand While Protecting Natural Resources and Improving Environmental Quality. *Ann. Rev. Environ. Res.* **28**:315–358. [3]

Casson, A. 2000. The Hesitant Boom: Indonesia's Palm Oil Sub Sector in an Era of Economic Crisis and Political Change. CIFOR Occasional Paper 29. http://www.cifor.cgiar.org/publications/pdf_files/OccPapers/OP-029.pdf. (accessed Dec. 8, 2013). [9]

Castells, M. 1989. The Informational City: Information Technology, Economic Restructuring and the Urban-Regional Process. Oxford: Basil Blackwell. [14]

———. 1996. The Rise of the Network Society. Oxford: Blackwell. [14]

Castells, M. 2010. Globalisation, Networking, Urbanisation: Reflections on the Spatial Dynamics of the Information Age. *Urban Stud.* **47**:2737–2745. [14]

Center for Human Rights Global Justice. 2010. Foreign Land Deals and Human Rights: Case Studies on Agricultural and Biofuel Investment. New York: NYU School of Law. [1]

Chan, K. M. A., A. D. Guerry, P. Balvanera, et al. 2012. Where Are Cultural and Social in Ecosystem Services? A Framework for Constructive Engagement. *Bioscience* **62**:744–756. [13]

Chan, K. M. A., M. R. Shaw, D. R. Cameron, U. E. C., and G. C. Daily. 2006. Conservation Planning for Ecosystem Services. *PLoS Biol.* **4**:e379. [4]

Chape, S., J. Harrison, M. Spalding, and I. Lysenko. 2005. Measuring the Extent and Effectiveness of Protected Areas as an Indicator for Meeting Global Biodiversity Targets. *Phil. Trans. R. Soc. B.* **360**:443–455. [13]

Chen, Y., and Y. Zhou. 2006. Reinterpreting Central Place Networks Using Ideas from Fractals and Self-Organized Criticality. *Environ. Plann. B* **33**:345–364. [14]

Cheshire, P., and S. Sheppard. 2002. The Welfare Economics of Land Use Planning. *J. Urban Econ.* **52**:242–269. [4]

Choudhury, K., and L. J. M. Jansen. 1998. Terminology for Integrated Resource Planning and Management. Rome: FAO. [6]

Christaller, W. 1933/1966. Die Zentralen Orte in Suddeutschland (translated by Charlisle W. Baskin as "Central Places in Southern Germany"). Jena/Upper Saddle River: Gustav Fischer/Prentice Hall. [14]

Chum, H., A. Faaij, J. Moreira, et al. 2012. Bioenergy. In: IPCC Special Report on Renewable Energy Sources and Climate Change Mitigation, ed. O. Edenhofer et al., pp. 209–332. Cambridge: Cambridge Univ. Press. [4]

Cinzano, P., F. Falchi, and C. D. Elvidge. 2001. The First World Atlas of the Artificial Night Sky Brightness. *Mon. Not. R. Astron. Soc.* **328**:689–707. [15]

Clapp, J. 1998. The Privatization of Global Environmental Governance: ISO 14000 and the Developing World. *Global Governance* **4**:295–316. [12]

Clough, Y., J. Barkmann, J. Juhrbandt, et al. 2011. Combining High Biodiversity with High Yields in Tropical Agroforests. *PNAS* **108**:8311–8316. [4]

Coelho, S., O. Agbenyega, A. Agostini, et al. 2012. Land and Water: Linkages to Bioenergy. In: Global Energy Assessment: Toward a Sustainable Future, ed. T. Johansson et al., pp. 1459–1525. Cambridge/Laxenburg: Cambridge Univ. Press/ Intl. Institute of Applied Systems Analysis (IIASA). [4]

Colchester, M. 2011. Palm Oil and Indigenous Peoples in South East Asia. International Land Coalition. http://www.forestpeoples.org/sites/fpp/files/publication/2010/08/palmoilindigenouspeoplesoutheastasiafinalmceng_0.pdf. (accessed Jan. 9, 2013). [9]

Combes, P.-P., G. Duranton, L. Gobillon, and S. Roux. 2010. Estimating Agglomeration Effects with History, Geology, and Worker Fixed-Effects. In: Agglomeration Economics, ed. E. L. Glaeser. Chicago: Univ. of Chicago Press. [14]

Connors, J. P., C. S. Galletti, and W. T. L. Chow. 2013. Landscape Configuration and Urban Heat Island Effects: Assessing the Relationship between Landscape Characteristics and Land Surface Temperature in Phoenix, Arizona. *Landsc. Ecol.* **28**:271–283. [4]

Conroy, M. E. 2006. Branded: How the "Certification Revolution" Is Transforming Global Corporations. Gabriola Island, BC: New Society Publishers. [12]

Corbera, E., and H. Schroeder. 2011. Governing and Implementing REDD+. *Environ. Sci. Policy* **14**:89–99. [10]

Corson, C., and K. MacDonald. 2012. Enclosing the Global Commons: The Convention on Biological Diversity and Green Grabbing. *J. Peasant Stud.* **39**:263–283. [13]

Costanza, R., and H. E. Daly. 1992. Natural Capital and Sustainable Development. *Conserv. Biol.* **6**:37–46. [16]

Cotula, L. 2011. Land Deals in Africa: What Is in the Contracts? London: Intl. Institute for Environment and Development. [10]

———. 2012. The International Political Economy of the Global Land Rush: A Critical Appraisal of Trends, Scale, Geography and Drivers. *J. Peasant Stud.* **39**:649–680. [7, 10, 13]

Cotula, L., S. Vermeulen, R. Leonard, and J. Keeley. 2009. Land Grab or Development Opportunity? Agricultural Investment and International Land Deals in Africa. London/Rome: IIED/FAO/IFAD. [10]

Couch, C., J. Karecha, H. Nuissl, and D. Rink. 2005. Decline and Sprawl: An Evolving Type of Urban Development: Observed in Liverpool and Leipzig. *Eur. Plann. Stud.* **13**:117–136. [15]

Cowie, A., U. A. Schneider, and L. Montanarella. 2007. Potential Synergies between Existing Multilateral Environmental Agreements in the Implementation of Land Use, Land-Use Change and Forestry Activities. *Environ. Sci. Policy* **10**:335–352. [13]

Cowling, R. M., B. Egoh, A. T. Knight, et al. 2008. An Operational Model for Mainstreaming Ecosystem Services for Implementation. *PNAS* **105**:9483–9488. [15]

Cox, S., and B. Searle. 2009. The State of Ecosystem Services. The Bridgespan Group. http://www.bridgespan.org/Publications-and-Tools/Environment/The-State-of-Ecosystem-Services.aspx#.Umjwjvl3auk (accessed Oct. 24, 2013). [15]

Cramb, R. A., C. J. P. Colfer, W. Dressler, et al. 2009. Swidden Transformations and Rural Livelihoods in Southeast Asia. *Human Ecol.* **37**:323–346. [2]

Crooks, J. A., and M. E. Soule. 1999. Lag Times in Population Explosions of Invasive Species: Causes and Implications. In: Invasive Species and Biodiversity Management, ed. O. T. Sandlund et al., pp. 103–125. Dordrecht: Kluwer Academic Publishers. [7]

Crosby, A. W. 1986. Ecological Imperialism: The Biological Expansion of Europe, 900–1900. Cambridge: Cambridge Univ. Press. [4]

Cudahy, B. J. 2006. The Containership Revolution. *TR News* **246**: 5–9. [14]

Cutler, A. C. 1997. Artifice, Ideology and Paradox: The Public/Private Distinction in International Law. *Rev. Intl. Polit. Econ.* **4**:261–285. [12]

Cutler, A. C., V. Haufler, and T. Porter. 1999a. The Contours and Significance of Private Authority in International Affairs. In: Private Authority and International Affairs, ed. A. C. Cutler et al., J. N. Rosenau, series ed., pp. 333–376. Albany: SUNY Press. [12]

———. 1999b. Private Authority in International Politics. In: SUNY Series in Global Politics, ed. J. N. Rosenau. Albany: SUNY Press. [10, 12]

D'Antonio, C. M., and P. M. Vitousek. 1992. Biological Invasions by Exotic Grasses, the Grass/Fire Cycle, and Global Change. *Annu. Rev. Ecol. System.* **23**:63–87. [7]

Daily, G., ed. 1997. Nature's Services: Societal Dependence on Natural Ecosystems. Washington, D.C.: Island Press. [14]

Daniel, S. 2012. Situating Private Equity Capital in the Land Grab Debate. *J. Peasant Stud.* **39**:703–729. [10]

Daniel, S., and A. Mittal. 2010. (Mis)Investment in Agriculture. The Role of the International Finance Corporation in Global Land Grabs. Oakland: Oakland Institute. [7]

Daniel, T. C., A. Muhar, A. Arnberger, et al. 2012. Contributions of Cultural Services to the Ecosystem Services Agenda. *PNAS* **109**:8812–8819. [13]

Danielsen, F., H. Beukema, N. D. Burgess, et al. 2009. Biofuel Plantations on Forested Lands: Double Jeopardy for Biodiversity and Climate. *Conserv. Biol.* **23**:348–358. [4, 7]

Dauvergne, P., and J. Lister. 2012. Big Brand Sustainability: Governance Prospects and Environmental Limits. *Global Environ. Change* **22**:36–45. [2]

Davies, J. B., S. Sandstrom, A. Shorrocks, and E. N. Wolff. 2006. The World Distribution of Household Wealth. UNU-WIDER Research Paper. Helsinki: U.N. Univ., World Institute for Development Economics Research. [6]

Decressin, J., and A. Fatas. 1995. Regional Labor Market Dynamics in Europe. *Eur. Econ. Rev.* **39**:1627–1655. [14]

DeFries, R. S., J. A. Foley, and G. P. Asner. 2004. Land-Use Choices: Balancing Human Needs and Ecosystem Function. *Front. Ecol. Environ.* **2**:249–256. [6]

DeFries, R. S., T. K. Rudel, M. Uriarte, and M. Hansen. 2010. Deforestation Driven by Urban Population Growth and Agricultural Trade in the Twenty-First Century. *Nature Geosci.* **3**:178–181. [2, 6, 7]

Degnan, F. H. 1997. The Food Label and the Right-to-Know. *Food Drug Law J.* **52**:49. [12]

Deininger, K., D. Byerlee, J. Lindsay, et al. 2011. Rising Global Interest in Farmland: Can It Yield Sustainable and Equitable Benefits? Washington, D.C.: World Bank. [6, 7, 10, 11]

Deininger, K., and G. Feder. 2001. Land Institutions and Land Markets. In: Handbook of Agricultural Economics, ed. B. L. Gardner and G. C. Rausser, vol. 1, Part A, pp. 288–331. Amsterdam: Elsevier. [4]

Dekle, R., and J. Eaton. 1999. Agglomeration and Land Rents: Evidence from the Prefectures. *J. Urban Econ.* **46**:200–214. [14]

De la Cruz, R. J. G. 2012. Land Title to the Tiller: Why It's Not Enough and How It's Sometimes Worse. ISS Working Paper No. 534. The Hague: Erasmus Univ. Rotterdam. [11]

Deller, S. C., T. Tsung-Hsiu, D. W. Marcouiller, and D. B. K. English. 2001. The Role of Amenities and Quality of Life in Rural Economic Growth. *Am. J. Agric. Econ.* **83**:352–365. [4]

Delzeit, R., and K. Holm-Müller. 2009. Steps to Discern Sustainability Criteria for a Certification Scheme of Bioethanol in Brazil: Approach and Difficulties. *Energy* **34**:662–668. [12]

Deng, X. Z., J. K. Huang, S. Rozelle, and E. Uchida. 2006. Cultivated Land Conversion and Potential Agricultural Productivity in China. *Land Use Policy* **23**:372–384. [5]

———. 2008. Growth, Population and Industrialization, and Urban Land Expansion of China. *J. Urban Econ.* **63**:96–115. [5]

De Schutter, O. 2010a. Access to Land and the Right to Food. Report of the Special Rapporteur on the Right to Food (A/65/281). http://www.srfood.org/images/stories/pdf/officialreports/20101021_access-to-land-report_en.pdf. (accessed Oct. 3, 2012). [13]

———. 2010b. Large-Scale Land Acquisitions and Leases: A Set of Core Principles and Measures to Address the Human Rights Challenge. Report of the Special Rapporteur on the Right to Food (A/HRC/13/33/Add.2). http://www.srfood.org/images/stories/pdf/officialreports/20100305_a-hrc-13-33-add2_land-principles_en.pdf. (accessed Oct. 3, 2012). [13]

———. 2011. Agroecology and the Right to Food. Report Submitted by the Special Rapporteur on the Right to Food (A/HRC/16/49). http://www.srfood.org/images/stories/pdf/officialreports/20110308_a-hrc-16-49_agroecology_en.pdf. (accessed Oct. 3, 2012). [13]

Desrochers, P., and H. Shimizu. 2012. The Locavore's Dilemma: In Praise of the 10,000-Mile Diet. New York: PublicAffairs. [7]

Dicken, D. 2003. Global Shift: Reshaping the Global Economic Map in the 21st Century, 4th edition. New York: Guilford Press. [6, 8]

Dickenson, N. 1993. Catering for the Ethical Shopper: A Look at a Growing Consumer Trend. In: Management (Marketing and Advertising) Section, P. 19. *Financial Times London* http://web.lexis-nexis.com. (accessed April 15, 2013). [12]

Dieleman, F., and M. Wegener. 2004. Compact City and Urban Sprawl. *Build. Environ.* **30**:308–323. [15]

Dingwerth, K. 2007. The New Transnationalism: Transnational Governance and Democratic Legitimacy. Transformations of the State. Basingstoke: Palgrave Macmillan. [12]

Dingwerth, K., and P. Pattberg. 2009. World Politics and Organizational Fields: The Case of Transnational Sustainability Governance. *Eur. J. Intl. Relat.* **15**:707–743. [12]

Dittmarr, H., and G. Ohland. 2004. The New Transit Town: Best Practices in Transit. Oriented Development. Washington, D.C.: Island Press. [14]

Dixon, R. K., S. Brown, R. A. Houghton, et al. 1994. Carbon Pools and Flux of Global Forest Ecosystems. *Science* **263**:185. [13]

Doriye, E. 2010. The Next Stage of Sovereign Wealth Investment: China Buys Africa. *J. Financ. Reg. Compl.* **18**:23–31. [1]

Douglas, I. 2006. Peri-Urban Ecosystems and Societies Transitional Zones and Contrasting Values. In: Peri-Urban Interface: Approaches to Sustainable Natural and Human Resource Use, ed. D. McGregor et al., pp. 18–29. London: Earthscan. [15]

Duany, A., E. Plater-Zyberk, and J. Speck. 2001. Suburban Nation: The Rise of Sprawl and the Decline of the American Dream. New York: North Point Press. [14]

Ducruet, C., and T. Notteboom. 2012. The Worldwide Maritime Network of Container Shipping: Spatial Structure and Regional Dynamics. *Global Networks* **12**:395–423. [14]

Duhon, M., H. McDonald, and S. Kerr. 2012. Nitrogen Trading in Lake Taupo: An Analysis and Evaluation of an Innovative Water Management Policy. Wellington: Motu Economic and Public Policy Research. [4]

Dunlap, R. E., and W. R. J. Catton. 2002. Which Function(s) of the Environment Do We Study? A Comparison of Environmental and Natural Resource Sociology. *Soc. Nat. Res.* **15**:239–249. [4]

Dunning, J. 1988. Explaining International Production. London: Unwin Hyman. [6]

Duranton, G., and D. Puga. 2001. Nursery Cities: Urban Diversity, Process Innovation, and the Life Cycle of Products. *Am. Econ. Rev.* **91**:1454–1477. [14]

Dyer, N., and S. Counsell. 2010. McREDD: How Mckinsey "Cost-Curves" Are Distorting REDD. Climate and Forests Policy Brief November 2010. London: Rainforest Foundation. [13]

Eakin, H., H. Bohle, A. Izac, et al. 2010. Food, Violence and Human Rights. In: Food Security and Global Environmental Change, ed. J. Ingram et al., pp. 66–77. London: Earthscan. [3, 8]

Eakin, H., A. Winkles, and J. Sendzimir. 2009. Nested Vulnerability: Exploring Cross-Scale Linkages and Vulnerability Teleconnections in Mexican and Vietnamese Coffee Systems. *Environ. Sci. Policy* **4**:398–412. [7, 8]

Ebeling, J., and M. Yasué. 2009. The Effectiveness of Market-Based Conservation in the Tropics: Forest Certification in Ecuador and Bolivia. *J. Environ. Manag.* **90**:1145–1153. [12]

EC. 2012a. New Commission Proposal to Minimise the Impacts of Biofuel Production. Brussels: European Commission. [13]

———. 2012b. Proposal for a Directive of the European Parliament and of the Council Amending Directive 98/70/EC Relating to the Quality of Petrol and Diesel Fuels and Amending Directive 2009/28/EC on the Promotion of the Use of Energy from Renewable Sources. http://ec.europa.eu/energy/renewables/biofuels/doc/biofuels/com_2012_0595_en.pdf. (accessed Dec. 8, 2013). [8]

Eichengreen, B. 2008. Globalizing Capital: A History of the International Monetary System 2nd edition. Princeton: Princeton Univ. Press. [6]

Elgert, L. 2011. Certified Discourse? The Politics of Developing Soy Certification Standards. *Geoforum* **43**:295–304. [12]

Eliasch, J. 2008. Climate Change: Financing Global Forests, the Eliasch Review. London: Crown. [13]

Ellis, E. C. 2011. Anthropogenic Transformation of the Terrestrial Biosphere. *Phil. Trans. R. Soc. A* **369**:1010–1035. [4]

Enkvist, P.-A., T. Naucler, and J. Rosander. 2007. A Cost Curve for Greenhouse Gas Reduction. *McKinsey Q.* **1**:35–45. [13]

Erb, K.-H., H. Haberl, and C. Plutzar. 2012. Dependency of Global Primary Bioenergy Crop Potentials in 2050 on Food Systems, Yields, Biodiversity Conservation and Political Stability. *Energy Policy* **47**:260–269. [4]

Erb, K.-H., F. Krausmann, W. Lucht, and H. Haberl. 2009. Embodied HANPP: Mapping the Spatial Disconnect between Global Biomass Production and Consumption. *Ecol. Econ.* **69**:328–334. [4]

Ericksen, P. J. 2008. Conceptualizing Food Systems for Global Environmental Change Research. *Global Environ. Change* **18**:234–245. [3]

Espach, R. 2005. Private Regulation Amid Public Disarray: An Analysis of Two Private Environmental Regulatory Programs in Argentina. *Business Polit.* **7**:1–36. [12]

———. 2006. When Is Sustainable Forestry Sustainable? The Forest Stewardship Council in Argentina and Brazil. *Global Environ. Polit.* **6**:55–84. [12]

Esquinas-Alcázar, J. 2005. Protecting Crop Genetic Diversity for Food Security: Political, Ethical and Technical Challenges. *Nature Rev. Gen.* **6**:946–953. [13]

European Biofuels Technology Platform. 2012. EBTP Steering Committee. http://www.biofuelstp.eu/steering.html#mems. (accessed Dec. 18, 2012). [11]

European Union. 2010. Press Release (IP/10/711): Commission Sets up System for Certifying Sustainable Biofuels. Brussels: European Union. [13]

Evans, B., M. Joas, S. Sunback, and K. Theobald. 2004. Governing Sustainable Cities. London: Earthscan. [4]

Evans, L. T. 1998. Feeding the Ten Billion: Plants and Population Growth. Cambridge: Cambridge Univ. Press. [3]

Ewers, R. M., J. P. W. Scharlemann, A. Balmford, and R. E. Green. 2009. Do Increases in Agricultural Yield Spare Land for Nature? *Global Change Biol.* **15**:1716–1726. [3]

Ewing, R., and R. Cervero. 2001. Travel and the Built Environment: A Synthesis. *J. Transp. Res. Board* **1780**:87–114. [14]

Fairhead, J., M. Leach, and I. Scoones. 2012. Green Grabbing: A New Appropriation of Nature? *J. Peasant Stud.* **39**:237–261. [10, 11, 13]

Fan, S., and A. Ramirez. 2012. Achieving Food Security While Switching to Low Carbon Agriculture. *J. Renew. Sust. Energy* **4**:041405. [10]

FAO. 1980. Tropical Forest Resources Assessment Project: Forest Resources of Tropical Asia. http://www.fao.org/docrep/007/ad908e/AD908E00.htm#TOC. (accessed Dec. 8, 2013). [13]

———. 1981. Tropical Forest Resources Assessment Project: Forest Resources of Tropical Africa, vol. 1. http://www.fao.org/docrep/007/ad909e/AD909E00.HTM. (accessed Dec. 8, 2013). [13]

———. 2004. Voluntary Guidelines to Support the Progressive Realization of the Right to Adequate Food in the Context of National Food Security. Rome: FAO. [13]

———. 2006. World Agriculture: Towards 2030/2050. Interim Report: Prospects for Food, Nutrition, Agriculture and Major Commodity Groups. Rome: FAO. [4]

———. 2008. Livestock's Long Shadow: Environmental Issues and Options. Rome: FAO. [4]

———. 2009a. The State of Agricultural Commodity Market: High Food Prices and the Food Crisis, Experiences and Lessons Learned. Rome: FAO. [4]

———. 2009b. State of Food Insecurity in the World. Rome: FAO. [13]

———. 2010. Zero Draft: Voluntary Guidelines on the Responsible Governance of Tenure of Land and Other Natural Resources. Rome: FAO. [10]

———. 2011a. Constitutional and Legal Protection of the Right to Food around the World. Rome: FAO. [13]

———. 2011b. Global Livestock Production Systems. Rome: FAO. [13]

———. 2011c. State of Food Insecurity in the World. Rome: FAO. [13]

———. 2012a. FAOstat. Food and Agriculture Organization of the United Nations. http://faostat.fao.org/. (accessed Dec. 8, 2013). [6, 11]

———. 2012b. Voluntary Guidelines on the Responsible Governance of Tenure of Land, Fisheries and Forests in the Context of National Food Security. Rome: FAO. [10]

———. 2013. Food Outlook: Biannual Report on Global Food Markets. Rome: FAO. http://www.globefish.org/upl/Publications/Food%20Outlook%20June%202013.pdf (accessed Nov. 7, 2013). [8]

FAO, IFAD, UNCTAD Secretariat, and World Bank Group. 2010. Principles for Responsible Agricultural Investment That Respects Rights, Livelihoods and Resources. http://unctad.org/en/Pages/DIAE/G-20/PRAI.aspx. (accessed Dec. 8, 2013). [13]

FAOSTAT. 2012. FAO Statistical Database. http://faostat.fao.org/. (accessed Dec. 6, 2013). [9]

———. 2013. FAO Statistical Database. http://faostat.fao.org/. (accessed Dec. 6, 2013). [9]

Fargione, J., J. Hill, D. Tilman, S. Polasky, and P. Hawthorne. 2008. Land Clearing and the Biofuel Carbon Debt. *Science* **319**:1235–1238. [13]

Feng, S. Y., N. Heerink, R. Ruben, and F. T. Qu. 2010. Land Rental Market, Off-Farm Employment and Agricultural Production in Southeast China: A Plot-Level Case Study. *China Econ. Rev.* **21**:598–606. [5]

Fensholt, R., K. Rasmussen, T. T. Nielsen, and C. Mbow. 2009. Evaluation of Earth Observation Based Long Term Vegetation Trends: Intercomparing NDVI Time Series Trend Analysis Consistency of Sahel from AVHRR GIMMS, Terra MODIS and SPOT VGT Data. *Remote Sens. Environ.* **113**:1886–1898. [4]

Fertner, C. 2012. Urbanization, Urban Growth and Planning in the Copenhagen Metropolitan Region with Reference Studies from Europe and the USA. Forest and Landscape Research No. 54-2012. Frederiksberg: Forest & Landscape. [14]

Filer, C. 2012. Why Green Grabs Don't Work in Papua New Guinea. *J. Peasant Stud.* **39**:599–617. [10]

Finger-Stitch, A., and M. Finger. 2003. State versus Participation: Natural Resources Management in Europe. London: IIED and IDS. [13]

Fischel, W. 1985. The Economics of Zoning Laws. Baltimore: Johns Hopkins Univ. Press. [4]

Fischer, G., M. Shah, F. N. Tubiello, and H. van Velhuizen. 2005. Socio-Economic and Climate Change Impacts on Agriculture: An Integrated Assessment, 1980–2080. *Phil. Trans. R. Soc. B.* **360**:2067–2083. [3]

Fischer, J., B. Brosi, G. C. Daily, et al. 2008. Should Agricultural Policies Encourage Land Sparing or Wildlife-Friendly Farming? *Front. Ecol. Environ.* **6**:380–385. [4]

Fischer-Kowalski, M., and H. Haberl. 2007. Socioecological Transitions and Global Change. Trajectories of Social Metabolism and Land Use. In: Advances in Ecological Economics, ed. J. van den Bergh. Cheltenham: Edward Elgar. [4]

Foley, J. A., R. S. DeFries, G. P. Asner, et al. 2005. Global Consequence of Land Use. *Science* **309**:570–574. [6]

Foley, J. A., N. Ramankutty, K. A. Brauman, et al. 2011. Solutions for a Cultivated Planet. *Nature* **478**:337–342. [4]

Foresight. 2011. The Future of Food and Farming. Final Project Report. London: Government Office of Science. [3]

Fortin, E. 2005. Reforming Land Rights: The World Bank and the Globalization of Agriculture. *Soc. Leg. Stud.* **14**:147–177. [11]

Foster, K. R., P. Vecchia, and M. H. Repacholi. 2000. Risk Management: Science and the Precautionary Principle. *Science* **288**:979–981. [16]

Fox, J. 1993. The Politics of Food in Mexico: State Power and Social Mobilization. Ithaca: Cornell Univ. Press. [11]

Franco, J., D. Carranza, and J. Fernandez. 2011. New Biofuel Project in Isabela, Philippines: Boon or Bane for Local People? Amsterdam: Transnational Institute. [11]

Frank, A. G. 1998. ReOrient: Global Economy in the Asian Age. Berkeley: Univ. of California Press. [7]

Fransen, L. W., and A. Kolk. 2007. Global Rule-Setting for Business: A Critical Analysis of Multi-Stakeholder Standards. *Organization* **14**:667–684. [12]

Frenkel, A., and M. Ashkenazi. 2008. Measuring Urban Sprawl: How Can We Deal with It? *Environ. Plann. B* **35**:56–79. [15]

Freobel, F., J. Heinrichs, and O. Kreye. 1980. The New International Division of Labor: Structural Unemployment in Industrialized Countries and Industrialization in Developing Countries. Cambridge: Cambridge Univ. Press. [6]

Fridell, G. 2007. Fair Trade Coffee: The Prospects and Pitfalls of Market-Driven Social Justice. Studies in Comparative Political Economy and Public Policy. Toronto: Univ. of Toronto Press. [12]

Friis, C., and A. Reenberg. 2010. Land Grab in Africa: Emerging Land System Drivers in a Teleconnected World. GLP Report No. 1. http://farmlandgrab.org/14816. (accessed Nov. 24, 2013). [6, 7, 10]

Fuchs, D., A. Kalfagianni, and M. Arentsen. 2009. Retail Power, Private Standards, and Sustainability in the Global Food System. In: Corporate Power in Global Agrifood Governance, ed. J. Clapp and D. Fuchs, R. Gottlieb, series ed., pp. 29–59. Cambridge, MA: MIT Press. [12]

Fuerst, F., and P. McAllister. 2011. Eco-Labeling in Commercial Office Markets: Do Leed and Energy Star Offices Obtain Multiple Premiums? *Ecol. Econ.* **70**:1220–1230. [12]

Fujita, M., P. Krugman, and A. J. Venables. 1999. The Spatial Economy: Cities, Regions, and International Trade. Cambridge, MA: MIT Press. [14]

Fuller, R. A., and K. J. Gaston. 2009. The Scaling of Green Space Coverage in European Cities. *Biol. Lett.* **5**:352–355. [15]

G-8. 2009. Responsible Leadership for a Sustainable Future. Declaration of the L'aquila Summit, July 8, 2009. http://www.g8italia2009.it/static/G8_Allegato/G8_Declaration_08_07_09_final,0.pdf. (accessed June 17, 2013). [10]

Galey, J. 1979. Industrialist in the Wilderness: Henry Ford's Amazon Venture. *J. Interamer. Stud. World Affairs* **21**:261–289. [6]

Garcia-Johnson, R. 2001. Multinational Corporations and Certification Institutions: Moving First to Shape a Green Global Production Context. Paper Presented at the International Studies Association Convention, Chicago, February 20–24, 2001. http://www.nicholas.duke.edu/solutions/documents/rgj_isa_mncs.pdf. (accessed Aug. 9, 2013). [12]

Garreau, J. 1991. Edge Cities. New York: Doubleday. [14]

Garrity, D. P., M. Soekardi, M. van Noordwijk, et al. 1997. The Imperata Grasslands of Tropical Asia: Area, Distribution, and Typology. *Agrofor. Syst.* **36**:3–29. [9]

GEA. 2012. Global Energy Assessment: Toward a Sustainable Future, ed. T. Johansson et al. Cambridge/Laxenburg: Cambridge Univ. Press/Intl. Institute of Applied Systems Analysis (IIASA). [4]

Geist, H., and E. F. Lambin. 2002. Proximate Causes and Underlying Driving Forces of Tropical Deforestation. *BioScience* **52**:143–150. [6]

Geist, H., W. McConnell, E. F. Lambin, et al. 2006. Causes and Trajectories of Land-Use/Cover Change. In: Land-Use and Land-Cover Change, ed. E. F. Lambin and H. Geist. Heidelberg: Springer. [6]

Gerber, J.-F. 2011. Conflicts over Industrial Tree Plantations in the South: Who, How and Why? *Global Environ. Change* **21**:165–176. [4]

Giannini, A., M. Biasutti, I. Held, M., and A. H. Sobel. 2008. A Global Perspective on African Climate. *Clim. Change* **90**:359–383. [4]

Gibbs, H. K., M. Johnston, J. A. Foley, et al. 2008. Carbon Payback Times for Crop-Based Biofuel Expansion in the Tropics: The Effects of Changing Yield and Technology. *Environ. Res. Lett.* **3**:34001–34011. [9]

Gibbs, H. K., A. S. Ruesch, F. Achard, et al. 2010. Tropical Forests Were the Primary Sources of New Agricultural Land in the 1980s and 1990s. *PNAS* **107**:16732–16737. [2]

Gibson, R. B. 2006. Sustainability Assessment: Basic Components of a Practical Approach. *Impact Assess. Proj. Appr.* **24**:170–182. [16]

Glaeser, E. L., and J. D. Gottlieb. 2009. The Wealth of Cities: Agglomeration Economies and Spatial Equilibrium in the United States. *J. Econ. Lit.* **47**:983–1028. [14]

Glaeser, E. L., H. D. Kallal, J. A. Scheinkman, and A. Shleifer. 1992. Growth in Cities. *J. Polit. Econ.* **100**:1126–1152. [14]

Glaeser, E. L., and J. E. Kohlhase. 2004. Cities, Regions and the Decline of Transport Costs. *Papers Reg. Sci.* **83**:197–228. [14]

Glaeser, E. L., and G. A. M. Ponzetto. 2007. Did the Death of Distance Hurt Detroit and Help New York? NBER Working Papers 13710. Cambridge, MA: National Bureau of Economic Research. [14]

Glaeser, E. L., and B. A. Ward. 2009. The Causes and Consequences of Land Use Regulation, Evidence from Greater Boston. *J. Urban Econ.* **65**:265–278. [4]

Glantz, M. H., R. W. Katz, and N. Nicholls. 1991. Teleconnections Linking Worldwide Climate Anomalies. Cambridge: Cambridge Univ. Press. [7]

Gleeson, T., Y. Wada, M. F. P. Bierkens, and L. P. H. v. Beek. 2012. Water Balance of Global Aquifers Revealed by Groundwater Footprint. *Nature Lett.* **488**:197–200. [4]

Godfray, H. C. J., J. R. Beddington, I. R. Crute, et al. 2010. Food Security: The Challenge of Feeding 9 Billion People. *Science* **327**:812–818. [3, 4]

Goldewijk, K. K., and N. Ramankutty. 2001. Land Use Changes During the Past 300 Years. Encyclopedia of Life Support Systems (EOLSS). Paris: UNESCO. [6]

Goldstein, J. H., G. Caldarone, T. K. Duarte, et al. 2012. Integrating Ecosystem-Service Tradeoffs into Land-Use Decisions. *PNAS* **109**:7565–7570. [4]

Gonzalez, P., C. J. Tucker, and H. Sy. 2012. Tree Density and Species Decline in the African Sahel Attributable to Climate. *J. Arid Environ.* **78**:55–64. [4]

Gopal, D. 2011. Flora in Slums of Bangalore, India: Ecological and Socio-Cultural Perspectives. M. Sci. thesis, Institute of Botany and Landscape Ecology, Ernst Moritz Arndt University, Greifswald, Germany. [16]

Gordon, D. R. 1998. Effects of Invasive, Non-Indigenous Plant Species on Ecosystem Processes: Lessons from Florida. *Ecol. Appl.* **8**:975–989. [7]

Gottmann, J. 1961. Megalopolis: The Urbanized Northeastern Seaboard of the United States. New York: The Twentieth Century Fund. [14]

Goulder, L. H., and I. W. H. Parry. 2008. Instrument Choice in Environmental Policy. Resources for the Future Discussion Paper RFF-DP 08-07. http://www.rff.org/documents/RFF-DP-08-07.pdf. (accessed Dec. 8, 2013). [12]

Grafton, R. Q. 2011. Economic Instruments for Water Management. Working Party on Biodiversity, Water and Ecosystems (October 27–28, 2011, OECD Conference Centre, Paris). http://www.kysq.org/docs/Grafton_Instruments.pdf. (accessed Aug. 30, 2013). [4]

Graham, A., S. Aubry, R. Künnemann, and S. M. Suárez. 2010. Land Grab Study. Heidelberg: FIAN. [10]

GRAIN. 2008. Seized! The 2008 Land Grab for Food and Financial Security. Grain Briefing. http://www.grain.org/article/entries/93-seized-the-2008-landgrab-for-food-and-financial-security. (accessed Dec. 8, 2013). [10, 11]

———. 2012. Grain Releases Data Set with over 400 Global Land Grabs. http://www.grain.org/article/entries/4479-grain-releases-data-set-with-over-400-global-land-grabs. (accessed Dec. 8, 2013). [6]

Green, D. P. 1992. The Price Elasticity of Mass Preferences. *Am. Polit. Sci. Rev.* **86**:128–148. [12]

Greenland, D. J., P. J. Gregory, and P. H. Nye. 1998. Land Resources and Constraints to Crop Production. In: Feeding a World Population of More Than Eight Billion People: A Challenge to Science, ed. J. C. Waterlow et al., pp. 39–55. Oxford: Oxford Univ. Press. [3]

Gregory, P. J., and T. S. George. 2011. Feeding Nine Billion: The Challenge to Sustainable Crop Production. *J. Exp. Bot.* **62**:5233–5239. [3]

Gregory, P. J., J. S. I. Ingram, R. Andersson, et al. 2002. Environmental Consequences of Alternative Practices for Intensifying Crop Production. *Agric. Ecosys. Environ.* **88**:279–290. [3]

Gregory, P. J., and B. Marshall. 2012. Attribution of Climate Change: A Methodology to Estimate the Potential Contribution to Increases in Potato Yield in Scotland since 1960. *Global Change Biol.* **18**:1372–1388. [3]

Grimm, N. B., S. H. Faeth, N. E. Golubiewski, et al. 2008. Global Change and the Ecology of Cities. *Science* **319**:756–760. [15]

Grubler, A. 2003. Technology and Global Change. Cambridge: Cambridge Univ. Press. [6]

GTZ. 2009. Foreign Direct Investment (FDI) in Land in Developing Countries. Echbern: Ministry of Economic Cooperation and Development. [10]

Gulbrandsen, L. H. 2005. Explaining Different Approaches to Voluntary Standards: A Study of Forest Certification Choices in Norway and Sweden. *J. Environ. Policy Plan.* **7**:43–59. [12]

———. 2006. Creating Markets for Eco-Labelling: Are Consumers Insignificant? *Intl. J. Consumer Stud.* **30**:477–489. [12]

———. 2010. Transnational Environmental Governance: The Emergence and Effects of the Certification of Forests and Fisheries. Cheltenham: Edward Elgar. [12]

Gullison, R. E. 2003. Does Forest Certification Conserve Biodiversity? *Oryx* **37**:153–165. [12]

Gunningham, N., R. A. Kagan, and D. Thornton. 2003. Shades of Green: Business, Regulation and Environment. Stanford: Stanford Univ. Press. [12]

Guo, B. 2011. Relationship between Urbanization and Food Crisis in China (in Chinese). *Reform Econ. Syst.* **1**:32–35. [5]

Guo, J. X. 2004. Coordination between Urbanization and Grain Production Security. *Res. Agric. Modern.* **25**:279–282. [5]

Guthman, J. 2007. The Polanyian Way? Voluntary Food Labels as Neoliberal Governance. *Antipode* **39**:456–478. [12]

Haase, D. 2008. Urban Ecology of Shrinking Cities: An Unrecognised Opportunity? *Nature Cult.* **3**:1–8. [15]

———. 2012a. The Importance of Ecosystem Services for Urban Areas: Valuation and Modelling Approaches. *UGEC Viewpoints* **7**:4–7. [15]

———. 2012b. Processes and Impacts of Urban Shrinkage and Response by Planning. In: Encyclopedia of Sustainability Science and Technology, ed. R. A. Meyers. New York: Springer. [15]

Haase, D., N. Kabisch, A. Haase, S. Kabisch, and D. Rink. 2012. Actors and Factors in Land Use Simulation: The Challenge of Urban Shrinkage. *J. Env. Mod. Softw.* **35**:92–103. [15]

Haase, D., and H. Nuissl. 2010. The Urban-to-Rural Gradient of Land Use Change and Impervious Cover: A Long-Term Trajectory for the City of Leipzig. *Land Use Sci.* **5**:123–142. [15]

Haase, D., R. Seppelt, and A. Haase. 2007. Land Use Impacts of Demographic Change: Lessons from Eastern German Urban Regions. In: Use of Landscape Sciences for the Assessment of Environmental Security, ed. I. Petrosillo et al., pp. 329–344. Amsterdam: Springer. [15]

Haberl, H., T. Beringer, S. C. Bhattacharya, K.-H. Erb, and M. Hoogwijk. 2010. The Global Technical Potential of Bio-Energy in 2050 Considering Sustainability Constraints. *Curr. Opin. Environ. Sustain.* **2**:394–403. [2, 4]

Haberl, H., K.-H. Erb, F. Krausmann, et al. 2009. Using Embodied HANPP to Analyze Teleconnections in the Global Land System: Conceptual Considerations. *Geografisk Tidsskrift* **109**:119–130. [7, 8]

Haberl, H., K.-H. Erb, F. Krausmann, et al. 2011a. Global Bioenergy Potentials from Agricultural Land in 2050: Sensitivity to Climate Change, Diets and Yields. *Biomass Bioenergy* **35**:4753–4769. [4]

Haberl, H., M. Fischer-Kowalski, F. Krausmann, J. Martinez-Alier, and V. Winiwarter. 2011b. A Socio-Metabolic Transition Towards Sustainability? Challenges for Another Great Transformation. *Sustain. Devel.* **19**:1–14. [4]

Haberl, H., and S. Geissler. 2000. Cascade Utilization of Biomass: Strategies for a More Efficient Use of a Scarce Resource. *Econ. Engineer.* **16**:S111–S121. [4]

Haberl, H., J. K. Steinberger, C. Plutzar, et al. 2012. Natural and Socioeconomic Determinants of the Embodied Human Appropriation of Net Primary Production and Its Relation to Other Resource Use Indicators. *Ecol. Ind.* **23**:222–231. [4]

Hall, D. 2011. Land Grabs, Land Control, and Southeast Asian Crop Booms. *J. Peasant Stud.* **38**:837–857. [10]

Hall, J., S. Matos, L. Severino, and N. Beltrao. 2009. Brazilian Biofuels and Social Exclusion: Established and Concentrated Ethanol versus Emerging and Dispersed Biodiesel. *J. Cleaner Prod.* **17**:S77–S85. [13]

Hall, P. 2009. Looking Backward, Looking Forward: The City Region of the Mid-21st Century. *Reg. Stud.* **43**:803–817. [14]

Hall, P., and N. Green. 2005. POLYNET Action 1.1: Commuting and the Definition of Functional Urban Regions: South East England. London: Institute of Community Studies, The Young Foundation and POLYNET Partners. [14]

Hall, P., and K. Pain. 2006. The Polycentric Metropolis: Learning from Mega-City Regions in Europe. London: Earthscan. [14]

Hall, R. B., and T. J. Biersteker. 2002. The Emergence of Private Authority in Global Governance. In: Cambridge Studies in International Relations, vol. 85. Cambridge: Cambridge Univ. Press. [12]

Hamilton, K., R. Bayon, G. Turner, and D. Higgins. 2007. State of the Voluntary Carbon Markets 2007: Picking up Steam. http://ecosystemmarketplace.com/documents/acrobat/StateoftheVoluntaryCarbonMarket18July_Final.pdf. (accessed July 18, 2013). [13]

Hamilton, K., M. Sjardin, T. Marcello, and G. Xu. 2008. Forging a Frontier: State of the Voluntary Carbon Markets 2008. http://www.ecosystemmarketplace.com/documents/cms_documents/2008_StateofVoluntaryCarbonMarket2.pdf. (accessed May 8, 2013). [13]

Hamilton, K., M. Sjardin, M. Peters-Stanley, and T. Marcello. 2010. Building Bridges: State of the Voluntary Carbon Markets 2010. http://www.forest-trends.org/documents/files/doc_2434.pdf. (accessed June 14, 2013). [13]

Hamilton, K., M. Sjardin, A. Shapiro, and T. Marcello. 2009. Fortifying the Foundation: State of the Voluntary Carbon Markets 2009. http://ecosystemmarketplace.com/documents/cms_documents/StateOfTheVoluntaryCarbonMarkets_2009.pdf. (accessed May 20, 2013). [13]

Hao, P., R. V. Sliuzas, and S. Geertman. 2011. The Development and Redevelopment of Urban Villages in Shenzhen. *Habitat Intl.* **35**:214–224. [5]

Hardin, G. 1968. The Tragedy of the Commons. *Science* **162**:1243–1248. [4]

Harper, R. J., S. J. Sochacki, K. R. J. Smettem, and N. Robinson. 2009. Bioenergy Feedstock Potential from Short-Rotation Woody Crops in a Dryland Environment. *Energy Fuels* **24**:225–231. [4]

Harvey, D. 1990. The Condition of Postmodernity: An Enquiry into the Origins of Cultural Change. Cambridge, MA: Blackwell. [4]

———. 2003. The New Imperialism. Oxford: Oxford Univ. Press. [13]

———. 2006. Spaces of Global Capitalism: A Theory of Uneven Geographical Development. New York: Verso Books. [15]

Hasselmann, F., E. Csaplovics, I. Falconer, M. Bürgi, and A. M. Hersperger. 2010. Technological Driving Forces of Lucc: Conceptualization, Quantification, and the Example of Urban Power Distribution Networks. *Land Use Policy* **27**:628–637. [15]

Havlík, P., H. Valin, A. Mosnier, et al. 2013. Crop Productivity and the Global Livestock Sector: Implications for Land Use Change and Greenhouse Gas Emissions. *Am. J. Agric. Econ.* **95**:442–448. [4]

He, R. W., S. Q. Liu, and Y. W. Liu. 2011. Application of SD Model in Analyzing the Cultivated Land Carrying Capacity: A Case Study in Bijie Prefecture, Guizhuo Province, China. *Proc. Environ. Sci.* **10**:1985–1991. [5]

Headrick, D. R. 1991 The Invisible Weapon: Telecommunications and International Politics 1851–1945. Oxford: Oxford Univ. Press. [7]

Held, D., A. McGrew, D. Goldblatt, and J. Perraton. 1998. Global Transformations, Politics, Economics, Culture. Stanford: Stanford Univ. Press. [6]

———. 1999. Global Transformations, Politics, Economic and Culture. Stanford: Stanford Univ. Press. [8]

Herold, M., J. Scepan, and K. C. Clarke. 2002. The Use of Remote Sensing and Landscape Metrics to Describe Structures and Changes in Urban Land Uses. *Environ. Plann. A* **34**:1443–1458. [15]

Hiernaux, P., E. Mougin, L. Diarra, et al. 2009. Sahelian Rangeland Response to Changes in Rainfall over Two Decades in the Gourma Region, Mali. *J. Hydrol.* **375**:114–127. [4]

Hilson, G. 2002. Land Use Competition between Small- and Large-Scale Miners: A Case Study of Ghana. *Land Use Policy* **19**:149–156. [4]

Hirst, P., and G. Thompson. 1996. Globalization in Question: The International Economy and the Possibilities of Governance. Cambridge: Policy Press. [6]

HLPE. 2011. Land Tenure and International Investments in Agriculture. A Report by the High Level Panel of Experts on Food Security and Nutrition of the Committee on World Food Security. Rome: FAO. [13]

Hofman, I., and P. Ho. 2012. China's "Developmental Outsourcing": A Critical Examination of Chinese Global "Land Grabs" Discourse. *J. Peasant Stud.* **39**:1–48. [11]

Hopkins, A. G. 2002. Globalization in World History. New York: W. W. Norton. [6]

Horne, R. E. 2009. Limits to Labels: The Role of Eco-Labels in the Assessment of Product Sustainability and Routes to Sustainable Consumption. *Intl. J. Consumer Stud.* **33**:175–182. [12]

Houghton, R. A. 1994. The Worldwide Extent of Land-Use Change. *BioScience* **44**:305–313. [6]

Houghton, R. A., and J. L. Hackler. 2000. Changes in Terrestrial Carbon Storage in the United States. 1: The Roles of Agriculture and Forestry. *Global Ecology and Biogeography* **9**(2):125–144. [6]

Howarth, R. W., S. Bringezu, M. Bekunda, et al. 2009. Rapid Assessment on Biofuels and Environment: Overview and Key Findings. In: Biofuels: Environmental Consequences and Interactions with Changing Land Use, ed. S. Bringezu and R. W. Howarth, pp. 1–13. Ithaca: Cornell Univ. [13]

Howden, S. M., S. J. Crimp, and C. J. Stokes. 2010. Australian Agriculture in a Climate Change. In: Managing Climate Change: Papers from Greenhouse 2009 Conference, ed. I. Jubb et al., pp. 101–112. Melbourne: CSIRO Publishing. [3]

Hsu, W.-T. 2012. Central Place Theory and City Size Distribution. *Econ. Journal* **122**:903–932. [14]

Huang, J., X. X. Lu, and J. M. Sellers. 2007a. A Global Comparative Analysis of Urban Form: Applying Spatial Metrics and Remote Sensing. *Landsc. Urban Plan* **82**:184–197. [15]

Huang, J., L. Zhu, and X. Deng. 2007b. Regional Differences and Determinants of Built-up Area Expansion in China. *Science in China D* **50**:1853–1843. [4]

Hulme, P. E. 2009. Trade, Transport and Trouble: Managing Invasive Species Pathways in an Era of Globalization. *J. Appl. Ecol.* **46**:10–18. [7]

Hummels, D. 2007. Transportation Costs and International Trade in the Second Era of Globalization. *J. Econ. Persp.* **21**:131–154. [14]

Humphreys, D. 2006. Logjam: Deforestation and the Crisis of Global Governance. The Earthscan Forestry Library. London: Earthscan. [13]

Huntsinger, L., M. Johnson, M. Stafford, and J. Fried. 2010. Hardwood Rangeland Landowners in California from 1985 to 2004: Production, Ecosystem Services, and Permanence. *Rangeland Ecol. Manag.* **63**:324–334. [13]

Hurtt, G. C., L. P. Chini, S. Frolking, et al. 2011. Harmonization of Land-Use Scenarios for the Period 1500–2100: 600 Years of Global Gridded Annual Land-Use Transitions, Wood Harvest, and Resulting Secondary Lands. *Clim. Change* **109**:117–161. [4]

Hutton, J. M., and N. Leader-Williams. 2003. Sustainable Use and Incentive-Driven Conservation: Realigning Human and Conservation Interests. *Oryx* **37**:215–226. [13]

IAASTD. 2009. Agriculture at a Crossroads. International Assessment of Agricultural Knowledge, Science and Technology for Development (IAASTD), Global Report. Washington, D.C.: Island Press. [4]

IIASA/FAO. 2012. Global Agroecological Zones (GAEZ v3.0). Laxenburg/Rome: IIASA/FAO. [2]

IMF. 2012. Statistics Department COFER Database: International Financial Statistics. Washington, D.C.: International Monetary Fund. [6]

IPCC. 2007. Climate Change 2007, Synthesis Report. Contribution of Working Groups I, II and III to the Fourth Assessment Report of the Intergovernmental Panel on Climate Change. Geneva: Intergovernmental Panel on Climate Change. [4]

Irwin, E. G., K. P. Bell, N. E. Bockstael, et al. 2009. The Economics of Urban-Rural Space. *Annu. Rev. Res. Econ.* **1**:435–459. [4]

Irwin, E. G., and D. Wrenn. 2013. An Assessment of Empirical Methods for Modeling Land Use. In: The Handbook of Land Economics, ed. J. M. Duke and J. Wu. Oxford: Oxford Univ. Press. [4]

Isard, W. 1956. Location and Space Economy: A General Theory Relating to Industrial Location, Market Areas, Land Use, Trade, and Urban Structure. Cambridge, MA: MIT Press. [14]

Ishagi, N., S. Ossiya, L. Aliguma, and C. Aisu. 2003. Urban and Peri-Urban Livestock Keeping among the Poor in Kampala City. Kampala, Uganda: Ibaren Konsultants. [16]

ISRIC. 1991. Global Assessment of Human-Induced Land Degradation (GLASOD). Wageningen: ISRIC. [3]

———. 2008. Global Assessment of Land Degradation and Improvement (GLADA). Wageningen: ISRIC. [3]

Jabareen, Y. R. 2006. Sustainable Urban Forms: Their Typologies, Models, and Concepts. *J. Plan. Educ. Res.* **26**:38–52. [15]

Jabbour, J., F. Keita-Ouane, C. Hunsberger, et al. 2012. Internationally Agreed Environmental Goals: A Critical Evaluation of Progress. *Environ. Devel. Sustain.* **3**:5–24. [10]

Jack, B. K., C. Kousky, and K. R. E. Sims. 2008. Designing Payments for Ecosystem Services: Lessons Form Previous Experience with Incentive-Based Mechanisms. *PNAS* **105**:9465–9470. [4]

Jackson, T. 2009. Prosperity without Growth: Economics for a Finite Planet. London: Earthscan. [4]

Jacobs, J. 1969. The Economy of Cities. New York: Random House. [14]

Jaggard, K. W., A. Qi, and E. S. Ober. 2010. Possible Changes to Arable Crop Yields by 2050. *Phil. Trans. R. Soc. B.* **365**:2835–2851. [3]

Jenkins, P. T., and H. A. Mooney. 2006. The United States, China, and Invasive Species: Present Status and Future Prospects. *Biol. Inv.* **8**:1589–1593. [7]

Jenks, M., and N. Dempsey. 2005. Future Forms and Design for Sustainable Cities. Oxford: Architectural Press. [15]

Jones, P. S. 2011. Powering up the People? The Politics of Indigenous Rights Implementation: International Labour Organisation Convention 169 and Hydroelectric Power in Nepal. *Intl. J. Human Rights* **16**:624–647. [10]

Joppa, L. N., and A. Pfaff. 2009. High and Far: Biases in the Location of Protected Areas. *PLoS ONE* **4**:e8273. [4]

———. 2011. Global Protected Area Impacts. *Proc. Roy. Soc. B* **278**:1633–1638. [4]

Jostock, C. 2008. Ludwig, Daniel Keith (1897–1992). In: Encyclopedia of Latin American History and Culture, 2nd edition, ed. J. Kinsbruner and E. D. Langer, vol. 4, p. 284. Detroit: Charles Scribner's Sons. [6]

Kabisch, N., and D. Haase. 2011. Diversifying European Agglomerations: Evidence of Urban Population Trends for the 21st Century. *Pop. Space Place* **17**:236–253. [15]

Kabisch, N., D. Haase, and A. Haase. 2010. Evolving Reurbanisation? Spatio-Temporal Dynamics Exemplified at the Eastern German City of Leipzig. *Urban Stud.* **47**:967–990. [15]

———. 2012. Urban Population Development in Europe, 1991–2008: The Examples of Poland and the UK. *Intl. J. Urban Reg. Res.* **36**:1326–1348. [15]

Kahn, B., D. Zaks, M. Fulton, et al. 2009. Investing in Agriculture: Far-Reaching Challenge, Significant Opportunity. An Asset Management Perspective. Frankfurt: DB Climate Change Advisors, Deutsche Bank Group. http://www.dbcca.com/dbcca/EN/_media/Investing_in_Agriculture_July_13_2009.pdf. (accessed Dec. 6, 2013). [4]

Kareiva, P., S. Watts, R. McDonald, and T. Boucher. 2007. Domesticated Nature: Shaping Landscapes and Ecosystems for Human Welfare. *Science* **316**:1866–1869. [1]

Kasanko, M., J. I. Barredo, C. Lavalle, et al. 2006. Are European Cities Becoming Dispersed? A Comparative Analysis of 15 European Urban Areas. *Landsc. Urban Plan* **77**:111–130. [15]

Kastner, T., M. J. I. Rivas, W. Koch, and S. Nonhebel. 2012. Global Changes in Diets and the Consequences for Land Requirements for Food. *PNAS* **109**:6868–6872. [1, 9]

Kauppi, P. E., J. H. Ausubel, J. Fang, et al. 2006. Returning Forests Analyzed with the Forest Identity. *PNAS* **103**(46):17,574–17,579. [6]

Kaur, A. 2010. Labour Migration in Southeast Asia: Migration Policies, Labour Exploitation and Regulation. *J. Asia Pacific Econ.* **15**:6–19. [9]

Keene, D. 2012. Medieval London and Its Supply Hinterlands. *Reg. Environ. Change* **12**:263–281. [14]

Kent, G. 2008. Global Obligations for the Right to Food. Plymouth: Rowman and Littlefield Publishers. [13]

Kenwood, A. G., and A. L. Lougheed. 1999. The Growth of the International Economy, 4th edition. London: Routledge. [6]

Kenworthy, J. R. 2006. The Eco-City: Ten Key Transport and Planning Dimensions for Sustainable City Development. *Environ. Urban.* **18**:67–85. [14]

Keohane, N. O., and S. M. Olmstead. 2007. Markets and the Environment. Foundations of Contemporary Environmental Studies. Washington, D.C.: Island Press. [12]

Kerr, I. M. 1984. A History of the Eurobond Market: The First 21 Years. London: Euromoney Publications Ltd. [6]

Kerr, S. 2013. The Economics of International Policy Agreements to Reduce Emissions from Deforestation and Degradation. *Rev. Environ. Econ. Policy* **7**:47–66. [4]

Khai, N. M., P. Q. Ha, and I. Öborn. 2007. Nutrient Flows in Small-Scale Peri-Urban Vegetable Farming Syatems in Southeast Asia: A Case Study in Hanoi. *Agric. Ecosys. Environ.* **122**:192–202. [3]

Kindleberger, C. P., and R. Aliber. 2011. Manias, Panics and Crashes, a History of Financial Crisis 6th edition. New York: Palgrave Macmillan. [6]

King, A. A., and M. J. Lenox. 2000. Industry Self-Regulation without Sanctions: The Chemical Industry's Responsible Care Program. *Acad. Manag. J.* **43**:698–716. [12]

Kinzig, A. P., C. Perrings, F. S. Chapin, et al. 2011. Paying for Ecosystem Services: Promise and Peril. *Science* **334**:603–604. [4]

Kirby, M. G., and M. J. Blyth. 1987. Economic Aspects of Land Degradation in Australia. *Austral. J. Agricul. Res. Econ.* **31**:154–174. [4]

Klemmedson, J. O., and J. G. Smith. 1964. Cheatgrass (*Bromus Tectorum* L.). *Bot. Rev.* **30**:226–262. [7]

Klooster, D. 2010. Standardizing Sustainable Development? The Forest Stewardship Council's Plantation Policy Review Process as Neoliberal Environmental Governance. *Geoforum* **41**:117–129. [12]

Knapp, P. A. 1996. Cheatgrass (*Bromus Tectorum* L.) Dominance in the Great Basin Desert: History, Persistence, and Influences to Human Activities. *Global Environ. Change* **6**:37–52. [7]

Knight, J. 1992. Institutions and Social Conflict. Political Economy of Institutions and Decisions. Cambridge: Cambridge Univ. Press. [12]

———. 1995. Models, Interpretations, and Theories: Constructing Explanations of Institutional Emergence and Change. In: Explaining Social Institutions, ed. J. Knight and I. Sened, pp. 95–119. Ann Arbor: Univ. of Michigan Press. [12]
Knill, C., and D. Lehmkuhl. 2002. Private Actors and the State: Internationalization and Changing Patterns of Governance. *Governance* **15**:41–63. [12]
Kogut, B. 2001. Multinational Corporations. In: International Encyclopedia of Social and Behavioral Sciences, ed. N. J. Smelser and P. B. Baltes. Oxford: Elsevier. [14]
Koh, L. P., P. Levang, and J. Ghazoul. 2009. Designer Landscapes for Sustainable Biofuels. *Trends Ecol. Evol.* **24**:431–438. [4]
Koh, L. P., J. Miettinen, S. C. Liew, and J. Ghazoul. 2011. Remotely Sensed Evidence of Tropical Peatland Conversion to Oil Palm. *PNAS* **108**:5127–5132. [9]
Koh, L. P., and D. S. Wilcove. 2008. Is Oil Palm Agriculture Really Destroying Tropical Biodiversity? *Conserv. Lett.* **1**:60–64. [9]
Konar, M., C. Dalin, S. Suweis, et al. 2011. Water for Food: The Global Virtual Water Trade Network. *Water Resources Research* **47**:W05520. [7]
Kondratieff, N. D. 1926. The Long Waves of Economic Life. *Rev. Intl. Polit. Econ.* **11**:519–562. [6]
Korotayev, A. V., and S. V. A. Tsirei. 2010. Special Analysis of World GDP Dynamics: Kondratieff Waves, Kuznets Swings, Juglar and Kitchin Cycles in Global Economic Development 2008–2009 Economic Crisis. *Struct. Dynam.* **4**:13–57. [6]
Kose, M. A., and E. S. Prasad. 2010. Emerging Markets: Resilience and Growth Amid Global Turmoil. Washington, D.C.: Brookings Institution Press. [14]
Kraas, F. 2007. Megacities and Global Change: Key Priorities. *Geogr. J.* **173**:79–82. [15]
Krasner, S. D. 1991. Global Communications and National Power: Life on the Pareto Frontier. *World Politics* **43**:336–366. [12]
Krausmann, F., K.-H. Erb, S. Gingrich, C. Lauk, and H. Haberl. 2008. Global Patterns of Socioeconomic Biomass Flows in the Year 2000: A Comprehensive Assessment of Supply, Consumption and Constraints. *Ecol. Econ.* **65**:471–487. [4]
Krausmann, F., S. Gingrich, H. Haberl, et al. 2012. Long-Term Trajectories of the Human Appropriation of Net Primary Production: Lessons from Six National Case Studies. *Ecol. Econ.* **77**:129–138. [4]
Krausmann, F., H. Haberl, K.-H. Erb, et al. 2009. What Determines Geographical Patterns of the Global Human Appropriation of Net Primary Production? *J. Land Use Sci.* **4**:15–33. [4]
Krausmann, F., H. Haberl, N. B. Schulz, et al. 2003. Land-Use Change and Socio-Economic Metabolism in Austria. Part I: Driving Forces of Land-Use Change: 1950–1995. *Land Use Policy* **20**:1–20. [4]
Kröger, M. 2012. Global Tree Plantation Expansion: A Review. ICAS Review Paper Series No. 3. Rotterdam: ICAS. [11]
Krugman, P. 1991. Increasing Returns and Economic. *Geogr. J. Polit. Econ.* **49**:137–150. [14]
———. 1995. Growing World Trade: Causes and Consequences. *Brookings Papers Econ. Activity* **26**:327–377. [6]
———. 2011. The New Economic Geography, Now Middle-Aged. *Reg. Stud.* **45**:1–7. [14]
Kühn, I., R. Brandl, and S. Klotz. 2004. The Flora of German Cities Is Naturally Species Rich. *Evol. Ecol. Res.* **6**:749–764. [15]

Kull, C. A., C. K. Ibrahim, and T. C. Meredith. 2007. Tropical Forest Transitions and Globalization: Neo-Liberalism, Migration, Tourism, and International Conservation Agendas. *Soc. Nat. Res.* **20**:723–737. [4]

Kupfer, F., H. Meersman, E. Onghena, and V. Voorde. 2011. Air Freight and Merchandise Trade: Towards a Disaggregated Analysis. *Ed. Adv. Board* **2**:28. [14]

Kuznets, S. 1967. Quantitative Aspects of the Economic Growth of Nations: X. Level and Structure of Foreign Trade: Long-Term Trends. *Econ. Dev. Cult. Change* **15**:1–140. [6]

Laborde, D. 2011. Asessing the Land Use Change Consequences of European Biofuels Policies. International Food Policy Research Institute. http://trade.ec.europa.eu/doclib/docs/2011/october/tradoc_148289.pdf. (accessed Dec. 8, 2013). [9]

Lambin, E. F. 2012. Global Land Availability: Malthus versus Ricardo. *Global Food Secur.* **1**:83–87. [2, 4]

Lambin, E. F., and H. J. Geist, eds. 2006. Land-Use and Land-Cover Change: Local Processes and Global Impacts. Global Change: IGBP Series, vol. 18. Berlin: Springer. [4, 15]

Lambin, E. F., H. J. Geist, and E. Lepers. 2003. Dynamics of Land-Use and Land-Cover Change in Tropical Regions. *Annu. Rev. Environ. Res.* **28**:205–241. [7]

Lambin, E. F., H. Gibbs, L. Ferraira, et al. 2013. Estimating the World's Potentially Available Cropland Using a Bottom-up Approach. *Global Environ. Change* **23**:892–901. [2]

Lambin, E. F., and P. Meyfroidt. 2010. Land Use Transitions: Socio-Ecological Feedback versus Socio-Economic Change. *Land Use Policy* **27**:108–118. [6]

———. 2011. Global Land Use Change, Economic Globalization, and the Looming Land Scarcity. *PNAS* **108**:3465–3472. [2, 4, 6–9]

Lambin, E. F., B. L. Turner, Jr., H. J. Geist, et al. 2001. The Causes of Land-Use and Land-Cover Change: Moving Beyond the Myths. *Global Environ. Change* **11**:261–269. [6]

Lang, R. 2003. Edgeless Cities: Exploring the Elusive Metropolis. Washington, D.C.: Brookings Institution Press. [14]

Lang, R., and D. Dhavale. 2005. Beyond Megalopolis: Exploring America's New "Megapolitan" Geography. Alexandria: Metropolitan Institute at Virginia Tech. [14]

Lang, R., and P. K. Knox. 2009. The New Metropolis: Rethinking Megalopolis. *Reg. Stud.* **43**:789–802. [14]

Langholz, J. A., and J. P. Lassoie. 2001. Perils and Promises of Privately Owned Protected Areas. *BioScience* **51**:1079–1085. [4]

Lapola, D. M., R. Schaldach, J. Alcamo, et al. 2010. Indirect Land-Use Changes Can Overcome Carbon Savings from Biofuels in Brazil. *PNAS* **107**:3388–3393. [4]

Leichenko, R., and W. Solecki. 2005. Exporting the American Dream: The Globalization of Suburban Consumption Landscapes. *Reg. Stud.* **39**:241–253. [6]

Lemaitre, S. 2011. Indigenous Peoples' Land Rights and REDD: A Case Study. *Rev. EC Intl. Environ. Law* **20**:150–162. [7]

Lenton, T. M., H. Held, E. Kriegler, et al. 2008. Tipping Elements in the Earth's Climate System. *PNAS* **105**:1786–1793. [4]

Leopold, A. 1949. A Sand County Almanac. New York: Oxford Univ. Press. [16]

Le Polain, Y., and E. F. Lambin. 2013. Niche Commodities and Rural Poverty Alleviation: Contextualizing the Contribution of Argan Oil to Rural Livelihoods in Morocco. *Ann. Assoc. Am. Geograph.* **103**:589–607. [2]

Levi-Faur, D. 2005. The Global Diffusion of Regulatory Capitalism. *Ann. Am. Acad. Polit. Soc. Sci.* **598**:12–32. [10]

Levinson, M. 2008. The Box: How the Shipping Container Made the World Smaller and the World Economy Bigger. Princeton: Princeton Univ. Press. [14]

Levitt, T. 1983. The Globalization of Markets. *Harvard Bus. Rev.* **61**:92–102. [7]

Li, L. X. 2011. The Incentive Role of Creating "Cities" in China. *China Econ. Rev.* **22**:172–181. [5]

Li, W., T. T. Feng, and J. M. Hao. 2009. The Evolving Concepts of Land Administration in China: Cultivated Land Protection Perspective. *Land Use Policy* **26**:262–272. [5]

Li, X., W. Zhou, Z. Ouyang, W. Xu, and H. Zheng. 2012. Spatial Pattern of Greenspace Affects Land Surface Temperature: Evidence from the Heavily Urbanized Beijing Metropolitan Area, China. *Landsc. Ecol.* **27**:887–898. [4]

Liang, S. M. 2005. Forecasting Arable Land in Mid and Long Run under the Background of Urbanization in China. *Iss. Agric. Econ.* **S1**:101–107. [5]

Lieser, K., and A. P. Groh. 2011. The Attractiveness of 66 Countries for Institutional Real Estate Investments. *J. Real Estate Port. Manag.* **17**:191–211. [1]

Lindquist, E. J., R. D'Annunzio, A. Gerrand, et al. 2012. Global Forest Land-Use Change 1990–2005. Rome: FAO. [2]

Linnaeus, C. 1749/1964. Systema Naturae, Sive Regna Tria Naturae Systematice Proposita Per Classes, Ordines, Genera, and Species (trans. M. S. J. Engel-Ledeboer and H. Engel). Leiden: B. De Graff. [7]

Lippman, T. W. 2010. Saudi Arabia's Quest for Food Security. *Middle East Policy* **17**:90–98. [1]

Lipschutz, R. D., and C. Fogel. 2002. Regulations for the Rest of Us? Global Civil Society and the Privatization of Transnational Regulation. In: The Emergence of Private Authority in Global Governance, ed. R. B. Hall and T. J. Biersteker, pp. 115–140. Cambridge: Cambridge Univ. Press. [12]

Liu, J., T. Dietz, S. R. Carpenter, et al. 2007a. Complexity of Coupled Human and Natural Systems. *Science* **317**:1513–1516. [7]

Liu, J., T. Dietz, S. R. Carpenter, et al. 2007b. Coupled Human and Natural Systems. *Ambio* **36**:639–649. [7]

Liu, J., V. Hull, M. Batistella, et al. 2013. Framing Sustainability in a Telecoupled World. *Ecol. Soc.* **18**:26. [7]

Liu, Y. G., G. W. Yin, and L. J. C. Ma. 2012a. Local State and Administrative Urbanization in Post-Reform China: A Case Study of Hebi City, Henan Province. *Cities* **29**:107–117. [5]

Liu, Z. F., C. Y. He, Q. F. Zhang, Q. X. Huang, and Y. Yang. 2012b. Extracting the Dynamics of Urban Expansion in China Using DMSP-OLS Night Time Light Data from 1992 to 2008. *Landsc. Urban Plan* **106**:62–72. [5]

Llavador, H., J. E. Roemer, and J. Silvestre. 2011. A Dynamic Analysis of Human Welfare in a Warming Planer. *J. Public Econ.* **95**:1607–1620. [16]

Lo, F.-C., and P. J. Marcotullio. 2000. Globalization and Urban Transformations in the Asia Pacific Region: A Review. *Urban Stud.* **37**:77–111. [6]

———. 2001. Globalization and the Sustainability of Cities in the Asia Pacific Region. Tokyo: United Nations Univ. Press. [6]

Lobell, D. B., K. G. Cassman, and C. B. Field. 2009. Crop Yield Gaps: Their Importance, Magnitudes, and Causes. *Ann. Rev. Environ. Res.* **34**:179–204. [3]

Lopez, R. A., F. A. Shah, and M. A. Altobello. 1994. Amenity Benefits and the Optimal Allocation of Land. *Land Econ.* **70**:53–62. [1]

Lorance Rall, E. D., and D. Haase. 2011. Creative Intervention in a Dynamic City: A Sustainability Assessment of an Interim Use Strategy for Brownfields in Leipzig, Germany. *Landsc. Urban Plan* **100**:189–201. [15]

Losada, H., H. Martinez, J. Vieyra, et al. 1998. Urban Agriculture in the Metropolitan Zone of Mexico City: Changes over Time in Urban, Suburban and Peri-Urban Areas. *Environ. Urban.* **10**:37–54. [16]

Lösch, A. 1940/1954. The Economics of Location (English Translation). New Haven: Yale Univ. Press. [14]

Loureiro, M. L., J. J. McCluskey, and R. C. Mittelhammer. 2001. Assessing Consumer Preferences for Organic, Eco-Labeled, and Regular Apples. *J. Agricul. Res. Econ.* **26**:404–416. [12]

Lowry, I. S. 1990. World Urbanization in Perspective. *Pop. Dev. Rev.* **16**:148–176. [6]

Lu, Q. S., F. Y. Liang, X. L. Bi, R. Duffy, and Z. P. Zhao. 2011. Effects of Urbanization and Industrialization on Agricultural Land Use in Shandong Peninsula of China. *Ecol. Ind.* **11**:1710–1714. [5]

Lubowski, R., and S. K. Rose. 2013. The Potential of REDD+: Economic Modelling Insights and Issues. *Rev. Environ. Econ. Policy* **7**:67–90. [4]

Lyon, T. P., and J. W. Maxwell. 2007. Environmental Public Voluntary Programs Reconsidered. *Policy Stud. J.* **35**:723–750. [12]

Ma, T., C. H. Zhou, T. Pei, S. Haynie, and J. F. Fan. 2012. Quantitative Estimation of Urbanization Dynamics Using Time Series of Dmsp/Ols Nighttime Light Data: A Comparative Case Study from China's Cities. *Remote Sens. Environ.* **124**:99–107. [5]

Ma, Y. H., and W. Y. Niu. 2009. Forecasting on Grain Demand and Availability of Cultivated Land Resources Based on Grain Safety in China. *China Soft Sci.* **3**:11–16. [5]

MacDonald, G. M. 2010. Water, Climate Change and Sustainability in the Southwest. *PNAS* **107**:21256–21262. [4]

Macedo, M. N., R. S. DeFries, D. C. Morton, et al. 2012. Decoupling of Deforestation and Soy Production in the Southern Amazon During the Late 2000s. *PNAS* **109**:1341–1346. [2, 4, 6]

Mack, R. N. 1981. Invasion of *Bromus Tectorum* L. Into Western North America: An Ecological Chronicle. *Agro-Ecosyst.* **7**:145–165. [7]

MacMillan, T. 2012. Eating Globally. *Nature* **486**:30. [7]

Maddison, A. 1991. Dynamic Forces in Capitalist Development, a Long-Run Comparative View. Oxford: Oxford Univ. Press. [6]

———. 2001. The World Economy: A Millennial Perspective. Paris: OECD. [6]

Malhi, Y., J. Timmons Roberts, R. A. Betts, et al. 2007. Climate Change, Deforestation and the Fate of the Amazon. *Science* **319**:169–172. [4]

Marcotullio, P. J. 2003. Globalization, Urban Form and Environmental Conditions in Asia Pacific Cities. *Urban Stud.* **40**:219–248. [6]

———. 2005. Time-Space Telescoping and Urban Environmental Transitions in the Asia Pacific. Yokohama: UNU-IAS. [6]

Margulis, M. E. 2011. Hunger in a Globalizing World: International Organizations and Contestation in the Global Governance of Food Security. Ph.D. dissertation, Political Science, McMaster Univ., Hamilton. [13]

———. 2012. Global Food Governance: The Committee for World Food Security, G8/G20 and the Comprehensive Framework for Action. In: The Challenge of Food Security, ed. R. Rayfuse and N. Wiesfelt, pp. 231–254. Cheltenham: Edward Elgar. [13]

Margulis, M. E., N. McKeon, and S. M. Borras. 2013. Land Grabbing and Global Governance: Critical Perspectives. *Globalizations* **10**:1–23. [10, 11, 13]

Margulis, M. E., and T. Porter. 2013. Governing the Global Land Grab: Multipolarity, Ideas and Complexity in Transnational Governance. *Globalizations* **10**:65–86. [10]

Marsh, G. P. 1864/1965. Man and Nature: Physical Geography as Modified by Human Action. Cambridge, MA: Belknap Press. [7]

Martin, R., and P. Sunley. 1996. Paul Krugman's Geographical Economics and Its Implications for Regional Development Theory: A Critical Assessment. *Econ. Geogr.* **72**:259–292. [14]

Marx, K. 1852/2008. The 18th Brumaire of Louis Bonaparte. Rockville: Wildside Press. [11]

Massey, D. B. 1984. Spatial Divisions of Labor: Social Structures and the Geography of Production. London: Macmillan. [6]

Mather, A. S. 1992. The Forest Transition. *Area* **24**:367–379. [6]

———. 2007. Recent Asian Forest Transitions in Relation to Forest-Transition Theory. *Intl. Forest. Rev.* **9**:491–502. [6]

Mather, A. S., G. Hill, and M. Nijnik. 2006. Post-Productivism and Rural Land Use: Cul de Sac or Challenge for Theoretization? *J. Rural Stud.* **22**:441–455. [13]

Mather, A. S., and C. L. Needle. 1998. The Forest Transition: A Theoretical Basis. *Area* **30**:117–124. [6]

Mather, A. S., C. L. Needle, and J. Fairbairn. 1999. Environmental Kuznets Curves and Forest Trends. *Geography* **84**:55–65. [6]

Matson, P. A., and P. M. Vitousek. 2006. Agricultural Intensification: Will Land Spared from Farming Be Land Spared for Nature? *Conserv. Biol.* **20**:709–710. [2]

Matthews, R., and G. Dyer. 2011. Evaluating the Impacts of REDD+ at Subnational Scales: Are Our Frameworks and Models Good Enough? *Carbon Manag.* **2**:517–527. [4]

Mattli, W., and T. Büthe. 2003. Setting International Standards: Technological Rationality or Primacy of Power? *World Politics* **56**:1–42. [12]

Maxted, N., S. Kell, B. Ford-Lloyd, E. Dulloo, and Á. Toledo. 2012. Toward the Systematic Conservation of Global Crop Wild Relative Diversity. *Crop Sci.* **52**:774–785. [13]

Mayo, J. H., T. J. Straka, and D. S. Leonard. 2003. The Cost of Slowing the Spread of the Gypsy Moth (Lepidoptera: Lymantriidae). *J. Econ. Entomol.* **96**:1448–1454. [7]

Mbow, C. 2010. Africa's Risky Gamble. *Global Change IGBP* **75**:20–23. [4]

Mbow, C., M. S. S. Smith, and P. Leadley. 2010. Appendix 4, West Africa: The Sahara, Sahel, and and Guinean Region. In: Biodiversity Scenarios: Projections of 21st Century Change in Biodiversity and Associated Ecosystem Services, ed. P. Leadley et al., pp. 78–86. Montreal: Secretariat of the Convention on Biological Diversity. [4]

McCarney, P. L., and R. E. Stren. 2008. Metropolitan Governance: Governing in a City of Cities in State of the World's Cities Report. Nairobi: UN-HABITAT. [14]

McCarthy, J. F. 2010. Processes of Inclusion and Adverse Incorporation: Oil Palm and Agrarian Change in Sumatra, Indonesia. *J. Peasant Stud.* **37**:821–850. [11]

McCarthy, J. F., J. A. C. Vel, and S. Afiff. 2012. Trajectories of Land Acquisition and Enclosure: Development Schemes, Virtual Land Grabs, and Green Acquisitions in Indonesia's Outer Islands. *J. Peasant Stud.* **39**:521–549. [11]

McConnell, W. J., J. D. A. Millington, N. J. Reo, et al. 2011. Research on Coupled Human and Natural Systems (CHANS): Approach, Challenges, and Strategies. *Bull. Ecol. Soc. Am.* **92**:218–228. [7]

McCullough, E. B., P. L. Pingali, and K. G. Stamoulis. 2008. The Transformation of Agri-Food Systems: Globalization, Supply Chains and Smallholder Farmers. Rome/London: FAO/EarthScan. [4]

McDermott, C. L., E. Noah, and B. Cashore. 2008. Differences That "Matter"? A Framework for Comparing Environmental Certification Standards and Government Policies. *J. Environ. Policy Plan.* **10**:47–70. [12]

McGranahan, D. A. 2008. Landscape Influence on Recent Rural Migration in the U.S. 2008. *Landsc. Urban Plan* **85**:228–240. [4]

McHale, M. R., D. N. Bunn, S. T. A. Pickett, and W. Twine. 2013. Urban Ecology in a Developing World: Why Advanced Socioecological Theory Needs Africa. *Front. Ecol. Environ.* **11**:e1–e8. [16]

McLaughlin Mitchell, S., and P. R. Hensel. 2007. International Institutions and Compliance with Agreements. *Am. J. Polit. Sci.* **51**:721–737. [4]

McMichael, P. 2012. The Land Grab and Corporate Food Regime Restructuring. *J. Peasant Stud.* **39**:681–701. [11]

McNeill, J. R. 2000. Something New under the Sun, an Environmental History of the Twenteith-Century World. New York: W. W. Norton. [6]

McNew, K., and D. Griffith. 2005. Measuring the Impact of Ethanol Plants on Local Grain Prices. *Appl. Econ. Persp. Policy* **27**:164–180. [4]

McShane, T. O., P. D. Hirsch, T. C. Trung, et al. 2011. Hard Choices: Making Trade-Offs between Biodiversity Conservation and Human Well-Being. *Biol. Conserv.* **144**:966–972. [16]

MEA. 2005. Millennium Ecosystem Assessment. Ecosystems and Human Well-Being: Synthesis. Washington, D.C.: Island Press. [3, 4, 6, 13]

Mehta, L., G. J. Veldwisch, and J. C. Franco. 2012. Introduction to the Special Issue: Water Grabbing? Focus on the (Re)Appropriation of Finite Water Resources. *Water Alternatives* **5**:193–207. [10, 11, 13]

Meidinger, E. E. 2006. The Administrative Law of Global Private-Public Regulation: The Case of Forestry. *Eur. J. Intl. Law* **17**:47–87. [12]

Mellaart, J. 1965. Earliest Civilizations of the near East. New York: McGraw-Hill. [14]

Menakis, J. P., D. M. Osborne, and M. M. 2003. Mapping the Cheatgrass-Caused Departure from Historical Natural Fire Regimes in the Great Basin, USA. In: Fire, Fuel Treatments, and Ecological Restoration, ed. P. N. Omi and Joycek L. A., pp. 281–287. Fort Collins: USDA Forest Service, Rocky Mountain Research Station. [7]

Mendez, V. E. 2008. Farmers' Livelihoods and Biodiversity Conservation in a Coffee Landscape of El Salvador. In: Confronting the Coffee Crisis: Fair Trade, Sustainable Livelihoods and Ecosystems in Mexico and Central America, ed. C. M. Bacon et al., R. Gottlieb, series ed., pp. 207–234. Cambridge, MA: MIT Press. [12]

Mertz, O. 2009. Trends in Shifting Cultivation and the REDD Mechanism. *Curr. Opin. Environ. Sustain.* **1**:156–160. [4]

Meyfroidt, P., and E. F. Lambin. 2011. Global Forest Transition: Prospects for an End to Deforestation. *Annu. Rev. Environ. Res.* **36**:343–371. [2, 4, 6]

Meyfroidt, P., T. K. Rudel, and E. F. Lambin. 2010. Forest Transitions, Trade and the Global Displacement of Land Use. *PNAS* **107**:20917–20922. [4, 6]

Micklin, P. P. 1988. Dessication of the Aral Sea: A Water Management Disaster in the Soviet Union. *Science* **241**:1170–1176. [4]

Miettinen, J., A. Hooijer, C. Shi, et al. 2012. Extent of Industrial Plantations on Southeast Asian Peatlands in 2010 with Analysis of Historical Expansion and Future Projections. *Global Change Biol. Bioenergy* **4**:908–918. [9]

Mills, E. S. 1967. An Aggregative Model of Resource Allocation in a Metropolitan Area. *Am. Econ. Rev.* **57**:197–210. [14]

———. 1972. Studies in the Structure of the Urban Economy. Baltimore: Johns Hopkins Press. [14]

Montgomery, M. R. 2008. The Urban Transformation of the Developing World. *Science* **319**:761–764. [1]

Mooney, H., and E. E. Cleland. 2001. The Evolutionary Impact of Invasive Species. *PNAS* **98**:5446–5451. [7]

Mooney, H., A. Larigauderie, M. Cesario, et al. 2009. Biodiversity, Climate Change, and Ecosystem Services. *Curr. Opin. Environ. Sustain.* **1**:46–54. [4]

Moran, E. F. 2010. Environmental Social Science: Human-Environment Interactions and Sustainability. Hoboken, NJ: Wiley-Blackwell. [7]

Morton, D. C., R. S. DeFries, Y. E. Shimabukuro, et al. 2006. Cropland Expansion Changes Deforestation Dynamics in the Southern Brazilian Amazon. *PNAS* **103**:14637–14641. [4]

Moss, T. 2008. "Cold Spots" of Urban Infrastructure: "Shrinking" Processes in Eastern Germany and the Modern Infrastructural Ideal. *Intl. J. Urban Reg. Res.* **32**:436–451. [15]

MPOB. 2013. Economics and Industry Development Division: Statistics. Malaysian Palm Oil Board. http://bepi.mpob.gov.my/. (accessed Jan. 9, 2013). [9]

Müller, B. 2004. Demographic Change and Its Consequences for Cities: Introduction and Overview. *German J. Urban Stud.* **44**: [15]

Mulligan, G. F. 1984. Agglomeration and Central Place Theory: A Review of the Literature. *Intl. Reg. Sci. Rev.* **9**:1–42. [14]

Mulligan, G. F., M. D. Partridge, and J. I. Carruthers. 2012. Central Place Theory and Its Reemergence in Regional Science. *Ann. Reg. Sci.* **48**:405–431. [14]

Murdiyarso, D., S. Dewi, D. Lawrence, and F. Seymour. 2011. Indonesia's Forest Moratorium: A Stepping Stone to Better Forest Governance? CIFOR Working Paper 76. http://www.cifor.org/publications/pdf_files/WPapers/WP-76Murdiyarso.pdf. (accessed Dec. 8, 2013). [9]

Murmis, M., and M. R. Murmis. 2011. Dinámica del Mercado de la Tierra en América Latina y el Caribe: El Caso de Argentina. http:// www.rlc.fao.org/fileadmin/content/events/semtierras/acaparamiento.pdf. (accessed Dec. 8, 2013). [11]

———. 2012. Land Concentration and Foreign Land Ownership in Argetina in the Context of Global Land Grabbing. *Can. J. Dev. Stud.* **33**:490–508. [11]

Murray, W. E. 2006. Geographies of Globalization. London: Routledge. [6]

Mutersbaugh, T. 2005. Fighting Standards with Standards: Harmonization, Rents, and Social Accountability in Certified Agrofood Networks. *Environ. Plann. A* **37**:2033–2051. [12]

Muth, R. F. 1969. Cities and Housing: The Spatial Pattern of Urban Residential Land Use. Chicago: Univ. of Chicago Press. [14]

Nair, J. 2005. The Promise of the Metropolis: Bangalore's Twentieth Century. New Delhi: Oxford Univ. Press. [16]

Nakhooda, S., A. Caravani, and L. Schalatek. 2011. Climate Finance Fundamentals. Berlin: Heinrich Boll Stiftung. [13]

National Bureau of Statistics of China. 1988–2011a. China City Statistical Yearbook. Beijing. Beijing: China Statistics Press. [5]

———. 1988–2011b. China Statistical Yearbook. Beijing: China Statistics Press. [5]

National Research Academy. 1999. Our Common Journey, a Transition toward Sustainability. Washington, D.C.: National Academy Press. [6]

Naylor, R., A. J. Liska, M. B. Burke, et al. 2007. The Ripple Effect: Biofuels, Food Security, and the Environment. *Environment* **49**:30–43. [3]

Naylor, R., H. Steinfeld, W. Falcon, et al. 2005. Losing the Links between Livestock and the Land. *Science* **310**:1621–1622. [2]

Nelson, G. C., M. W. Rosegrant, A. Palazzo, et al. 2010. Food Security, Farming, and Climate Change to 2050: Scenarios, Results, Policy Options. Washington, D.C.: IFPRI. [10]

Nepstad, D. C., C. M. Stickler, and O. T. Almeida. 2006. Globalization of the Amazon Soy and Beef Industries: Opportunities for Conservation. *Conserv. Biol.* **20**:1595–1603. [7]

Niasse, M. 2011. Access to Land and Water for the Rural Poor in a Context of Growing Resource Scarcity. IFAD Conference on New Directions for Smallholder Agriculture, January 24–25, 2011. http://www.ifad.org/events/agriculture/doc/papers/niasse.pdf. (accessed June 17, 2013). [10]

Nigh, R. 1997. Organic Agriculture and Globalization: A Maya Associative Corporation in Chiapas, Mexico. *Human Organiz.* **56**:427. [12]

Njenga, M., S. Kimani, D. Romney, and N. Karanja. 2007. Nutrient Recovery from Solid Waste and Linkage to Urban and Peri-Urban Agriculture in Nairobi, Kenya. In: Advances in Integrated Soil Fertility Management in Sub-Saharan Africa: Challenges and Opportunities, ed. A. Bationo et al., pp. 487–491. Dordrecht: Springer. [3]

NLGN. 2006. Seeing the Light: Next Steps for City-Regions. London: New Local Government Network. [15]

Noah, L. 1994. Imperative to Warn: Disentangling the Right to Know from the Need to Know About Consumer Product Hazards. *Yale J. Regul.* **11**:293. [12]

Nordregio et al. 2005. Espon 111: Potentials for Polycentric Development in Europe, Project Report. Luxembourg: ESPON Monitoring Committee. [14]

Norgaard, R. B. 2010. Ecosystem Services: From Eye-Opening Metaphor to Complexity Blinder. *Ecol. Econ.* **69**:1219–1227. [10]

Nowak, D. J., R. A. Rowntree, E. G. McPherson, et al. 1996. Measuring and Analyzing Urban Tree Cover. *Landsc. Urban Plan* **36**:49–57. [15]

NRC. 2009. Driving and the Built Environment: The Effects of Compact Development on Motorized Travel, Energy Use, and CO_2 Emissions. Special Report 298. Washington, D.C.: National Academies Press. [14]

Nuissl, H., D. Haase, H. Wittmer, and M. Lanzendorf. 2008. Impact Assessment of Land Use Transition in Urban Areas: An Integrated Approach from an Environmental Perspective. *Land Use Policy* **26**:414–424. [15]

Nuissl, H., and D. Rink. 2005. The "Production" of Urban Sprawl in Eastern Germany as a Phenomenon of Post-Socialist Transformation. *Cities* **22**:123–134. [15]

Nunn, N., and N. Qian. 2010. The Columbian Exchange: A History of Disease, Food, and Ideas. *J. Econ. Perspect.* **24**:163–188. [7]

Obidzinski, K., R. Andriani, H. Komarudin, and A. Andrianto. 2012. Environmental and Social Impacts of Oil Palm Plantations and Their Implications for Biofuel Production in Indonesia. *Ecol. Soc.* **17**:25. [9]

O'Brien, K., B. Hayward, and F. Berkes. 2009. Rethinking Social Contracts: Building Resilience in a Changing Climate. *Ecol. Soc.* **14**:12. [8]

Obstfeld, M., and A. M. Taylor. 2003. Globalization and Capital Markets. In: Globalization in Historical Perspective, ed. M. D. Bordo et al., pp. 121–187. Chicago: Univ. of Chicago Press. [6]

OECD. 1996. International Captial Market Statistics 1950–1995. Paris: OECD. [6]

———. 2011. OECD Guidelines for Multinational Enterprises: Recommendations for Responsible Business Conduct in a Global Context. http://www.oecd.org/corporate/mne/48004323.pdf. (accessed Dec. 8, 2013). [9]

———. 2012. Compact City Policies: A Comparative Assessment, OECD Green Growth Studies. Paris: OECD Publishing. [14]

Oldeman, L. R. 1994. The Global Extent of Soil Degradation. In: Soil Resilience and Sustainable Land Use, ed. D. J. Greenland and I. Szabolcs, pp. 99–118. Wallingford: CAB Intl. [3]

Olesen, J. E., and M. Bindi. 2002. Consequences of Climate Change for European Agricultural Productivity, Land Use and Policy. *Eur. J. Agron.* **16**:239–262. [4]

Ostrom, E. 1990. Governing the Commons: The Evolution of Institutions for Collective Action. New York: Cambridge Univ. Press. [4]

———. 1999. Coping with Tragedies of the Commons. *Annu. Rev. Polit. Sci.* **2**:493–535. [4]

Overdevest, C. 2010. Comparing Forest Certification Schemes: The Case of Ratcheting Standards in the Forest Sector. *Socio-Econ. Rev.* **8**:47–76. [12]

Overton, J., and J. Heitger. 2008. Maps, Markets and Merlot: The Making of an Antipodean Wine Appellation. *J. Rural Stud.* **24**:440–449. [13]

Oxfam. 2012. Land and Power: The Growing Scandal Surrounding the New Wave of Investments in Land. http://www.oxfam.org/sites/www.oxfam.org/files/bp151-land-power-rights-acquisitions-220911-en.pdf. (accessed Oct. 23, 2013). [10]

Painter, J. 2000. Localization. In: The Dictionary of Human Geography, ed. R. J. Johnston et al. Oxford: Blackwell. [6]

Palmer, D., S. Fricska, and B. Wehrmann. 2009. Towards Improved Land Governance. Rome: FAO. ftp://ftp.fao.org/docrep/fao/012/ak999e/ak999e00.pdf (accessed Dec. 6, 2013). [4]

Partridge, M. D., and D. S. Rickman. 2013. Integrating Regional Economic Development and Land Use Economics. In: The Handbook of Land Economics, ed. J. M. Duke and J. Wu. Oxford: Oxford Univ. Press. [4]

Partridge, M. D., D. S. Rickman, K. Ali, and M. R. Olfert. 2008. Lost in Space: Population Growth in the American Hinterlands and Small Cities. *J. Econ. Geogr.* **8**:727–757. [14]

Pattberg, P. 2006. Private Governance and the South: Lessons from Global Forest Politics. *Third World Q.* **27**:579–593. [12]

———. 2007. Private Institutions and Global Governance: The New Politics of Environmental Sustainability. Cheltenham: Edward Elgar. [12]

Pazarbaşioğlu, C., M. Goswami, and J. Ree. 2007. The Changing Face of Investors. *Finan. Dev.* **44**:28–31. [1]

Pelletier, N., and P. Tyedmers. 2010. Forecasting Potential Global Environmental Costs of Livestock Production 2000–2050. *PNAS* **107**:18371–18374. [4]

Peng, S., R. C. Laza, R. M. Visperas, et al. 2000. Grain Yield of Rice Cultivars and Lines Developed in the Philippines since 1966. *Crop Sci.* **40**:307–314. [4]

Perfecto, I., and J. Vandermeer. 2010. The Agroecological Matrix as Alternative to the Land-Sparing/Agriculture Intensification Model. *PNAS* **107**:5786–5791. [4]

Perlin, J. 2005. A Forest Journey: The Story of Wood and Civilization. Woodstock, VT: The Countryman Press. [13]

Perrine, B., G. Mathilde, A. R. Rivo, and R. Raphael. 2011. From International Land Deals to Local Informal Agreements: Regulations of and Local Reactions to Agricultural Investments in Madagascar. Paper Presented at Intl. Conf. on Global Land Grabbing, April 6–8, 2011, Univ. of Sussex. http://www.iss.nl/fileadmin/ASSETS/iss/Documents/Conference_papers/LDPI/35_Perrine_Mathilde_Rivo_and_Raphael.pdf. (accessed Nov. 23, 2013). [7]

Peterson, G. 2000. Political Ecology and Ecological Resilience: An Integration of Human and Ecological Dynamics. *Ecol. Econ.* **35**:323–336. [7]

Peters-Stanley, M., and K. Hamilton. 2012. Developing Dimensions: State of Voluntary Carbon Markets 2012. Washington, D.C.: Ecosystem Marketplace and Bloomberg New Energy Finance. [13]

Peters-Stanley, M., K. Hamilton, T. Marcello, and M. Sjardin. 2011. Back to the Future: State of the Voluntary Carbon Markets 2011. New York: Bloomberg New Energy Finance. [13]

Pettorelli, N., J. O. Vik, A. Mysterud, et al. 2005. Using the Satellite-Derived NDVI to Assess Ecological Responses to Environmental Change. *Trends Ecol. Evol.* **20**:503–510. [4]

Phalan, B., M. Onial, A. Balmford, and R. E. Green. 2011. Reconciling Food Production and Biodiversity Conservation: Land Sharing and Land Sparing Compared. *Science* **333**:1289–1291. [4]

Phelps, J., E. L. Webb, and A. Agrawal. 2010. Does REDD+ Threaten to Recentralize Forest Governance? *Science* **328**:312–313. [13]

Pielke, R. A., G. Marland, R. A. Betts, et al. 2002. The Influence of Land-Use Change and Landscape Dynamics on the Climate System: Relevance to Climate-Change Beyond the Radiative Effect of Greenhouse Gases. *Phil. Trans. R. Soc. A* **360**:1705–1719. [4]

Pimentel, D. 2002. Biological Invasions: Economic and Environmental Costs of Alien Plant, Animal, and Microbe Species. Boca Raton: CRC Press. [7]

Pimentel, D., W. Dazhong, and M. Giampietro. 1990. Technological Changes in Energy Use in U.S. Agricultural Production. In: Agroecology, Researching the Ecological Basis for Sustainable Agriculture, ed. S. R. Gliessmann, pp. 305–321. New York: Springer. [4]

Pimentel, D., M. Pimentel, and A. Wilson. 2007. Plant, Animal, and Microbe Invasive Species in the United States and World. In: Biological Invasions, ed. W. Nentwig, pp. 315–330. Berlin: Springer. [7]

Pimentel, D., R. Zuniga, and D. Morrison. 2005. Update on the Environmental and Economic Costs Associated with Alien-Invasive Species in the United States. *Ecol. Econ.* **52**:273–288. [7]

Pimm, S. L. 2009. Climate Disruption and Biodiversity. *Curr. Biol.* **19**:R595–R601. [4]

Piorr, A., J. Ravetz, and I. Tosics. 2011. Periurbanisation in Europe: Towards a European Policy to Sustain Urban-Rural Futures. A Synthesis Report. Frederiksberg: Univ. of Copenhagen, Academic Books Life Sciences. [14]

Pires, M. 2005. Watershed Protection for a World City: The Case of New York. *Land Use Policy* **21**:161–175. [4]

Pivo, G. 1990. The Net of Mixed Beads Suburban Office Development in Six Metropolitan Regions. *J. Amer. Plan. Assoc.* **56**:457–469. [14]

Plieninger, T., S. Ferranto, L. Huntsinger, M. Kelly, and C. Getz. 2012. Appreciation, Use, and Management of Biodiversity and Ecosystem Services in California's Working Landscapes. *Environ. Manag.* **50**:427–440. [13]

Ploch, L., and N. Cook. 2012. Madagascar's Political Crisis. Congressional Research Service 7-5700. http://www.fas.org/sgp/crs/row/R40448.pdf. (accessed June 18, 2013). [11]

Polasky, S., E. Nelson, E. Lonsdorf, P. Fackler, and A. Starfield. 2005. Conserving Species in a Working Landscape: Land Use with Biological and Economic Objectives. *Ecol. Appl.* **15**:1387–1401. [4]

Ponte, S. 2008. Greener Than Thou: The Political Economy of Fish Ecolabeling and Its Local Manifestations in South Africa. *World Devel.* **36**:159–175. [12]

Popp, A., J. P. Dietrich, H. Lotze-Campen, et al. 2011. The Economic Potential of Bioenergy for Climate Change Mitigation with Special Attention Given to Implications for the Land System. *Environ. Res. Lett.* **6**:034017. [4]

Popp, A., H. Lotze-Campen, and B. Bodirsky. 2010. Food Consumption, Diet Shifts and Associated Non-CO_2 Greenhouse Gases from Agricultural Production. *Global Environ. Change* **20**:451–462. [4]

Potapov, P. V., S. A. Turubanova, M. C. Hansen, et al. 2012. Quantifying Forest Cover Loss in Democratic Republic of the Congo, 2000–2010, with Landsat ETM+ Data. *Remote Sens. Environ.* **122**:106–116. [4]

Potschin, M. B., and R. H. Haines-Young. 2011. Ecosystem Services Exploring a Geographical Perspective. *Prog. Phys. Geogr.* **35**:575–594. [13]

Powlson, D. S., P. J. Gregory, W. R. Whalley, et al. 2011. Soil Management in Relation to Sustainable Agriculture and Ecosystem Services. *Food Policy* **36**:S72–S87. [3]

Prakash, A., and M. Potoski. 2006. The Voluntary Environmentalists: Green Clubs, ISO 14001, and Voluntary Regulations. Cambridge: Cambridge Univ. Press. [12]

Pratt, J. 2007. Food Values the Local and the Authentic. *Crit. Anthropol.* **27**:285–300. [13]

Pretty, J. 2008. Agricultural Sustainability: Concepts, Principles and Evidence. *Phil. Trans. R. Soc. B.* **363**:447–465. [3]

Primdahl, J., and S. Swaffield. 2010. Globalisation and the Sustainability of Agricultural Landscapes. In: Globalisation and Agricultural Landscapes. Change Patterns and Policy Trends in Developed Countries, ed. J. Primdahl and S. Swaffield, pp. 1–15. Cambridge: Cambridge Univ. Press. [13]

Puga, D. 2010. The Magnitude and Causes of Agglomeration Economies. *J. Reg. Sci.* **50**:203–219. [14]

Quinton, J. N., G. Govers, K. V. Oost, and R. D. Bardgett. 2010. The Impact of Agricultural Soil Erosion on Biogeochemical Cycling. *Nature Geosci.* **3**:311–314. [3]

Ramankutty, N. 2013. Agricultural Lands in the Year 2000 (M3-Cropland and M3-Pasture Data). http://www.geog.mcgill.ca/~nramankutty/Datasets/Datasets.html. (accessed Dec. 6, 2013). [2]

Ramankutty, N., J. A. Foley, J. Norman, and K. McSweeney. 2002. The Global Distribution of Cultivable Lands: Current Patterns and Sensitivity to Possible Climate Change. *Global Ecol. Biogeogr.* **11**:377–392. [2]

Ramankutty, N., L. Graumlich, F. Achard, et al. 2006. Global Land Cover Change: Recent Progress, Remaining Challenges. In: Land-Use and Land-Cover Change: Local Processes and Global Impacts, ed. E. F. Lambin and H. Geist. Heidelberg: Springer. [6]

Ratha, D., S. Mohapatra, and E. Scheja. 2011. Impact of Migration on Economic and Social Development: A Review of Evidence and Emerging Issues. In: World Bank Policy Research Working Paper, No. 5558. Washington, D.C.: World Bank. [6]

Ratha, D., S. Mohapatra, and A. Silwal. 2010. Outlook for Remittance Flows 2010–11. In: Migration and Development Brief, No. 12. Washington, D.C.: World Bank. [6]

Raustalia, K., and D. G. Victor. 2004. The Regime Complex for Plant Genetic Resources. *Intl. Org.* **58**:277–309. [10]

Ravetz, J. 2000. City Region 2020: Integrated Planning for a Sustainable Environment. London: Earthscan. [15]

Raynolds, L. T., D. Murray, and A. Heller. 2007. Regulating Sustainability in the Coffee Sector: A Comparative Analysis of Third-Party Environmental and Social Certification Initiatives. *Agric. Human Values* **24**:147–163. [12]

Redman, C. L. 1999. Human Impact on Ancient Environments. Tucson: Univ. of Arizona Press. [6]

Reenberg, A., and N. A. Fenger. 2011. Globalizing Land Use Transitions: The Soybean Acceleration. *Geografisk Tidsskrift* **111**:85–92. [6]

Reijnders, J. 1990. Long Waves in Economic Development. Aldershot: Edward Elgar. [6]

Rhodes, R. A. W. 1996. The New Governance: Governing without Government. *Political Stud.* **44**:652–667. [13]

Rickards, L., and S. M. Howden. 2012. Transformational Adaptation: Agriculture and Climate Change. *Crop Past. Sci.* **63**:240–250. [3]

Ricketts, T. H., B. Soares-Filho, G. A. B. Fonseca, D. Nepstad, and A. Pfaff. 2010. Indigenous Lands, Protected Areas, and Slowing Climate Change. *PLoS Biol.* **8**:e1000331. [4]

Rieniets, T. 2009. Shrinking Cities: Causes and Effects of Urban Population Losses in the Twentieth Century. *Nature Cult.* **4**:231–254. [15]

Riisgaard, L. 2009. Global Value Chains, Labor Organization and Private Social Standards: Lessons from East African Cut Flower Industries. *World Devel.* **37**:326–340. [12]

Rink, D. 2009. Wilderness: The Nature of Urban Shrinkage? The Debate on Urban Restructuring and Restoration in Eastern Germany. *Nature Cult.* **3**:275–292. [15]

Ritzer, G. 2009. Globalization: A Basic Text. Oxford: Blackwell. [6]

Rivera, J. 2002. Assessing a Voluntary Environmental Initiative in the Developing World: The Costa Rican Certification for Sustainable Tourism. *Policy Sci.* **35**:333–360. [12]

Rivera, J., and P. de Leon. 2004. Is Greener Whiter? Voluntary Environmental Performance of Western Ski Areas. *Policy Stud. J.* **32**:417–437. [12]

Rivera, J., P. de Leon, and C. Koerber. 2006. Is Greener Whiter Yet? The Sustainable Slopes Program after Five Years. *Policy Stud. J.* **34**:195–221. [12]

Roback, J. 1982. Wages, Rents, and the Quality of Life. *J. Polit. Econ.* **90**:1257–1278. [14]

Robalino, J., A. S. P. Pfaff, G. A. Sanchez-Azofeifa, et al. 2008. Changing the Deforestation Impacts of Ecopayments: Evolution (2000–2005) in Costa Rica's PSA Program. Durham: Duke University, Sanford School of Public Policy. [4]

Roberts, M. J., and R. N. Lubowski. 2007. Enduring Impacts of Land Retirement Policies: Evidence from the Conservation Reserve Program. *Land Econ.* **83**:4. [4]

Robertson, B., and P. Pinstrup-Andersen. 2010. Global Land Acquisition: Neo-Colonialism or Development Opportunity? *Food Security* **2**:271–283. [10]

Robertson, G. P., and S. M. Swinton. 2005. Reconciling Agricultural Productivity and Environmental Integrity: A Grand Challenge for Agriculture. *Front. Ecol. Environ.* **3**:38–46. [13]

Robertson, R. 1992. Globalization: Social Theory and Global Culture. London: Sage. [6]
Rodrigues, A. S. L., H. R. Akcakaya, S. J. Andelman, et al. 2004. Global Gap Analysis: Priority Regions for Expanding the Global Protected-Area Network. *BioScience* **54**:1092–1100. [13]
Romero, S. 2012. In Brazil, Violence Hits Tribes in Scramble for Land. *New York Times* June 9 2012. [1]
Rosen, S. 1979. Wage-Based Indexes of Urban Quality of Life. In: Current Issues in Urban Economics, ed. P. Mieszkowski and M. Straszheim. Baltimore: Johns Hopkins Univ. Press. [14]
Rosenau, J. N. 1995. Governance in the Twenty-First Century. *Global Governance* **6**:13–43. [10, 13]
Rosendal, G. K., and S. Andresen. 2011. Institutional Design for Improved Forest Governance through REDD: Lessons from the Global Environment Facility. *Ecol. Econ.* **70**:1908–1915. [4]
Rosenthal, S. S., and W. Strange. 2004. Evidence on the Nature and Sources of Agglomeration Economies. In: Handbook of Regional and Urban Economics, ed. V. Henderson and J.-F. Thisse, vol. 4, pp. 2119–2171. Amsterdam: North-Holland. [14]
Rosset, P. M. 2006. Food Is Different. Halifax: Fernwood Publishing. [13]
RSPO. 2007. RSPO Principles and Criteria for Sustainable Palm Oil Production. Including Indicators and Guidance. Kuala Lumpur: Roundtable for Sustainable Palm Oil. http://www.rspo.org/files/resource_centre/RSPO%20Principles%20 &%20Criteria%20Document.pdf. (accessed Dec. 8, 2013). [9]
Ruddiman, W. F. 2003. The Anthropogenic Greenhouse Era Began Thousands of Years Ago. *Clim. Change* **62**:261–293. [6]
Rudel, T. K. 1998. Is There a Forest Transition? Deforestation, Reforestation and Development. *Rural Sociology* **63**(4):533–552. [6]
Rudel, T. K., O. T. Coomes, E. Moran, et al. 2005. Forest Transitions: Towards a Global Understanding of Land Use Change. *Global Environ. Change* **15**:23–31. [2, 6]
Rudel, T. K., R. DeFries, G. P. Asner, and W. F. Laurance. 2009a. Changing Drivers of Deforestation and New Opportunitieis for Conservation. *Conserv. Biol.* **23**:1396–1405. [6]
Rudel, T. K., L. Schneider, M. Uriarte, et al. 2009b. Agricultural Intensification and Changes in Cultivated Areas, 1970–2005. *PNAS* **106**:20675–20680. [4]
Rulli, M. C., A. Saviori, and P. D'Odorico. 2013. Global Land and Water Grabbing. *PNAS* **110**:892–897. [10]
Sachs, J. D., and A. Warner. 1995. Economic Reform and the Process of Global Integration. *Brookings Papers Econ. Activity* **26**:1–118. [6]
Sahakian, M., and J. K. Steinberger. 2011. Energy Reduction through a Deeper Understanding of Household Consumption: Staying Cool in Metro Manila. *J. Indust. Ecol.* **1**:31–48. [15]
SAIP. 2012. Sustainable Agriculture Initiative: About Us. http://www.saiplatform.org/. (accessed Oct. 3, 2012). [13]
Salter, L. 1999. The Standards Regime for Communication and Information Technologies. In: Private Authority and International Affairs, ed. A. C. Cutler et al., J. N. Rosenau, series ed., pp. 97–128. Albany: SUNY Press. [12]
Sánchez-Azofeifa, G. A., A. Pfaff, J. A. Robalino, and J. Boomhower. 2007. Costa Rica's Payment for Environmental Services Program: Intention, Implementation, and Impact. *Conserv. Biol.* **21**:1165–1173. [4]

Sanders, N. J., N. J. Gotelli, N. E. Heller, and D. M. Gordon. 2003. Community Disassembly by an Invasive Species. *PNAS* **100**:2474–2477. [7]

Sassen, S. 2009. Global Inter-City Networks and Commodity Chains: Any Intersections? *Global Networks* **10**:150–163. [15]

———. 2011. Global Challenges and the City. *Arena* http://www.arena.org.au/2011/09/global-challenges-and-the-city/. (accessed Oct. 24, 2013). [15]

———. 2013. Land Grabs Today: Feeding the Disassembling of National Territory. *Globalizations* **10**:26–46. [10]

Sasser, E. N. 2003. Gaining Leverage: NGO Influence on Certification Institutions in the Forest Products Sector. In: Forest Policy for Private Forestry, ed. L. Teeter et al., pp. 229–244. Oxon: CAB Intl. [12]

Sasser, E. N., A. Prakash, B. Cashore, and G. Auld. 2006. Direct Targeting as an NGO Political Strategy: Examining Private Authority Regimes in the Forestry Sector. *Business Polit.* **8**:1–32. [12]

Satterthwaite, D. 2007. The Transition to a Predominantly Urban World and Its Underpinnings. Human Settlements Discussion Paper Series. Human Settlements Discussion Paper Series. London: Intl. Institute for Environment and Development. [15]

———. 2009. The Implications of Population Growth and Urbanization for Climate Change. *Environ. Urban.* **21**:545. [15]

Satterthwaite, D., G. McGranahan, and C. Tacoli. 2010. Urbanization and Its Implications for Food and Farming. *Phil. Trans. R. Soc. B.* **365**:2809–2820. [3]

Saunders, C., A. Barber, and G. Taylor. 2006. Food Miles: Comparative Energy/Emissions Performance of the NZ Agricultural Industry. AERU Research Report, No. 285. Christchurch: Lincoln Univ. [8]

Savills PLC. 2012. International Farmland Focus. www.savills.co.uk/research. (accessed Oct. 3, 2012). [13]

Scherer, C. 2012. Foreign Investment in Agricultural Land Down from 2009 Peak. Washington, D.C.: Worldwatch Institute. [7]

Schiller, G., and S. Siedentop. 2005. Follow-up Costs of Settlement Development for Infrastructure under Conditions of Shrinkage. *DISP Journal* **160**:83–93. [15]

Schneider, A., and C. E. Woodcock. 2008. Compact, Dispersed, Fragmented, Extensive? A Comparison of Urban Growth in Twenty-Five Global Cities Using Remotely Sensed Data, Pattern Metrics and Census Information. *Urban Stud.* **45**:659–692. [15]

Scholte, J. A. 2005. Globalization: A Critical Introduction, 2nd edition. New York: Palgrave Macmillan. [6]

Schouten, G., and P. Glasbergen. 2011. Creating Legitimacy in Global Private Governance: The Case of the Roundtable on Sustainable Palm Oil. *Ecol. Econ.* **70**:1891–1899. [13]

Schueler, V., T. Kuemmerle, and H. Schroeder. 2011. Impacts of Surface Gold Mining on Land Use Systems in Western Ghana. *Ambio* **40**:528–539. [4]

Schwarz, G. 2006. Enabling Global Trade above the Clouds: Restructuring Processes and Information Technology in the Transatlantic Air-Cargo Industry. *Environ. Plann. A* **38**:1463–1485. [14]

Schwarz, N. 2010. Urban Form Revisited: Selecting Indicators for Characterising European Cities. *Landsc. Urban Plan* **96**:29–47. [15]

Schwarz, N., and D. Haase. 2010. Urban Shrinkage: A Vicious Circle for Residents and Infrastructure? Coupling Agent-Based Models on Residential Location Choice and Urban Infrastructure Development. *Intl. Congress on Environmental Modelling and Software Modelling for Environment's Sake* http://www.iemss.org/iemss2010/index.php?n=Main.Proceedings. (accessed Dec. 8, 2013). [15]

Schwarz, N., D. Haase, and R. Seppelt. 2010. Omnipresent Sprawl? A Review of Urban Simulation Models with Respect to Urban Shrinkage. *Environ. Plann. B* **37**:265–283. [15]

Scurrah, M. 2006. Defendiendo Derechos y Promoviendo Cambiois: el Estado, Las Empresas Extractivas y Las Comunidades Locales en el Perú. Lima: Instituto de Estudios Peruanos. [13]

Searchinger, T. 2010. Biofuels and the Need for Additional Carbon. *Environ. Res. Lett.* **5**:024007. [4]

Searchinger, T., R. Heimlich, R. A. Houghton, et al. 2008. Use of U.S. Croplands for Biofuels Increases Greenhouse Gases through Emissions from Land-Use Change. *Science* **319**:1238–1240. [2, 8, 13]

Sedjo, R. A., and D. Botkin. 1997. Using Forest Plantations to Spare Natural Forests. *Environment* **39**:14–20 30. [2, 4]

Sellers, E., A. Simpson, and S. Curd-Hetrick. 2010. List of Invasive Alien Species Online Information Systems. A "Living Document" Based on a Preliminary Draft Document, Prepared for the Experts Meeting Towards the Implementation of a Global Invasive Species Information Network, Baltimore, Maryland, April 6–8, 2004. http://ibis.colostate.edu/cwis438/Websites/GISINDirectory/DatabaseDirectory_Table.php. (accessed Dec. 7, 2013). [7]

Seto, K. C., R. de Groot, S. Bringezu, et al. 2009. Stocks, Flows, and Prospects of Land. In: Linkages of Sustainability, ed. T. E. Graedel and E. van der Voet, Strüngmann Forum Reports, J. Lupp, series ed., vol. 4. Cambridge, MA: MIT Press. [16]

Seto, K. C., M. Fragkias, B. Güneralp, and M. K. Reilley. 2011. A Meta-Analysis of Global Urban Land Expansion. *PLoS ONE* **6**:e23777. [1, 2, 4, 6, 14, 15]

Seto, K. C., B. Güneralp, and L. Hutyra. 2012a. Global Forecasts of Urban Expansion to 2030 and Direct Impacts on Biodiversity and Carbon Pools. *PNAS* **109**:16083–16088. [5]

Seto, K. C., A. Reenberg, C. Boone, et al. 2012b. Urban Land Teleconnections and Sustainability. *PNAS* **109**:7687–7692. [1, 2, 4, 6–8, 14–16]

Seto, K. C., R. Sanchez-Rodrıguez, and M. Fragkias. 2010. The New Geography of Contemporary Urbanization and the Environment. *Annu. Rev. Environ. Res.* **35**:167–194. [8, 9]

Seto, K. C., C. E. Woodcock, C. Song, et al. 2002. Monitoring Land Use Change in the Pearl River Delta Using Landsat Tm. *Intl. J. Remote Sens.* **23**:1985–2004. [4]

Seufert, P. 2013. The FAO Voluntary Guidelines on the Responsible Governance of Tenure of Land, Fisheries and Forests. *Globalizations* **10**:181–186. [10]

Seufert, V., N. Ramankutty, and J. A. Foley. 2012. Comparing the Yields of Organic and Conventional Agriculture. *Nature* **485**:229–234. [4]

Shearman, P., J. Bryan, and W. F. Laurance. 2012. Are We Approaching "Peak Timber" in the Tropics? *Biol. Conserv.* **151**:17–21. [4]

Shortle, J. S., and R. D. Horan. 2008. The Economics of Water Quality Trading. *Intl. Rev. Environ. Res. Econ.* **2**:101–133. [4]

Siciliano, G. 2012. Urbanization Strategies, Rural Development and Land Use Changes in China: A Multiple-Level Integrated Assessment. *Land Use Policy* **29**:165–178. [5]

Sikor, T. 2012a. Public and Private in Natural Resource Governance: A False Dichotomy? London: Routledge. [12]

———. 2012b. Tree Plantations, Politics of Possession and the Absence of Land Grabs in Vietnam. *J. Peasant Stud.* **39**:1077–1101. [11]

Sikor, T., and C. Lund. 2009. Access and Property: A Question of Power and Authority. *Dev. Change* **40**:1–22. [13]

Sim, S., M. Barry, R. Clift, and S. J. Cowell. 2007. The Relative Importance of Transport in Determining an Appropriate Sustainability Strategy for Food Sourcing. *Intl. J. Life Cycle Assess.* **12**:422–431. [14]

Sime Darby. 2013. Sime Darby Core Businesses. http://www.simedarby.com/. (accessed Dec. 8, 2013). [9]

Simmons, A. J., J. M. Wallace, and G. W. Branstator. 1983. Barotropic Wave Propagation and Instability, and Atmospheric Teleconnection Patterns. *J. Atmos. Sci.* **40**:1363–1392. [8]

Sinclair, F. L., and L. Joshi. 2001. Taking Local Knowledge About Trees Seriously. In: Forestry, Forest Users and Research: New Ways of Learning, ed. A. Lawrence, pp. 45–61. Wageningen: ETFRN. [13]

Smaling, E. M. A., L. O. Fresco, and A. de Jager. 1996. Classifying, Monitoring and Improving Soil Nutrient Stocks and Flows in African Agriculture. *Ambio* **25**:492–496. [3]

Smaller, C., and H. Mann. 2009. A Thirst for Distant Land. Foreign Investment in Agricultural Land and Water. Manitoba: Intl. Institute for Sustainable Development. [7]

Smalley, R., and E. Corbera. 2012. Large-Scale Land Deals from the inside Out: Findings from Kenya's Tana Delta. *J. Peasant Stud.* **39**:1039–1075. [11]

Smeets, E. M. W., A. P. C. Faaij, I. M. Lewandowski, and W. C. Turkenburg. 2007. A Bottom-up Assessment and Review of Global Bioenergy Potentials to 2050. *Prog. Energy Comb. Sci.* **33**:56–106. [4]

Smith, J. 2007. Social Movements for Global Democracy. Baltimore: John Hopkins Univ. Press. [10]

Smith, P., P. J. Gregory, D. van Vuuren, et al. 2010. Competition for Land. *Phil. Trans. R. Soc. B.* **365**:2941–2957. [2–4, 13]

Smith, P., H. Haberl, A. Popp, et al. 2013. How Much Land Based Greenhouse Gas Mitigation Can Be Achieved without Compromising Food Security and Environmental Goals? *Global Change Biol.* **19**:2285–2302. [4]

Smith, P., and E. Wollenberg. 2012. Acheiving Mitigation through Synergies with Adapatation. In: Climate Change Mitigation and Agriculture, ed. E. Wollenberg et al., pp. 50–57. London: ICRAF-CIAT. [4]

Smith, R. C., I. Walter, and G. DeLong. 2012. Global Banking, 3rd edition. Oxford: Oxford Univ. Press. [6]

Snidal, D. 1985. Coordination Versus Prisoners Dilemma: Implications for International-Cooperation and Regimes. *Am. Polit. Sci. Rev.* **79**:923–942. [12]

———. 1996. Political Economy and International Institutions. *Intl. Rev. Law Econ.* **16**:121–137. [12]

Soares-Filho, B., P. Moutinho, D. Nepstad, et al. 2010. Role of Brazilian Amazon Protected Areas in Climate Change Mitigation. *PNAS* **107**:10821–10826. [4]

Soja, E. W. 2010. Cities and States in Geohistory. *Theor. Soc.* **39**:361–376. [14]

Song, Y., and Y. Zenou. 2012. Urban Villages and Housing Values in China. *Reg. Sci. Urban Econ.* **42**:495–505. [5]

Squires, G. D. 2002. Urban Sprawl: Causes, Consequences and Policy Responses. Washington, D.C.: The Urban Institute Press. [15]

Srinivas, S. 2004. Landscapes of Urban Memory: The Sacred and the Civic in India's High-Tech City. Hyderabad: Orient Longman. [16]

Stavins, R. N., N. O. Keohane, and R. L. Revesz. 1998. The Choice of Regulatory Instruments in Environmental Policy. *Harvard Environ. Law Rev.* **22**:313–367. [4]

Steering Committee of the State-of-Knowledge Assessment of Standards Certification. 2012. Toward Sustainability: The Roles and Limitations of Certification. Washington, D.C.: RESOLVE, Inc. [8]

Stehfest, E., L. Bouwman, D. P. Vuuren, et al. 2009. Climate Benefits of Changing Diet. *Clim. Change* **95**:83–102. [4]

Steinfeld, H., P. Gerber, T. Wassenaar, et al. 2006. Livestock's Long Shadow: Environmental Issues and Options. Rome: FAO. [2, 4, 13]

Stenger, A. J. 2011. Advances in Information Technology Applications for Supply Chain Management. *Transport. J.* **50**:37–52. [14]

Stern, N. 2006. The Economis of Climate Change; the Stern Review. Cambridge: Cambridge Univ. Press. [13]

Stevens, S. 2010. Implementing the UN Declaration on the Rights of Indigenous Peoples and International Human Rights Law through Recognition of Iccas. *Policy Matters* **17**: [13]

Su, S. L., Z. L. Jiang, Q. Zhang, and Y. Zhang. 2011. Transformation of Agricultural Landscapes under Rapid Urbanization: A Threat to Sustainability in Hang-Jia-Hu Region, China. *Appl. Geog.* **31**:439–449. [5]

Sudhira, H. S., T. V. Ramachandra, and M. H. B. Subrahmanya. 2007. City Profile Bangalore. *Cities* **24**:379–390. [16]

Sunstein, C. R. 1993. Endogenous Preferences, Environmental Law. *J. Legal Stud.* **22**:217–254. [12]

Swaney, D., R. Santoro, R. W. Howarth, B. Hong, and K. Donaghy. 2012. Historical Changes in the Food and Water Supply Systems of the New York City Metropolitan Area. *Reg. Environ. Change* **12**:363–380. [14]

Taylor, P. L. 2005. In the Market but Not of It: Fair Trade Coffee and Forest Stewardship Council Certification as Market-Based Social Change. *World Devel.* **33**:129–147. [12]

TEEB. 2011. The Economics of Ecosystem Services and Biodiversity for International and National Policymakers. London: Earthscan. [15]

Teisl, M. F. 2003. What We May Have Is a Failure to Communicate: Labeling Environmentally Certified Forest Products. *Forest Sci.* **49**:668–680. [12]

Teisl, M. F., S. Peavey, F. Newman, J. Buono, and M. Hermann. 2002. Consumer Reactions to Environmental Labels for Forest Products: A Preliminary Look. *Forest Prod. J.* **52**:44–50. [12]

Teisl, M. F., B. Roe, and A. S. Levy. 1999. Ecocertification: Why It May Not Be a "Field of Dreams." *Am. J. Agric. Econ.* **81**:1066–1071. [12]

TheCityUK. 2012. Sovereign Wealth Funds 2012 Report. http://www.thecityuk.com/research/our-work/reports-list/sovereign-wealth-funds-2012/. (accessed June 2, 2012). [1]

Thornton, P. K., and P. K. Thornton. 2010. Livestock Production: Recent Trends, Future Prospects. *Phil. Trans. R. Soc. B.* **365**:2853–2867. [13]

Tilman, D. 1999. Global Environmental Impacts of Agricultural Expansion: The Need for Sustainable and Efficient Practices. *PNAS* **96**:5995–6000. [6]

Tilman, D., C. Balzer, J. Hill, and B. L. Befort. 2011. Global Food Demand and the Sustainable Intensification of Agriculture. *PNAS* **108**:20260–20264. [2–4]

Tilman, D., K. G. Cassman, P. A. Matson, R. Naylor, and S. Polasky. 2002. Agricultural Sustainability and Intensive Production Practices. *Nature* **418**:671–677. [4]

Toit, J. T., B. H. Walker, and B. M. Campbell. 2004. Conserving Tropical Nature: Current Challenges for Ecologists. *Trends Ecol. Evol.* **19**:12–17. [4]

Tollefson, C., F. P. Gale, and D. Haley. 2008. Setting the Standard: Certification, Governance, and the Forest Stewardship Council. Vancouver: UBC Press. [12]

Torell, L. A., N. R. Rimbey, O. A. Ramirez, and D. W. McCollum. 2005. Income Earning Potential versus Consumptive Amenities in Determining Ranchland Values. *J. Agricul. Res. Econ.* **30**:537–560. [13]

Touza, J., K. Dehnen-Schmutz, and G. Jones. 2007. Economic Analysis of Invasive Species Policies. In: Biological Invasions, ed. W. Nentwig, pp. 353–366. Berlin: Springer. [7]

Trenbreth, K. E., and J. W. Hurrell. 1994. Decadal Atmospheric-Ocean Variations in the Pacific. *Clim. Dynam.* **9**:303–319. [8]

Tsai, Y. 2005. Quantifying Urban Form: Compactness versus "Sprawl." *Urban Stud.* **42**:141–161. [15]

Tscharntke, T., Y. Clough, T. C. Wanger, et al. 2012. Global Food Security, Biodiversity Conservation and the Future of Agricultural Intensification. *Biol. Conserv.* **151**:53–59. [4]

Tsing, A. 2005. Friction: An Ethnography of Global Connection. Princeton: Princeton Univ. Press. [11]

Tukker, A., R. A. Goldbohm, A. de Koning, et al. 2011. Environmental Impacts of Changes to Healthier Diets in Europe. *Ecol. Econ.* **70**:1776–1788. [4]

Turner, B. L., II, and K. W. Butzer. 1992. The Columbian Encounter and Land Use Change. *Environment* **34**:16–44. [4]

Turner, B. L., II, A. Janetos, and P. Verburg. 2013. Land Systems Architecture: A Novel Strategy for Global Environmental Change and Sustainability Science and Policy. *Global Environ. Change* **23**:395–397. [4]

Turner, B. L., II, R. E. Kasperson, P. A. Matson, et al. 2003. A Framework for Vulnerability Analysis in Sustainability Science. *PNAS* **100**:8074–8079. [7]

Turner, B. L., II, E. Lambin, and A. Reenberg. 2007. The Emergence of Land Change Science for Global Environmental Change and Sustainability. *PNAS* **104**:20666–20671. [1, 7, 10]

Turner, J. 1997. The Institutional Order. New York: Longman. [4]

UBOS. 2012. Livestock Census Report. Kampala, Uganda: Uganda Bureau of Statistics. [16]

UN. 2010. World Urbanization Prospects, the 2010 Revision. New York: United Nations Dept. of Economic and Social Affairs, Population Division. [6]

———. 2012a. United Nations Commodity Trade Statistics Database, U.N. Statistics Division. http://comtrade.un.org/. (accessed Dec. 8, 2013). [6]

———. 2012b. World Urbanization Prospects, the 2011 Revision. Department of Economic and Social Affairs, Population Division. New York: United Nations. [14]

UN Committee on Economic, Social, and Cultural Rights. 1999. Substantive Issues Arising in the Implementation of the International Covenant on Economic, Social and Cultural Rights: General Comment 12: The Right to Adequate Food (Art. 11). http://www.unhchr.ch/tbs/doc.nsf/0/3d02758c707031d58025677f003b73b9. (accessed Dec. 8, 2013). [13]

UN Committee on Food Security. 2012. Voluntary Guidelines on the Responsible Governance of Tenure of Land, Fisheries and Forests in the Context of National Food Security. Rome: FAO. [13]

UN Committee on Human Rights. 2000. The Right to Food: Commission on Human Rights Resolution 2000/10. Geneva: OHCHR. [13]

UN-Habitat. 2009. Global Report on Human Settlements 2009, Planning Sustainable Cities. London: Earthscan. [6]

UN Human Rights Council. 2011a. Guiding Principles for Business and Human Rights. United Nations Human Rights Council. http://www.ohchr.org/documents/issues/business/A.HRC.17.31.pdf. (accessed Dec. 8, 2013). [9]

———. 2011b. Report of the Secretary-General: Role and Achievements of the UN High Commissioner for Human Rights in Assisting the Government and People of Cambodia in the Promotion and Protection of Human Rights (A/HRC/18/47). New York: U.N. General Assembly. [11]

———. 2012. Report of the Special Rapporteur on the Situation of Human Rights in Cambodia, Surya P. Subedi (A/HRC/21/63/Add.1). New York: U.N. General Assembly. [11]

UNCTAD. 1997. World Investment Report. New York: United Nations. [6]

———. 2007. The Universe of the Largest Transnational Corporations. New York: United Nations. [6]

———. 2010. World Investment Report. New York: United Nations. [6]

———. 2011. World Investment Report. New York: United Nations. [6]

UNCTADstat. 2012. Statistical Database. Intl. Trade in Goods and Services, Economics Trends, Martime Transport, etc. http://unctadstat.unctad.org/ReportFolders/reportFolders.aspx, (accessed Dec. 6, 2013). [6]

UNDP. 2006. National Human Development Report: International Trade and Human Development. Vientiane Capital: Committee for Planning and Investment National Statistics Centre. [7]

UNECE and FAO. 2012. Forest Products Annual Market Review 2011–2012. http://www.unece.org/fileadmin/DAM/timber/publications/FPAMR_2012.pdf. (accessed July 24, 2013). [12]

UNFCCC. 1992. United Nations Framework Convention on Climate Change. http://unfccc.int/resource/docs/convkp/conveng.pdf. (accessed Dec. 8, 2013). [13]

———. 2011. Appendix 1 to the UNFCCC Decision 1/CP.16: Guidance and Safeguards for Policy Approaches and Positive Incentives on Issues Relating to Reducing Emissions from Deforestation and Forest Degradation in Developing Countries; and the Role of Conservation, Sustainable Management of Forests and Enhancement of Forest Carbon Stocks in Developing Countries (UNFCCC/CP/2010/7/Add.1). Bonn: U.N. Framework Convention on Climate Change. [13]

Unmüssig, B., W. Sachs, and T. Fatheuer. 2012. Critique of the Green Economy. Toward Social and Environmental Equity. Berlin: Heinrich Böll Stiftung. [13]

UNOHCHR. 2004. Land Concessions for Economic Purposes in Cambodia: A Human Rights Perspective. Phnom Penh: U.N. Office of the High Comissioner for Human Rights in Cambodia. [11]

———. 2007. Economic Land Concessions in Cambodia: A Human Rights Perspective. Phnom Penh: U.N. Office of the High Comissioner for Human Rights in Cambodia. [11]

Urioste, M. 2012. Concentration and "Foreignisation" of Land in Bolivia. *Can. J. Dev. Stud.* **33**:439–457. [11]

USDA. 2009. Indonesia: Palm Oil Production Growth to Continue. Commodity Intelligence Report, U.S. Dept. of Agriculture, Foreign Agricultural Service. http://www.pecad.fas.usda.gov/highlights/2009/03/Indonesia/. (accessed Dec. 8, 2013). [9]

———. 2012. Oilseeds: World Markets and Trade. U.S. Dept. of Agriculture, Foreign Agricultural Service. http://www.fas.usda.gov/oilseeds/Current/. (accessed Dec. 8, 2013). [9]

van Asselen, S., and P. H. Verburg. 2012. A Land System Representation for Global Assessments and Land-Use Modeling. *Global Change Biol.* **18**:3125–3148. [4]

Vandergeest, P. 2007. Certification and Communities: Alternatives for Regulating the Environmental and Social Impacts of Shrimp Farming. *World Devel.* **35**:1152–1171. [12]

van der Horst, D., and S. Vermeylen. 2011. Spatial Scale and Social Impacts of Biofuel Production. *Biomass Bioenergy* **35**:2435–2443. [10]

van Gelder, J. W., and L. German. 2011. Biofuel Finance: Global Trends in Biofuel Finance in Forest-Rich Countries of Asia, Africa and Latin America and Implications for Governance. *CIFOR InfoBrief* http://www.cifor.org/publications/pdf_files/infobrief/3340-infobrief.pdf. (accessed Dec. 8, 2013). [9]

van Vliet, N., O. Mertz, A. Heinimann, et al. 2012. Trends, Drivers and Impacts of Changes in Swidden Cultivation in Tropical Forest-Agriculture Frontiers: A Global Assessment. *Global Environ. Change* **22**:418–429. [2]

van Vuuren, D. P., J. Edmonds, M. Kainuma, et al. 2011. The Representative Concentration Pathways: An Overview. *Clim. Change* **109**:5–31. [4]

van Vuuren, D. P., M. T. J. Kok, B. Girod, P. L. Lucas, and B. de Vries. 2012. Scenarios in Global Environmental Assessments: Key Characteristics and Lessons for Future Use. *Global Environ. Change* **22**:884–895. [4]

van Vuuren, D. P., J. van Vliet, and E. Stehfest. 2009. Future Bioenergy Potential under Various Natural Constraints. *Energy Policy* **37**:4220–4230. [4]

Verburg, P. H., B. Eickhout, and H. van Meijl. 2008. A Multi-Scale, Multi-Model Approach for Analyzing the Future Dynamics of European Land Use. *Ann. Reg. Sci.* **42**:57–77. [4]

Visser, O., and M. Spoor. 2011. Land Grabbing in Post-Soviet Eurasia: The World's Largest Agricultural Land Reserves at Stake. *J. Peasant Stud.* **38**:299–323. [10]

Vitousek, P. M. 1997. Human Domination of Earth's Ecosystems. *Science* **277**:494–499. [15]

Vitousek, P. M., C. M. D'Antonio, L. L. Loope, M. Rejmanek, and R. Westbrooks. 1997. Introduced Species: A Significant Component of Human-Caused Global Change. *NZ J. Ecol.* **21**:1–16. [7]

Vitousek, P. M., P. R. Ehrlich, A. H. Ehrlich, and P. A. Matson. 1986. Human Appropriation of the Products of Photosynthesis. *Bioscience* **36**:363–373. [4]

Vitousek, P. M., R. Naylor, T. Crews, et al. 2009. Nutrient Imbalances in Agricultural Development. *Science* **324**:1519–1520. [3]

Vogel, D. 2008a. Private Global Business Regulation. *Annu. Rev. Polit. Sci.* **11**:261–282. [12]

Vogel, G. 2008b. Upending the Traditional Farm. *Science* **319**:752–753. [15]

Von Maltitz, G., and W. Stafford. 2011. Assessing Opportunities and Constraints for Biofuel Development in Sub-Saharan Africa. *CIFOR Working Paper* http://www.cifor.org/publications/pdf_files/WPapers/WP58CIFOR.pdf. (accessed Dec. 8, 2013). [9]

von Thünen, J. H. 1826. Der Isolierte Staat in Beziehung Auf Die Landwirtschaft und National Ökonomie, Erster Teil. Jena: Gustav Fischer. [14]
Wade, R. H. 2001. The Rising Inequality of World Income Distribution. *Finan. Dev.* 38, http://www.imf.org/external/pubs/ft/fandd/2001/12/wade.htm. (accessed Dec. 8, 2013). [6]
Waha, K., L. G. J. van Bussel, C. Müller, and A. Bondeau. 2012. Climate-Driven Simulation of Global Crop Sowing Dates. *Global Ecol. Biogeogr.* **21**:247–259. [4]
Wakker, E. 2004. Greasy Palms: The Social and Ecological Impacts of Large-Scale Oil Palm Plantation Development in Southeast Asia. London: Friends of the Earth. [9]
Walker, B., C. S. Holling, S. R. Carpenter, and A. Kinzig. 2004. Resilience, Adaptability and Transformability in Social-Ecological Systems. *Ecol. Soc.* **9**:5. [7]
Walker, P. A. 2003. Reconsidering "Regional" Political Ecologies: Toward a Political Ecology of the Rural American West. *Prog. Human Geogr.* **27**:7–24. [13]
Walker, P. A., and L. Fortmann. 2003. Whose Landscape? A Political Ecology of the "Exurban" Sierra. *Cult. Geograph.* **10**:469–491. [13]
Wang, H. J., Q. Q. He, X. J. Liu, Y. H. Zhuang, and S. Hong. 2012a. Global Urbanization Research from 1991 to 2009: A Systematic Research Review. *Landsc. Urban Plan* **104**:299–309. [5]
Wang, J., Y. Q. Chen, X. M. Shao, Y. Y. Zhang, and Y. G. Cao. 2012b. Land-Use Changes and Policy Dimension Driving Forces in China: Present, Trend and Future. *Land Use Policy* **29**:737–749. [5]
Warren, R. 2011. The Role of Interactions in a World Implementing Adaptation and Mitigation Solutions to Climate Change. *Phil. Trans. R. Soc. A* **369**:217–241. [4]
WBGU. 2009. Future Bioenergy and Sustainable Land Use. London: Earthscan. [4]
WCED. 1987. World Commission on Environment and Development. Our Common Future. Oxford: Oxford Univ. Press. [13, 16]
Webb, M. C. 1999. Private and Public Management of International Mineral Markets. In: Private Authority and International Affairs, ed. A. C. Cutler et al., J. N. Rosenau, series ed., pp. 53–96. Albany: SUNY Press. [12]
Weber, C. L., and H. S. Matthews. 2008. Food-Miles and the Relative Climate Impacts of Food Choices in the United States. *Environ. Sci. Tech.* **42**:3508–3513. [4]
Weis, T. 2010. Our Ecological Hoofprint and the Population Bomb of Reverse Protein Factories. *Review* **33**:131–152. [13]
Weng, Q., H. Liu, and D. Lu. 2007. Assessing the Effects of Land Use and Land Cover Patterns on Thermal Conditions Using Landscape Metrics in City of Indianapolis, United States. *Urban Ecosyst.* **10**:203–219. [15]
Wessolek, G. 2008. Sealing of Soils. In: Urban Ecology, ed. J. Marzluff et al., pp. 161–179. New York: Springer. [15]
West, P. C., H. K. Gibbs, C. Monfreda, et al. 2010. Trading Carbon for Food: Global Comparison of Carbon Stocks vs. Crop Yields on Agricultural Land. *PNAS* **107**:19645–19648. [13]
Westerink, J., D. Haase, A. Bauer, et al. 2012. Expressions of the Compact City Paradigm in Peri-Urban Planning across European City Regions: How Do Planners Deal with Sustainability Trade-Offs? *Eur. Plann. Stud.* **25**:1–25. [15]
Westphal, M., M. Browne, K. MacKinnon, and I. Noble. 2008. The Link between International Trade and the Global Distribution of Invasive Alien Species. *Biol. Inv.* **10**:391–398. [7]
White, B., and A. Dasgupta. 2010. Agrofuels Capitalism: A View from Political Economy. *J. Peasant Stud.* **37**:593–607. [11]

Wicke, B., R. Sikkema, V. Dornburg, and A. Faaij. 2011. Exploring Land Use Changes and the Role of Palm Oil Production in Indonesia and Malaysia. *Land Use Policy* **28**:193–206. [9]

Wicke, B., P. Verweij, H. Van Meijl, D. P. Van Vuuren, and A. P. C. Faaij. 2012. Indirect Land Use Change: Review of Existing Models and Strategies for Mitigation. *Biofuels* **3**:87–100. [9]

Wilbanks, T. J., and R. W. Kates. 1999. Global Change in Local Places: How Scale Matters. *Clim. Change* **43**:601–628. [6]

Wilcove, D. S., D. Rothstein, D. Jason, A. Phillips, and E. Losos. 1998. Quantifying Threats to Imperiled Species in the United States. *Bioscience* **48**:607–615. [7]

Williams, P. 2002. Transnational Organized Crime and the State. In: The Emergence of Private Authority in Global Governance, ed. R. B. Hall and T. J. Biersteker, vol. 85, pp. 161–182. Cambridge: Cambridge Univ. Press. [12]

Wirsenius, S., C. Azar, and G. Berndes. 2010. How Much Land Is Needed for Global Food Production under Scenarios of Dietary Changes and Livestock Productivity Increases in 2030? *Agric. Syst.* **103**:621–638. [2, 4]

Wolford, W., S. M. Borras, Jr., R. Hall, I. Scoones, and B. White. 2013. Governing the Global Land Grab: The Role of the State in the Rush for Land. *Dev. Change* **44**:189–210. [11]

Woodhouse, P. 2012. New Investment, Old Challenges. Land Deals and the Water Constraint in African Agriculture. *J. Peasant Stud.* **39**:777–794. [11]

World Bank. 2003. Striking a Better Balance: Extractive Industries Review Reports. The World Bank Group and Extractive Industries: Final Report of the Extractive Industries Review. http://web.worldbank.org/WBSITE/EXTERNAL/TOPICS/EXTOGMC/0,,contentMDK:20306686~menuPK:592071~pagePK:148956~piPK:216618~theSitePK:336930,00.html. (accessed Dec. 8, 2013). [13]

———. 2009. Large-Scale Acquisition of Land Rights for Agricultural or Natural Resource-Based Use. Washington, D.C.: World Bank. [10]

———. 2010. Rising Global Interest in Farmland: Can It Yield Sustainable and Equitable Benefits? Washington, D.C.: The World Bank. [2]

———. 2012. World Bank Indicators. http://data.worldbank.org/indicator. (accessed Dec. 8, 2013). [6]

World Bank, FAO, IFAD, and UNCTAD. 2010. Principles for Responsible Agricultural Investment That Respects Rights, Livelihoods and Resources: Discussion Note Prepared to Contribute to an Ongoing Global Dialogue. http://siteresources.worldbank.org/INTARD/214574-1111138388661/22453321/Principles_Extended.pdf. (accessed Oct. 28, 2013). [13]

Worldwatch Institute. 2011. Biofuels Regain Momentum. Vital Signs Online. http://vitalsigns.worldwatch.org/vs-trend/biofuels-regain-momentum. (accessed Oct. 20, 2012). [13]

WTO. 2011. Trade Patterns and Global Value Chains in East Asia: From Trade in Goods to Trade in Tasks. http://www.wto.org/english/res_e/publications_e/stat_tradepat_globvalchains_e.htm. (accessed Nov. 24, 2013). [14]

Wu, K. Y., and H. Zhang. 2012. Land Use Dynamics, Built-up Land Expansion Patterns, and Driving Forces Analysis of the Fast-Growing Hangzhou Metropolitan Area, Eastern China (1978–2008). *Appl. Geog.* **34**:137–145. [5]

Wu, Y. Z., X. L. Zhang, and L. Y. Shen. 2011. The Impact of Urbanization Policy on Land Use Change: A Scenario Analysis. *Cities* **28**:147–159. [5]

Wunder, S., S. Engel, and S. Pagiola. 2008. Taking Stock: A Comparative Analysis of Payments for Environmental Services Programs in Developed and Developing Countries. *Ecol. Econ.* **65**:834–852. [4]

WWF. 2012. Roundtable on Sustainable Palm Oil. http://wwf.panda.org/what_we_do/footprint/agriculture/palm_oil/solutions/roundtable_on_sustainable_palm_oil/. (accessed May 15, 2013). [9]

Wycherly, R. E. 1962. How the Greeks Built Cities. New York: W. W. Norton. [14]

Xanthaki, A. 2010. Indigenous Rights in International Law over the Last 10 Years and Future Developments. *Melb. J. Intl. Law* **10**:3. [10]

Xiang, W. N., R. M. B. Stuber, and X. C. Meng. 2011. Meeting Critical Challenges and Striving for Urban Sustainability in China. *Landsc. Urban Plan* **100**:418–420. [5]

Xu, Y., Q. Tang, J. Fan, S. J. Bennett, and Y. Li. 2011. Assessing Construction Land Potential and Its Spatial Pattern in China. *Landsc. Urban Plan* **103**:207–216. [5]

Yang, H. 2004. Land Conservation Campaign in China: Integrated Management, Local Participation and Food Supply Option. *Geoforum* **35**:507–518. [5]

You, L. Z., M. Spoor, J. Ulimwengu, and S. M. Zhang. 2011. Land Use Change and Environmental Stress of Wheat, Rice and Corn Production in China. *China Econ. Rev.* **22**:461–473. [5]

Young, A. 1999a. Is There Really Spare Land? A Critique of Estimates of Available Cultivable Land in Developing Countries. *Environ. Devel. Sustain.* **1**:3–18. [2]

Young, O. R. 1999b. Governance in World Affairs. Ithaca: Cornell Univ. Press. [12]

———. 2002. The Institutional Dimensions of Environmental Change: Fit, Interplay, and Scale. Cambridge, MA: MIT Press. [10, 12]

Young, O. R., F. Berkhout, G. C. Gallopin, et al. 2006. The Globalization of Socio-Ecological Systems: An Agenda for Scientific Research. *Global Environ. Change* **16**:304–316. [1]

Yue, Y. H., Q. H. Ran, C. M. Sun, and D. T. Xie. 2010. Grey Predication of Cultivated Land Change and Analysis of Farmland Protection in China. *Geogr. Geo-Info. Sci.* **26**:56–59. [5]

Zadek, S. 2001. The Civil Corporation: The New Economy of Corporate Citizenship. London: Earthscan. [12]

Zak, M. R., M. Cabido, D. Cáceres, and S. Díaz. 2008. What Drives Accelertaed Land Cover Change in Central America? Synergistic Consequences of Climatic, Socioeconomic, and Technological Factors. *Environ. Manag.* **42**:181–189. [4]

Zevin, R. 1992. Are World Financial Markets More Open? If So, Why and with What Effects? In: Financial Openness and National Autonomy, ed. T. Banuri and J. B. Schor, pp. 43–83. Oxford: Clarendon Press. [6]

Zhang, J. K., F. R. Zhang, D. Zhang, et al. 2008. The Grain Potential of Cultivated Lands in Mainland China in 2004. *Land Use Policy* **26**:68–76. [5]

Zhang, J. Y., and S. F. Jia. 2005. Mechanism of Population Urbanization Influence on Cultivated Land Changes: A Comparative Study between Japan and China. *China Pop. Res. Environ.* **15**:26–31. [5]

Zhang, Q., J. H. Jin, K. Zhang, and Z. C. Xu. 2007. Empirical Analysis and Enlightenment of the Relationship between Land Use and Economic Development in Japan and Korea. *Resources Sci.* **29**:149–155. [5]

Zhao, M., and S. W. Running. 2010. Drought-Induced Reduction in Global Terrestrial Net Primary Production from 2000 through 2009. *Science* **329**:940–943. [4]

Zhao, P. J. 2011. Managing Urban Growth in a Transforming China: Evidence from Beijing. *Land Use Policy* **28**:96–109. [5]

Zhou, W., G. Huang, and M. L. Cadenasso. 2011. Does Spatial Configuration Matter? Understanding the Effects of Land Cover Pattern on Land Surface Temperature in Urban Landscapes. *Landsc. Urban Plan* **102**:54–63. [4]

Zhu, J., L.-L. Sim, and X.-Q. Zhang. 2006. Global Real Estate Investments and Local Cultural Capital in the Making of Shanghai's New Office Locations. *Habitat Intl.* **30**:462–448. [1]

Ziegler, A., J. Fox, and J. Xu. 2009. The Rubber Juggernaut. *Science* **324**:1024–1025. [4]

Zimmerer, K. S. 2010. Biological Diversity in Agriculture and Global Change. *Annu. Rev. Environ. Res.* **35**:137–166. [8]

Zomer, R. J., A. Trabucco, R. Coe, and F. Place. 2009. Trees on Farm: Analysis of Global Extent and Geographical Patterns of Agroforestry. ICRAF Working Paper No. 89. Nairobi: World Agroforestry Centre. [4]

Zoomers, A. 2010. Globalisation and the Foreignisation of Space: Seven Processes Driving the Current Global Land Grab. *J. Peasant Stud.* **37**:429–447. [10, 13]

Subject Index

ActionAid 188, 190
actor theory-based approach 155
agrarian-industrial transition 58, 60
agri-business 43, 44, 48, 129, 172, 335
agricultural labor 78, 79
  migration of 78, 80, 82–85
agriculture 11, 14, 16, 42, 61
  flex crops 206
  intensification of 11, 16, 17, 23, 24, 37, 50, 60, 113, 173
  investment in 193–195
  mechanized 16, 17, 19, 37, 43, 76, 79
  organic 61, 64, 67, 226, 257
  peri-urban 28, 49
  rainfed 12, 41, 130
  sustainable 31, 32, 129, 256, 257, 266
agrobiodiversity 148, 152
agroforestry 18, 63
air freight 287

Bangalore, India 322, 323
Better Sugarcane Initiative (BSI) 229, 230
biodiesel 145, 166, 171–173, 208
biodiversity 1, 12, 22, 24, 42, 45, 52, 63–65, 68, 246, 248, 259, 268, 307
  agro- 148, 152
  conservation 13, 26, 35–39, 249, 269, 337
  governance conflicts 265
  loss of 26, 44, 59, 120, 128, 130, 147, 164, 172, 173
bioenergy 11, 15, 18, 19, 35, 38, 40, 43–45, 48, 50, 60, 68
  ethanol 28, 40, 145, 206, 256
  palm oil-based 171, 174, 179, 180
  production 51, 52, 61–64, 197
biofuel 14, 18, 123, 170–173, 187, 189, 206, 208, 255, 261
  governance 206, 260–262
  mandates 22, 145, 146
  production 4, 13, 22, 28, 62, 67, 153
BREEAM 231, 232

brownfields 301, 304
built environment 42, 45, 277, 286, 294, 304–306, 317, 319, 325, 329
  density 275, 311, 315

carbon 12, 248
  forest 198, 255, 262–265, 337
  markets 41, 47, 151
  sequestration 1, 13, 19, 35, 38, 42, 62–64, 67, 68
  tax 164, 180
Castells' network society 275, 276, 291, 296, 297
central place theory (CPT) 275, 276, 279–285, 293, 295, 297
certification 160, 163, 174, 227, 231–233, 247, 261, 262, 336. *See also* Forest Stewardship Council
  eco- 21, 160, 246
  Program for the Endorsement of Forest Certification 229, 230, 233
  Utz Certified 229, 230
Chaco soybean 36, 40
cheatgrass 133–136
climate change 2, 14, 22, 27, 32, 33, 35, 43, 59, 135, 171, 269, 297, 298, 310, 337
  mitigating 36, 63–67, 203, 260, 261, 264
  role of agriculture 189, 190
coffee 44, 148, 229, 230, 232, 256
collective decision making 240, 243, 244, 269
communication 2, 144, 156, 285, 291, 296
compact city 275, 292–294, 297, 307, 308, 310
competition 11–22, 24–28, 35–70, 145, 239, 241, 247, 251, 255, 269, 337
  in China 71–86
  defined 37
  impact on sustainability 49–52
  managing 52–58

competition (continued)
  mapping 47–49
  trends 38–42
  types of 42–47, 68
connectivity 144, 314, 315, 318–322, 325–328, 330
conservation 15, 21, 39, 44–51, 128, 152, 268. *See also* biodiversity
  defined 42
  financing 266
  forest 1, 2, 16, 35, 38, 264
  land demand for 63–65
consumer 224–227
  behavior 149, 154, 158, 281
  demand 66, 142, 145, 179, 180
  responsibility 160, 175, 178, 179, 257
containerization 96, 286, 287
continuum of urbanity 318, 327, 329, 330
control grabbing 185, 202, 206
corn. *See* maize
cotton 73, 230
coupled human-natural systems 119–122, 138
cropland 12–15, 19, 22, 31, 32, 44, 58, 60, 145
  conversion 127
  expansion 12, 15, 23, 112
crop production 12, 18, 25–27, 32, 84, 112, 145, 230. *See also* food production
cultural ecosystem services 42, 253, 266

deforestation 2, 12, 16–20, 45, 50, 51, 65, 108, 110, 113, 114, 164, 172, 197
  mitigating 60, 137, 198, 263
  tropical 16, 19, 40, 163, 171, 179
deindustrialization 114, 283, 304
derivatives 104, 107, 108
diet 14, 58, 67, 170, 179, 301, 330
  meat consumption 1, 4, 40, 43, 56, 61, 171, 256, 322
  vegetarian 15, 22, 61
diffusion of technology 170, 171, 179
distal land connections 4, 119, 120, 163–180, 334–336
  defined 142, 143, 164

eco-certification 21, 160, 246
eco-labeling 110, 228, 230
ecological integrity 143, 257, 314–317, 325–330
ecological risk 189–191, 195, 198
economic development 28, 50, 51, 72, 78, 107, 199, 284, 306, 309
  phases of 91, 92
  role of trade in 95
  uneven 303
economic flows 90, 93, 95–107, 108, 160, 289, 336
  impact on land use 109–112
economic integration. *See* integration of economies
economic spatial theories 275, 279–285
ecosystem services 2, 25, 277, 297, 309, 310, 338
  cultural 253, 266
  demand for 14, 31, 36
  loss in 19, 36, 130, 145
  nonprovisioning 35, 38–45, 50, 51, 63, 68
  paradigm 268, 269
  payments for 21, 38, 41, 47, 53, 68, 125, 193, 249
edge cities 285, 286
edgeless cities 285, 286
emergent global land governance 183, 184, 190–193, 198–200
  policy instruments 193–198
energy 16, 28, 38, 152. *See also* bioenergy, fossil fuels
  consumption 171, 330
  demand 40, 61, 62, 120, 185, 204
  green 189
  nuclear 40, 61, 152
  renewable 18, 61, 67
  security 203, 205, 260, 261
  solar 40
equity 5, 35, 50, 54, 66, 104, 106, 224, 233, 236, 264, 314, 317, 325, 326, 329, 334, 338
  intragenerational 269
e-Stewards 217, 231
ethanol 19, 28, 40, 43, 145, 206, 256
ethics 251, 268, 269, 324–326
  new land 270, 313, 314, 329, 330

exclusion 188–191, 195, 198, 261
Extractive Industries Transparency
    Initiative 194, 195

fair trade 148, 191, 235, 246
farming. *See* agriculture
feed 11–14, 18–21, 37, 38, 42–48, 52, 173
fertilizers 16, 25, 29–32, 37, 43, 61, 83, 113
fire 133–135, 146, 164, 173
food crisis 30, 83, 127, 141, 185, 203, 204, 253, 256, 269
food miles 57, 129, 158
food production 1, 3, 16, 21, 23–34, 43–51, 57, 64, 67, 205, 257, 294
    in China 71–86
    governance conflicts 256–259
    impact of urbanization 80–84, 127
    local 18, 42
food security 3, 22, 23, 24, 28–30, 33, 51, 57, 58, 62, 71, 75, 80–83, 120, 131, 143, 148, 194–196, 203–205, 249, 255–257, 261, 267, 270
foreign direct investment (FDI) 3, 20, 93, 95, 99–103, 109, 110, 114, 205, 289
forest 11, 15–21, 31, 62, 63, 67, 114
    carbon 198, 255, 262–265, 337
    conservation 1, 2, 16, 35, 38, 44–46, 264
    degradation 17, 44
    global governance of 193, 197
    management 4, 72–74, 198, 199, 217, 240, 262
    reforestation 17, 21, 147, 263
forestry 28, 41, 199, 233
    agro- 18, 63
    industrial 13, 14, 73
    intensification 17, 18
    timber extraction 42–48, 51, 168
Forest Stewardship Council (FSC) 197, 217, 223, 226, 227, 229–231, 233, 235, 240
forest transition 17, 18, 40, 111–116
    defined 109

fossil fuels 35, 40, 43, 50, 61, 62, 67, 171, 260
future research 3, 21, 22, 53, 63, 64, 68, 138, 143, 151, 153, 160, 224, 242, 253, 255, 270–272

GlobalG.A.P. 230, 234
globalization 109–112, 121, 143–145, 168–173, 178, 179, 276, 335
    historic perspective 91–95
global land governance 5, 58, 163, 178, 180, 214, 247, 255, 258
    demand for 188–190
    emergent 183, 184, 190–200
    role of knowledge 250–253
global land-use estimates 12–16
governance 3, 7, 30, 37, 52, 66, 239–272, 275, 277, 295, 298, 310, 336, 337. *See also* global land governance
    actor involvement 149–151, 205, 245–247
    challenges 56–58, 298
    conceptual framework 242–248
    conflicts 256–269
    impact of urbanization 239–241
    nonstate market-driven 219, 220
    role in telecoupling 137, 143, 147
    role of economic instruments 41, 47
    role of private regulations 217–238
green economy 241, 249, 250
green grabs 185, 197, 266, 269, 270
greenhouse gas emissions 24, 32, 50, 56–58, 65, 164, 171, 173, 189
    reducing 31, 51, 60–63, 249, 260, 261, 297
green space 304, 306–308, 311

housing 1, 75, 83, 124, 127, 280, 293, 299, 304, 307, 308, 322
    residential density 294, 304
    second homes 67, 266, 301, 321
human-natural systems 119–122, 138
human rights 4, 30, 154, 193, 196, 250, 258, 265, 268. *See also* indigenous rights
    UN Guiding Principles for Business and 178

human well-being. *See* well-being
hydrocarbons 255, 258–260

Imperata grasslands 175, 176
income gap 75–78, 301
income growth 1, 11, 17, 20, 43, 50, 51, 61, 170, 171, 256
indigenous rights 154, 246, 264, 268–270
indirect land-use change 19, 22, 51, 145, 146, 173, 174, 261, 262
industrial production model 289, 290
information 144, 149, 156, 170, 174
 systems 159, 188
 technology 276, 281, 284, 286, 288, 289, 296, 323
innovation 50, 124, 145, 159, 281, 288, 291, 325, 326
institutions 3, 7, 211, 325, 330
 defined 52
 managing complexity 66, 67
 role in telecoupling 147
 types 53–56
integration of economies 1, 3, 4, 6, 95, 168–173, 178, 179, 218, 227, 333
International Labor Organization Convention 259, 268, 269
International Land Coalition 190, 255
investment in land 193–199, 203, 213, 241, 243, 260, 321, 326
 large-scale 205, 213
irrigation 25, 26, 29, 41, 83, 276

jatropha 208, 212
justice 5, 149, 188, 196, 264, 268, 269, 326, 338

knowledge 144, 156, 170–172, 176, 177, 271
 creation 145, 243, 326
 networks 159, 160, 253
 role in governance 250–253
 ways of knowing 243, 244, 255, 263, 268, 269
Kunming, China 76–79

labor 76–80, 226, 281
 division of 92, 114, 303
 migration 78, 80, 82–85, 169–171, 179
 rights 164, 173
Lake Taupo, New Zealand 54
land architecture 35, 50, 65, 68
land conversion 12, 13, 16, 19–22, 28, 78, 124, 145, 152, 297
land degradation 14, 21, 22, 24, 110, 131, 148
 global assessment 30, 31
land ethic 268, 270, 313, 314, 329, 330
land grabbing 4, 6, 43, 111, 114, 128, 183–187, 196–198, 201–206, 210–212, 243, 249–253, 256, 257
 characteristics 186–188
 defining 185, 202, 203, 254
land management 4, 15, 25, 52, 53, 61, 62, 65, 82, 93, 110, 111, 165, 178, 249, 264, 267, 337
Land Matrix Project 130, 186, 187
landscape amenities 42, 68, 239, 255, 266, 267, 281, 329
land sparing 6, 21, 26, 27, 49, 51
land teleconnections. *See* teleconnections
land tenure 110, 164, 173, 193–197, 198, 199, 208, 241, 262, 270
land-use competition. *See* competition
land-use governance. *See* governance
land-use intensification 17, 19, 21, 23, 32, 64, 112, 116, 148
land-use perforation 301, 305
land-use transitions 41, 89, 107–116
 defined 90
Laos 44, 129, 130
large-scale land acquisitions 196, 253, 261
 defined 253
large-scale land transactions 196, 201–216
LEED standard 223, 235
life cycle analysis 157, 158
lifestyle 1, 239, 241, 309, 314, 317–319, 322, 326, 330
 rural 51, 321, 323
 urban 17, 39, 157, 240, 313, 315
livelihood 1, 20, 39, 49–51, 128, 147, 194, 314–323, 326, 330

livestock production 14, 19, 28, 29, 38, 43, 126, 135, 176, 205, 209, 248, 257
local couplings 137, 138
location theory 279–281
logistics 276, 285–289, 296

maize 25–28, 40, 43, 44, 47, 67
Marine Stewardship Council (MSC) 223, 227
maritime shipping 96, 283, 286, 287
market-based regulations 217–235
market capitalization 103, 104, 108
meat consumption 1, 4, 40, 43, 56, 61, 171, 256, 322
mega-cities 39, 278, 291, 295–297, 300, 302, 306, 307
metropolitanization 285, 296
migration 79, 107, 119, 124, 125, 132, 146, 148, 152, 169, 173, 179, 267, 303, 317, 321
 rural to urban 2, 28, 39, 49, 80, 84, 85
mining 42, 45–47, 51, 126, 230, 267
 governance conflicts 258–260
multinational corporations 167, 172, 180, 244

network society 275, 276, 291, 296, 297
new economic geography (NEG) 283, 295, 297
new land ethic 270, 313, 314, 329, 330
New Urbanism 292, 293
NGOs 21, 38, 42, 45, 55, 57, 68, 108, 149–151, 167, 174, 183, 220, 225–227, 245, 264
 international 188–190
nitrogen 29–32, 113
nonprovisioning ecosystem services 35, 38–45, 50, 51, 63, 68
nuclear energy 40, 61, 152

official development assistance 104, 107, 108
organic agriculture 61, 64, 67, 226, 257
Oxfam 183, 188

palm oil 19, 40, 44, 67, 145, 163–180, 206, 209, 213, 256
 producer responsibility 175, 178
 production 98, 165–169, 172, 175
 yields 176–179
pasture expansion 14, 15, 19, 43, 50, 58
payments for ecosystem services (PES) 21, 38, 41, 47, 53, 68, 125, 193, 249
peatland degradation 164, 171, 172
peri-urban areas 29, 39, 46, 110, 255, 278, 279, 301, 304–307
 industry in 114
pesticides 16, 37, 43, 50, 61, 83, 124, 125, 132
phosphorus 26, 29, 30, 113
population 39
 aging 303, 304, 308, 311
 decline 84, 301–304
 density 127, 169, 188, 251, 275, 277, 295, 299, 302, 305, 306
population growth 2, 4, 11, 14, 17, 24–28, 67, 72, 82, 170, 179, 256, 257, 301
 link to GDP 58–60, 75, 91, 94, 308
 urban 17, 39, 46, 74, 76, 83, 112, 127, 277, 279, 300
precipitation 12, 27, 40, 62, 65, 252
Principles on Responsible Agriculture Investment (PRAI) 193–195
private market-based regulations 217–238, 243
 characteristics 219–228
Process tracing 155
production
 globalization of 290
 vertical integration of 289, 296
product labeling 110, 178, 191, 223, 224, 228, 230, 235
product life cycle 229
Program for the Endorsement of Forest Certification (PEFC) 229, 230, 233

quality of life 266, 281, 301–303, 307, 310
quinoa 154

rainfall. *See* precipitation

rapeseed oil 19, 145, 173
rebound effect 37, 50, 176–178, 180, 232
REDD 48, 137, 175, 197, 263
REDD+ 17, 21, 22, 38, 44, 45, 53, 128, 158, 197, 198, 247, 263, 264
reforestation 17, 21, 147, 263
remittances 20, 89, 95, 107–109, 148, 152, 319
renewable energy 18, 61, 67
  directive 145, 146, 153, 174
revalorization 239, 243, 244, 248–250, 256–259, 261–264, 267, 268, 271, 334, 337
Rio+20 268, 269
rubber 44, 98, 111, 127, 129
rural hinterland 275–277, 283–286, 293, 295, 297, 307, 309
rural-urban income gap 75–78

sacredness 255, 267–270
second homes 67, 266, 301, 321
Shanghai, China 77–80
shrinking cities 299, 301, 304, 305, 310
smoke pollution 164, 172
social-ecological systems 119, 120, 142, 146, 335
social equity. *See* equity
social networks 142, 146, 149, 150, 154, 155, 158–160, 245, 336
social values 239, 240, 257, 337
socioeconomic development 71, 72
soil 42
  degradation 2, 30, 31, 50, 59
  erosion 30, 42, 61, 72, 128, 152, 248
  fertility 26, 28, 32, 49, 248
  pollution 16, 46
  sealing 301, 306
solar power 40
sovereign wealth funds 3, 111, 187, 258
soybean 4, 19, 43, 44, 67, 112, 206
  Chaco 36, 40
spatial equilibrium theories 279–281
species invasion 119, 125, 131–133, 138
stewardship 31, 137, 249, 250, 267, 268, 317, 329, 330. *See also* Forest Stewardship Council

e-Stewards 217, 231
  Marine Stewardship Council 223, 227
suburbanization 6, 46, 285, 292, 296, 304
sugarcane 40, 43, 44, 145, 206
  Better Sugarcane Initiative 229, 230
sustainability 5, 6, 36, 194, 320, 324–328, 325, 334, 338, 339
  challenges 4, 59, 68, 69, 313
  compact city model 292–294
  impact of land-use competition 37, 49–52, 55, 66
  palm oil-based bioenergy 171, 174, 179, 180
  standards 21, 160, 174, 256, 261, 336
sustainable agriculture 31, 32, 129, 256, 257, 266
sustainable urbanity 297, 298, 324–329

teleconnections 2–6, 35, 40, 68, 117, 121, 149, 189, 192, 251, 314–317, 321, 322, 330, 335
  defined 142
  framework 121, 327
  impact on institutions 66, 67
  urban land 2, 4, 17, 90, 117, 301, 302, 309
telecoupling 39, 119, 121, 141–162, 245, 247, 269, 335, 336
  concept 144–146
  defined 142, 143
  drivers of 39
  examples of 124, 129, 134–137, 148
  framework 120–123, 146–153, 175
  palm oil system 166–175
  role of governance 239–241
  systems conditions 151–153
temperature 12, 26, 27, 40, 51, 62, 64, 65
territory 267–270, 324
  dismantling divisions 89, 104, 192, 239, 240, 334, 336
timber extraction 42–48, 51, 168
time and space interdependencies 334, 335
timing of land-use change 114–116

tourism 39, 124, 126, 148, 228, 231, 266, 309, 321, 322
  vacation homes 67, 266, 301, 321
trade 2, 28, 47, 54, 95–99, 108, 109, 119, 132, 133, 136, 172, 241, 261, 276, 281, 296
  barriers 93, 94, 288
  globalization of 286, 288, 290
  multilateral 21
  openness ratio 288
  palm oil 168, 169, 171
  virtual 20, 137, 188
tragedy of the commons 53
transit-oriented development (TOD) 292–294
transnational corporations 92, 94, 99, 100, 103, 107, 110, 114, 150, 180, 191, 193, 246, 256, 257
transnational economies 90, 92, 214
transnational land deals 16, 119, 129, 130, 138, 207, 208. *See also* large-scale land transactions
  telecoupling components 126–128
transportation 57, 92, 280, 285, 289, 291, 299
  advances in 113, 120, 288, 289
  air freight 287
  automobile 292, 293, 297
  costs 94, 96, 283, 284
  infrastructure 172, 276, 277, 283, 285, 296
  ocean freight 96, 283, 286, 287
  patterns 302, 307
  rail 135, 285, 286, 296
tree plantations 11, 15, 17, 18, 44, 46, 48, 51
tropical deforestation 16, 19, 40, 163, 171, 179

UN Framework Convention on Climate Change (UNFCCC) 244, 263, 264
UN Permanent Forum on Indigenous Issues (UNPFII) 268, 269
urban areas 302, 310
  conceptualizing 277, 278, 315–317
  spatial form 307–309
urban economics 279, 295

urban expansion 14, 18, 127, 148, 279, 300
  in China 75, 76–78, 83, 85
urban governance 295, 298, 310
urban growth 39, 110, 114, 279
  management 275, 276, 292
urbanity 45, 240, 247, 284, 313–330
  continuum of 318, 327, 329, 330
  defined 39, 314
  framework 315–318, 324
  sustainable 297, 298, 324–329
urbanization 1–4, 17, 35, 39, 45–49, 68, 71, 141, 142, 218, 322, 333–336. *See also* sustainable urbanity
  characteristics 275–277, 296
  in China 74–80
  defined 278, 295, 317
  framework 315–317
  future trajectories 64, 300, 311
  impact on food production 22, 23, 28–30, 80–84, 127
  impact on governance 239–241
  measuring 302, 303
  rates 112, 277, 300
  teleconnections 145, 301, 314, 317
urban land teleconnections 2, 4, 17, 90, 117, 301, 302, 309. *See also* teleconnections
urban land use 11, 39, 117, 299–312, 333
  characteristics 305–307
  patterns 275–298
urban planning 81, 292, 310
  examples of 76
urban policy 292, 295, 297
urban-rural distinction 278, 281, 283, 295, 313, 316, 317. *See also* lifestyle, livelihood
  income gap 75–78
urban shrinkage 301–305, 310
urban sprawl 1, 15, 21, 67, 279, 306
  defined 308
urban village 293
U.S. Green Building Council (USGBC) 231, 232, 234, 235
Utz Certified 229, 230

values 6, 52, 141, 144, 145, 152,
    154, 158, 160, 225, 324–326. *See
    also* revalorization
  social 239, 240, 257, 337
vegetarian diet 15, 22, 61
virtual trade 20, 137, 188
VisaNet 106
Voluntary Guidelines for Responsible
    Governance of Land, Fisheries and
    Forests in the Context of National
    Food Security 190, 195–197

water 3, 16, 26, 27, 47, 83, 125, 212,
    246, 248, 257
  infrastructures 276, 296, 305
  irrigation 25, 26, 29, 41

  power 40, 61
  protection 38, 42
  quality 29–31, 46, 54, 55, 326
  scarcity 64, 298
water grabbing 128, 185, 202, 203
well-being 1, 35, 36, 143, 147, 243, 268,
    269, 299, 309, 314, 317, 325–330,
    338
wind power 40, 61, 92
world city systems analysis 155
World Social Forum 150
World Wildlife Fund 149

Yibin, China 77–80

zoning 20, 21, 57, 62, 110, 163, 175, 179